Lecture Notes in Networks and Systems

Volume 405

The series "Lecture Notes in Networks and Systems" publishes the latest developments in Networks and Systems—quickly, informally and with high quality. Original research reported in proceedings and post-proceedings represents the core of LNNS.

Volumes published in LNNS embrace all aspects and subfields of, as well as new challenges in, Networks and Systems.

The series contains proceedings and edited volumes in systems and networks, spanning the areas of Cyber-Physical Systems, Autonomous Systems, Sensor Networks, Control Systems, Energy Systems, Automotive Systems, Biological Systems, Vehicular Networking and Connected Vehicles, Aerospace Systems, Automation, Manufacturing, Smart Grids, Nonlinear Systems, Power Systems, Robotics, Social Systems, Economic Systems and other. Of particular value to both the contributors and the readership are the short publication timeframe and the world-wide distribution and exposure which enable both a wide and rapid dissemination of research output.

The series covers the theory, applications, and perspectives on the state of the art and future developments relevant to systems and networks, decision making, control, complex processes and related areas, as embedded in the fields of interdisciplinary and applied sciences, engineering, computer science, physics, economics, social, and life sciences, as well as the paradigms and methodologies behind them.

Indexed by SCOPUS, INSPEC, WTI Frankfurt eG, zbMATH, SCImago.

All books published in the series are submitted for consideration in Web of Science.

For proposals from Asia please contact Aninda Bose (aninda.bose@springer.com).

More information about this series at https://link.springer.com/bookseries/15179

Miguel Botto-Tobar · Henry Cruz ·
Angela Díaz Cadena · Benjamin Durakovic
Editors

Emerging Research in Intelligent Systems

Proceedings of the CIT 2021 Volume 1

Springer

Editors
Miguel Botto-Tobar
Eindhoven University of Technology
Eindhoven, Noord-Brabant, The Netherlands

Henry Cruz
Universidad de las Fuerzas Armadas ESPE
Sangolquí, Ecuador

Angela Díaz Cadena
University of Valencia
Valencia, Spain

Benjamin Durakovic
International University of Sarajevo
Sarajevo, Bosnia and Herzegovina

ISSN 2367-3370 ISSN 2367-3389 (electronic)
Lecture Notes in Networks and Systems
ISBN 978-3-030-96042-1 ISBN 978-3-030-96043-8 (eBook)
https://doi.org/10.1007/978-3-030-96043-8

This Springer imprint is published by the registered company Springer Nature Switzerland AG
The registered company address is: Gewerbestrasse 11, 6330 Cham, Switzerland

Contents

Artificial Intelligence, Communications, Security and Cryptography, and Software Engineering

Lot Sizing Decisions Under Uncertain Demand Considering Skewness and Kurtosis

Fernando Rojas[1,2](✉)

[1] School of Nutrition and Dietetics, Universidad de Valparaíso, Valparaíso, Chile
fernando.rojas@uv.cl

[2] Center of Micro-Bioinnovation, Faculty of Pharmacy, Universidad de Valparaíso, Valparaíso, Chile

Abstract. The lot sizing model is useful for supply making decisions based on probabilistic modeling of demand, using two-stage stochastic programming, calculating the optimal costs of a supply model. In this paper, we study this model by using simulated scenarios subject to different degrees of skewness and kurtosis to model demand, considering univariate Weibull statistical distribution described by a generalized additive models of location, scale and shape (GAMLSS).

We carried out a simulation study of 10,000 different demand scenarios with different degrees of skewness and kurtosis, evaluating relationships between total costs, lot size decisions, expected stock and out of stock respect to coefficients of demand skewness and kurtosis.

In this study it has been shown that the coefficients of skewness and kurtosis impact on the total costs of supplying an item. The results also allow generating a predictive pattern of the first and second stage decisions, that is, the expected quantities in stock and shortages for the use of stochastic lot sizing. Our results indicate that the higher total cost of supply and greater shortage are related to demand patterns with more negative symmetry and lower kurtosis.

Keywords: GAMLSS · kurtosis · Lot sizing · Statistical moments · Skewness · Weibull statistical distribution

1 Introduction

An inventory management system is essential to give structure and direction to the decision-making of an organization regarding the supply, see [1]. In inventory management, the costs of purchasing, ordering, holding and shortages are determined, requiring an estimate of demands in a certain time whose restriction can be of different types: budgetary, by storage volume, service levels, among others [2].

Stochastic programming is an approximation to model optimization problems that involve uncertainty, see [3]. While deterministic optimization problems are formulated with known parameters, real-world problems almost invariably include parameters that are unknown at the time decisions must be made. When the parameters are uncertain, but they are assumed to be within a set of possible values, a solution could be sought that is feasible for all possible choices of parameters and optimizes a given objective function, see [4]. In this way, it is possible to have extensions of lot sizing models in probabilistic environments, through an approach based on two-stage stochastic programming

M. Botto-Tobar et al. (Eds.): CIT 2021, LNNS 405, pp. 3–14, 2022.
https://doi.org/10.1007/978-3-030-96043-8_1

(SP), obtaining first-stage decisions without yet knowing the realization of a random variable, such as demand per unit of weather. Subsequently, in a second stage, stock decisions and probable shortages are obtained, considering the generation of stochasticity scenarios for the demand of a product, see [2].

To the best of our knowledge, the lot sizing models has been mostly studied assuming independent and normally distributed demands per unit time (DPUT) of the items, but DPUTs are random variables (RVs) that may show any shape, see [5]. For the statistical distributions of RV independently and identically distributed (IID), such as the DPUTs, a statistical moment is a particular calculable dimension of the shape of its probability density functions (PDFs). The zero-th moment is always 1, the 1-th moment is the mean, the 2-th central moment is the variance, the 3-th standardized moment is the skewness, while the 4-th standardized moment is the kurtosis; see [6,7] and [8]. As we will see later in the background of this paper, all these moments are related and for this paper we will postulate that they can influence inventory lot sizing.

Often an RV data set has a joint statistical distribution based on marginal distributions with different skewness and kurtosis, as noted in [9] and [10]. In this context, it is essential to have excellent goodness of fit to actual data for theoretical description of the marginal statistical distributions, see [11] and [12].

Generalized Additive Model for Location, Scale and Shape (GAMLSS) is a semi-parametric regression type model introduced by [13] that allows great versatility in the modeling of RVs, see [14].

The main objective of this paper is to propose a new methodology for studying how the skewness and kurtosis of marginal distributions of DPUTs affect the total costs (TC) and inventory decisions under a stochastic lot sizing model in two-stage. The remainder of the paper is organized as follows. Section 2 presents the methodology proposed in this study built upon two pillars: (i) modelling of marginal statistical distributions of DPUTs with different skewness and kurtosis; and (ii) stochastic lot sizing model in two-stage. A simulation study is performed in Sect. 3 to analyze how the marginal statistical distribution of DPUTs with different skewness and kurtosis affect the TC and inventory decisions of stochastic lot sizing. In Sect. 4 finishing with a discussion and the conclusions of the results obtained in this research, along with their limitations and possible future research.

2 Methodology

As mentioned in the introduction, it is possible to use the GAMLSS model to model the marginal distributions of random variables. Unlike generalized linear models (GLM), the GAMLSS considers a family of generalized, discrete, or continuous statistical distributions, which can have varying degrees of skewness and kurtosis. Thanks to its formulation, it is possible to model any parameter of these statistical distributions for the response variable linearly or not in an additive parametric or non-parametric form of covariates with known or random values; see [15]. The advantage of using this type of statistical modeling is that GAMLSS is a regression toolbox appropriate for a big dataset of response variables that can consider linear or smoothing functions of predictive covariates to model any parameter of location, scale, or shape of the statistical distribution. The current packages available in R software, see [16], allow working with continuous (any type of skewness or kurtosis), discrete (including zero inflated data), and mixture statistical distributions. Models can be selected according to criteria of goodness of fit to the real data, as well as by generating random numbers with arbitrary distributions of interest for theoretical or empirical research, see [2, 17] and [18]. Our proposal differs from what has been addressed in the literature on the stochastic lot sizing of inventories. We describe IID DPUT as a marginal statistical distribution described by GAMLSS models, making it possible to generate different skewness and kurtosis to explore the TC and two-stage decisions that leads a stochastic lot sizing.

The new methodology proposed in this paper is subdivided into the following two sections:

- Sect. 2.1 shows how to describe the DPUT for an inventory item using GAMLSS, moments, skewness and kurtosis and their use with Weibull type 3 statistical distribution.
- Sect. 2.2 provides the elements of a probabilistic lot size inventory model using SP in two-stages.

2.1 How to Describe the DPUT for an Inventory Item

GAMLSS Formulation. Let Y be an IID RV corresponding to the DPUT. If Y be the DPUT of an inventory item, we considered that μ is the expected value of a response variable. Consider to d as a covariate. If $f(y|\boldsymbol{\theta})$ be a conditional PDF on parameters $\boldsymbol{\theta}$ ($F_{y|\theta}$ is the conditional cumulative distribution function (CDF)), where $\boldsymbol{\theta} = (\mu, \sigma, \nu, \tau)^\top = (\theta_1, \theta_2, \theta_3, \theta_4)^\top$ is a vector of four distribution parameters. In the GAMLSS formulation, only μ is a function of the covariates, μ and σ are location and scale parameters, and ν and τ are shape parameters. If $\{y_i\}, i = 1, \ldots, n$ is an $n \times 1$ vector of the response variable to model, considering $k = 1, 2, 3, 4$ as parameters. Then g_k is a link functions related to the k-th parameter $\boldsymbol{\theta}_k$ to covariates by additive models:

$$g_1(\boldsymbol{\mu}) = \boldsymbol{\eta}_1 = \boldsymbol{D}_1\boldsymbol{\beta}_1 + \sum_{j=1}^{J_1} h_{j1}(\boldsymbol{d}_{j1}), \tag{1}$$

$$g_2(\boldsymbol{\sigma}) = \boldsymbol{\eta}_2 = \sum_{j=1}^{J_2} h_{j2}(\boldsymbol{d}_{j2}), \tag{2}$$

$$g_3(\boldsymbol{\nu}) = \boldsymbol{\eta}_3 = \sum_{j=1}^{J_3} h_{j3}(\boldsymbol{d}_{j3}), \text{ and} \tag{3}$$

$$g_4(\boldsymbol{\tau}) = \boldsymbol{\eta}_4 = \sum_{j=1}^{J_4} h_{j4}(\boldsymbol{d}_{j4}), \tag{4}$$

where $\boldsymbol{\mu}, \boldsymbol{\sigma}, \boldsymbol{\nu}, \boldsymbol{\tau}, \boldsymbol{\eta}_t$ and $\boldsymbol{d}_{j1}$, for $j = 1, \ldots, J_k$ and $k = 1, 2, 3, 4$, are $n \times 1$ vectors. $\boldsymbol{D}_1$ is an $n \times J_1$ known matrix of variables and the regression coefficients $\boldsymbol{\beta}_1$ to be estimated is a $J_1 \times 1$ vector. h_{jk} is a semi-parametric additive function for the covariate D_{jk} evaluated at the vector $\boldsymbol{d}_{jk}$, which is assumed fixed and known.

For details of parameter estimate, diagnostic and good fit on the data see [13].

Moments, Skewness, and Kurtosis. If F is a CDF of any statistical distribution, considering the Riemann Stieltjes integral, see [19], the n-th moment of the statistical distribution is expressed by:

$$\mu_n' = \mathrm{E}\left[Y^n\right] = \int_{-\infty}^{\infty} y^n \, \mathrm{d}F(y)$$

with E as an expectation operator for the mean.
The zero-th moment of any PDF is 1.
The first raw moment is the mean:

$$\mu \equiv \mathrm{E}[Y].$$

The second central moment is the variance, and the square root of the variance is the SD:

$$SD \equiv \left(\mathrm{E}\left[(y-\mu)^2\right]\right)^{\frac{1}{2}}.$$

The normalized n-th central moment of the RV Y is

$$\frac{\mu_n}{SD^n} = \frac{\mathrm{E}\left[(Y-\mu)^n\right]}{SD^n},$$

and represents the distribution.

The normalized third central moment is called the skewness. A distribution that is skewed to the left has a negative skewness, and vice versa. Zero values indicate symmetry of the distribution. The Fisher coefficient of skewness (CSk), is defined as:

$$CSk = \frac{\mu_3'}{SD^3},$$

where μ_3' is the third moment centered.

The fourth central moment is a measure of outliers values far from the average distribution values and is denominated kurtosis. Statistical distributions with kurtosis less than 3 are said to be "platykurtic", while distributions with kurtosis greater than 3 are said to be "leptokurtic". The Fisher coefficient of skewness (CK), is defined as:

$$CK = \frac{\mu_4'}{SD^4},$$

where μ_4' is the fourth moment centered.

The parameterization of the probability density function (PDF) for Weibull type 3 statistical distribution (WEI3) in GAMLSS environment is given by

$f(y|\mu,\sigma) = (\sigma/\beta) * (y/\beta)^{(\sigma-1)} exp(-(y/\beta)^{\sigma})$

where $\beta = \mu/(\Gamma(1/\sigma)+1)$ for $y > 0, \mu > 0$ and $\sigma > 0$ are parameters of GAMLSS showed in Sect. 2.1, see [20]. We can also calculate the variance (sd_Y^2) as:

$\mathrm{sd}_Y^2 = \mu_Y^2\{\frac{\Gamma(\frac{2}{\sigma_Y+1})}{\Gamma(\frac{1}{\sigma_Y+1})^2} - 1\}.$

2.2 Probabilistic Lot Size Inventory Model Using SP in Two-Stages

Parameters		Variables	
t:	Period index of the decision stage in the planning time horizon ($t = 1, \ldots, T$)	Z_t:	Binary variable indicating whether a purchase is carried out in period t or not
C_t:	Purchase budget in period t	Q_t:	Quantity of units to be purchased in period t
u_t:	Unitary cost of purchase in period t	I_t:	Stock level at the end of period t
o_t:	Fixed order cost in period t	I_0:	Initial stock level
h_t:	Holding cost at the end of period t	S_t:	Shortage level at the end of period t
s_t:	Shortage cost at the end of period t		
p_t^{ω}: Probability of occurrence of the scenario ω in period t of the decision stage.			

To obtain the observed values of a forecast DPUT y_t^{ω} of Y_t, their probabilities p_t^{ω}, and E(TC), we adapted method showed in [2] to GAMLSS environment. The corresponding SP framework used to minimize the expected TC-E(TC)-of the inventory model can be formulated as [4]

$$\min\{\mathrm{E(TC)}\} = \min\left\{\sum_{\omega\in\Omega}\sum_{t=T}^{T+1} o_t Z_t + u_t Q_t + p_t^{\omega}(h_t I_t^{\omega} + s_t S_t^{\omega})\right\}, \tag{5}$$

subject to

$$\begin{aligned} Q_t + (I_{t-1}^{\omega} - S_{t-1}^{\omega}) - (I_t^{\omega} - S_t^{\omega}) = y_t^{\omega}, \\ Q_t \le C_t Z_t, \end{aligned} \tag{6}$$

$$\forall t \in T,\ \forall \omega \in \Omega,\ Q_t \ge 0,\ I_t^{\omega} \ge 0,\ S_t^{\omega} \ge 0, y_t^{\omega} \ge 0,\ p_t^{\omega} \in [0, 1],\ Z_t \in \{0, 1\},$$

where Ω is the set of selected possible demand scenarios and ω is a specific scenario, with a fixed number of scenarios in each period of the decision stages. The objective function defined in Eq. (5) attains a solution that minimizes E(TC) over all scenarios. This minimization can be carried out through the addition of sharing cuts for feasibility and optimality at the resource function, whenever this function or its constraints contain stochastic coefficients in a multi-stage problem [21].

3 Simulation and Analysis of Results

The new methodology proposed in this paper is studied using simulation and corroborated by an illustrative actual case. Here, we consider the following three sections:

- Section 3.1 presents details of the computational framework utilized and describes the simulation scenarios, which is divided into two parts as indicated below.
- Section 3.2 provide the results of simulation study where we analyzed the performance of a new methodology of probabilistic lot sizing on inventory with different skewness and kurtosis for the demand.

3.1 Computational Framework and Simulation Scenarios

We implemented our proposal in a non-commercial software named R; see http://www.r-project.org. See [18,22,23] and [24] to visualize R applications in supply models.

The simulation of 10.000 scenarios establish different: (i) inventory policies, (ii) statistical models for the DPUT, and (iii) SP to minimize costs. First, we used probabilistic lot sizing model, see [2]. Second, the statistical modeling is based on IID DPUTs assuming a Weibull type 3 (WEI3) statistical distribution, see [20]. We assumed several structures of skewness and kurtosis with WEI3 marginal statistical distributions for the DPUT. The uniformly distributed parameters employed to build these scenarios are chosen from values found in selected papers; see Table 2 and Appendix C of [25] for a list of these values. Third, SP is performed to minimize the total cost of inventory by using an objective function showed in Subsect. 2.2.

We used the following indicators with WEI3 statistical distribution to generate these 10,000 scenarios and inventory policies obtained by SP in two-stage:

Statistical parameters

- $\mu_Y \sim \text{U}(300, 1200)$,
- $\sigma_Y \sim \text{U}(10, 60)$,

Inventory parameter

- $h_t \sim \text{U}(0.25, 0.33)$,
- $o_t \sim \text{U}(12000, 18000)$,
- $u_t \sim \text{U}(15000, 30000)$,
- $s_t \sim \text{U}(700, 2000)$,
- $C_t \sim \text{U}(10000000, 20000000)$,

The choice of these values is based on previous studies on the topic; see [26,27] and [25].

3.2 Simulation Study

Firstly, in the 10,000 scenarios proposed their skewness and kurtosis frameworks for the DPUTs, we analyzed to respect that probabilistic lot sizing that provide the minimum total cost, and their decisions in first and second stage. Table 1 show descriptive statistical indicators of 10000 scenarios of the simulation study

Table 1. Descriptive statistical indicators of 10000 scenarios of the simulation study

Indicator	μ_Y	σ_Y	sd	CSk	CK	Q	I	S	TC	o	h	s	u	C
Min	300.1	10.01	688	−1.7274	2.926	280.5	3.114	2.499	4791629	12003	0.2500	700.2	15133	100007582
1st Qu.	454.3	22.47	1764	−1.0365	3.958	441.4	10.106	9.899	40749989	13481	0.2707	1024.3	64455	129817770
Median	633.3	35.09	2468	−0.9286	4.376	617.4	15.756	15.667	72711355	14980	0.2911	1358.9	114840	156133623
Mean	671.9	35.11	2698	−0.9340	4.517	653.3	19.558	19.926	75374976	14989	0.2916	1357.9	125808	154022583
3rd Qu.	869.0	47.85	3437	−0.8231	4.899	844.3	24.362	25.270	104361290	16466	0.3125	1692.0	176845	178923427
Max.	1199.7	60.00	6362	−0.3879	17.451	1191.2	110.935	120.437	196036684	17998	0.3333	2000.0	299964	199984413

Subsequently, in Table 2, the correlation between variables of the simulation study was evaluated using Pearson's correlation coefficient matrix.

Table 2. Pearson's correlation coefficient matrix between variables of the simulation study

	C	h	I	CK	μ_Y	o	Q	s	S	sd	CSk	σ_Y	TC	u
C	1.00	−0.02	0.03	−0.00	0.06	0.02	0.06	−0.01	0.03	0.05	−0.01	0.01	0.22	0.14
h	−0.02	1.00	−0.00	0.00	−0.01	−0.00	−0.01	0.02	−0.01	−0.01	0.01	0.01	−0.01	−0.01
I	0.03	−0.00	1.00	−0.31	0.55	−0.01	0.54	0.00	0.73	0.05	0.43	−0.67	0.12	−0.22
CK	−0.00	0.00	−0.31	1.00	−0.02	−0.01	−0.01	0.01	−0.29	0.19	−0.25	0.39	−0.00	0.01
μ_Y	0.06	−0.01	0.55	−0.02	1.00	−0.01	1.00	0.01	0.55	0.83	0.03	−0.04	0.23	−0.38
o	0.02	−0.00	−0.01	−0.01	−0.01	1.00	−0.01	0.00	−0.02	−0.01	−0.01	−0.00	0.01	0.01
Q	0.06	−0.01	0.54	−0.01	1.00	−0.01	1.00	0.01	0.51	0.84	0.01	−0.01	0.24	−0.38
s	−0.01	0.02	0.00	0.01	0.01	0.00	0.01	1.00	−0.00	0.01	−0.00	0.00	−0.01	−0.01
S	0.03	−0.01	0.73	−0.29	0.55	−0.02	0.51	−0.00	1.00	0.06	0.43	−0.64	0.11	−0.20
sd	0.05	−0.01	0.05	0.19	0.83	−0.01	0.84	0.01	0.06	1.00	−0.27	0.49	0.21	−0.32
CSk	−0.01	0.01	0.43	−0.25	0.03	−0.01	0.01	−0.00	0.43	−0.27	1.00	−0.55	−0.00	−0.01
σ_Y	0.01	0.01	−0.67	0.39	−0.04	−0.00	−0.01	0.00	−0.64	0.49	−0.55	1.00	0.01	0.01
TC	0.22	−0.01	0.12	−0.00	0.23	0.01	0.24	−0.01	0.11	0.21	−0.00	0.01	1.00	0.75
u	0.14	−0.01	−0.22	0.01	−0.38	0.01	−0.38	−0.01	−0.20	−0.32	−0.01	0.01	0.75	1.00

An extract of correlations of interest for this study according to their significance given by p-value, are showed in Table 3.

Table 3. Extract of correlations of interest for this study according to their significance given by p-value

Relationship	Type	p-value
I vs CK	Inverse	<.0001
I vs μ_Y	Direct	<.0001
I vs S	Direct	<.0001
I vs CSk	Direct	<.0001
S vs CK	Inverse	<.0001
S vs μ_Y	Direct	<.0001
S vs CSk	Direct	<.0001

Table 4 show results of linear regression between variables with significant correlations. Note that the inventory stock (I) is lower when the coefficient of kurtosis (CK) takes more positive values. Also, the higher the mean DPUT (μ_Y), the more inventory stock (I) remains. For its part, the symmetry coefficient (CSk) is directly related to the inventory balance in stock (I), the latter being higher when the coefficient takes more positive values. On the other hand, more negative kurtosis coefficients (CK) generate more shortages (S). Along with increasing I, the larger μ_Y, the larger will also be S. This relationship is important to consider since according to the adjusted R-squared value shown in the Table 4, 29.82% of the shortage is explained by μ_Y. Finally, positive coefficients of CSk generate shortages. However, negative values of CSk result in negative degrees of shortage, which is interpreted as over-balance stock.

Table 4. Results of linear regression between variables with significant correlations.

Dependent variable	Independent variable	Intercept	Slope	Adjusted R-squared
I	CK	427.348	−51.289	0.0974
I	μ_Y	−6338260	0.03006	0.3048
I	CSk	527.786	355.662	0.1890
S	CK	432.582	−51.649	0.0869
S	μ_Y	−13.702.134	0.03169	0.2982
S	CSk	546.998	372.328	0.1827

In order to form groupings of scenarios that are homogeneous among themselves, but in turn heterogeneous among groups of scenarios, a hierarchical cluster analysis was carried out, which associated the simulated scenarios according to the similarity of multiple variables. This grouping was illustrated in a dendrogram presented in Fig. 1.

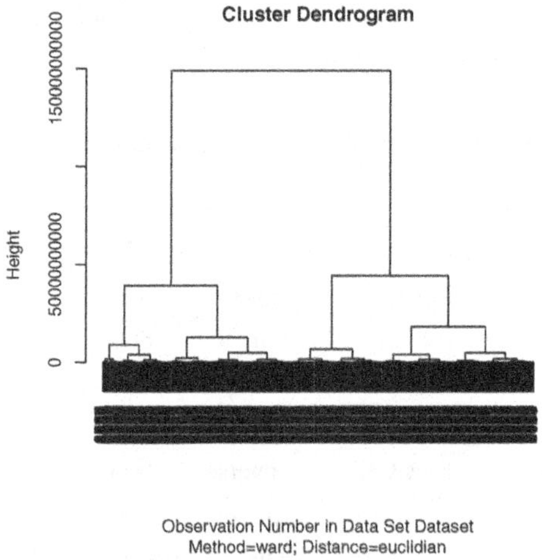

Fig. 1. Hierarchical cluster analysis of variables in the simulation study.

The dendrogram allows us to visualize the variables with which the scenarios were formed in 4 groups. Subsequently, a cluster analysis called "k-means" was carried out, which quantified the number of scenarios by groups, in addition to averaging their variables, thus facilitating data analysis. To confirm this, it is necessary to study the variability of the groups given by the sum of squares, resulting in the intragroup variability $9.948302e^{17}$ and the intergroup variability $1.099817e^{19}$. The intergroup variability is greater than the intragroup variability, which shows us that there are effectively 4 different groups of data. The mean of the variables of the groups of variables of conformation of the scenarios is shown in Table 5.

Table 5. Mean of the variables of the groups of variables of conformation of the scenarios.

Group	I	CK	μ_Y	Q	S	sd	CSk	TC
1	19.14	4.53	648.93	631.16	19.24	2597.06	−0.93	66597107
2	18.18	4.50	621.31	604.12	18.52	2493.29	−0.93	28205097
3	19.79	4.52	695.12	675.42	20.76	2796.15	−0.92	103328543
4	23.25	4.49	794.72	773.61	23.06	3209.21	−0.93	147193441

Table 5 shows an important finding when comparing the averages of the variables, group 4 is the one with the highest total cost, along with the most negative symmetry coefficient and the lowest kurtosis coefficient. It also has the largest shortage and inventory stock. It is also the group with the highest demand and quantity to buy. Group number 2, on the other hand, has a lower total cost, higher kurtosis coefficient, lower shortage, demand and therefore quantity to buy and higher symmetry coefficient. Finally, the probability that a scenario belongs to one of the groups was determined using a multinominal logistic regression model. For this, the values of the regression of the coefficients were tabulated in Table 6.

Table 6. Multinominal logistic regression model.

Group	Intercept	CK (slope)	CSk (slope)
2	0.1828067	−0.041315977	0.009013863
3	0.1222597	−0.003733192	0.252487023
4	−0.7170231	−0.059291893	−0.149842434

As a way to better interpret the multinomial logistic regression for the probability of belonging to groups 2 and 4, which presented lower and higher total cost respectively, then in Table 7 we show an evaluation of these probabilities for minimum values and CK and CSk maximums. Given the characteristics of higher total cost, inventory stock and shortage of group 4, the probability of belonging to this group increases when the kurtosis coefficient is higher, however, there is no significant increase in probability when taking a higher symmetry coefficient. Likewise, only the kurtosis coefficient also significantly determines an increase in group 2 membership when this coefficient takes higher values.

4 Discussion

This study provides a guide by using novel stochastic inventory optimization methods for policy-makers to mitigate inventory shortage of critical supplies while significantly improving their availability.

The scope of this document is open to both the academic world in the area of operations research and to the practical management of decisions carried out in the real

Table 7. Probability of belonging to the groups according to the coefficient of symmetry and kurtosis.

Group	Coefficient	valor	p (group)
	CK	2.92 (min)	0.4851
2		17.451 (max)	0.6319
	CSk	−1.72 (min)	0.4589
		−0.38 (max)	0.4559
	CK	2.92 (min)	0.7072
4		17.451 (max)	0.8506
	CSk	−1.72 (min)	0.6151
		−0.38 (max)	0.6585

world. In comparison with previous work, our results suggest new decision-making patterns such as considering the skewness and kurtosis of data related to demands per unit of time since it has shown its effect on the variance, which interacts with the best lot sizing, total cost and the balance or shortages of the items. More precisely, a predictive pattern of the first and second stage decisions, that is, the expected quantities in stock and shortages for the use of stochastic lot sizing. Our results indicate that the higher total cost of supply and greater shortage are related to demand patterns with more negative symmetry and lower kurtosis.

A selection of the main implications is provided next:

- In general, demand asymmetry and kurtosis seem not to affect first-stage lot sizing decisions.
- DPUT distributions with a more negative skewness (CSk < 0) and more leptokurtic, that is to say more pointed and with thicker tails than normal (CK > 3), favor to a greater extent obtaining high TC in probabilistic lot sizing systems. This conditions of symmetry and kurtosis of DPUT it is associated with an increase in the variances of this RV.
- Increases of 1 unit of C_K, that is more positive leptokurtic DPUT distributions, have a diffuse effect on the expected quantities stored (I) and shortage (S), causing decreases of 12% of I, with predictive power of 9%, and a 13% increase in S, with a slightly lower predictive power of 8.69 %.
- Increases of 1 unit of C_{Sk}, that is more positive skewness DPUT distributions, have a great effect on the expected quantities stored (I) and shortage (S), causing increases of 67.38% of I, with predictive power of 18.9%, and a 68.07% of increase in S, with a predictive power of 18.27 %, increasing the costs of second-stage decisions.

5 Conclusions, Limitations, and Future Research

In this paper we show relationships between the skewness and kurtosis of the DPUT distributions, with respect to the second stage decisions of a probabilistic lot sizing model. It appears that these coefficients that define the shape of the DPUT distribution would have no effect on first-stage lot size decisions.

Assuming the limitations of this work, which have to do with the non-consideration of heterogeneity of the DPUT and the time-dependent scenarios, this work is expandable to more general models with heteroscedasticity of variance, such as the widely known variants of generalized lineal models, which are very flexible allowing linear and non-linear functional structures. In addition, in line with this work, the methodology to model a time series, such as the generalized autoregressive and moving average (GARMA) model can be explored and its effects over lot sizing decisions. Furthermore, multivariate time series may be also considered. Some of these issues are being analyzed by the authors, whose findings will be reported in future articles.

The future research trend will focus more on cost saving, inventory availability, uninterrupted supply as well as dealing with unexpected/unpredicted change in the demand. Our study can be regarded as a pioneer in this research perspective by investigating the lot sizing decisions through considering the skewness and kurtosis of data.

Acknowledgment. This research was carried out thanks to the funding of the Fondecyt initiation project code: 11190004, Chile.

References

1. Sabet, E., Yazdani, B., Kian, R., Galanakis, K.: A strategic and global manufacturing capacity management optimisation model: a Scenario based multi-stage stochastic programming approach. Omega **93**, 102026 (2020)
2. Rojas, F., Leiva, V., Wanke, P., Lillo, C., Pascual, J.: Modeling lot-size with time-dependent demand based on stochastic programming and case study of drug supply in Chile. PloS One **14**(3), e0212768 (2019)
3. Shapiro, A., Dentcheva, D., Ruszczynski, A.: Lectures on Stochastic Programming: Modeling and Theory, vol. 16. SIAM, Philadelphia (2014)
4. Raa, B., Aghezzaf, H.: A robust dynamic planning strategy for lot-sizing problems with stochastic demands. J. Intell. Manufact. **16**, 207–213 (2005)
5. Sadeghi, J., Niaki, S.T.A.: Two parameter tuned multiobjective evolutionary algorithms for a bi-objective vendor managed inventory model with trapezoidal fuzzy demand. Appl. Soft Comput. **30**, 567–576 (2015)
6. Casella, G., Berger, R.L.: Statistical Inference, vol. 2. Duxbury, Pacific Grove (2002)
7. Deng, C., Miao, J., Ma, Y., Wei, B., Feng, Y.: Reliability analysis of chatter stability for milling process system with uncertainties based on neural network and fourth moment method. Int. J. Prod. Res. **58**(9), 2732–2750 (2020)
8. Lin, E.M.H., Sun, E.W., Yu, M.-T.: Behavioural data-driven analysis with Bayesian method for risk management of financial services. Int. J. Prod. Econ. **228**, 107737 (2020)
9. Escribano, A., Pfann, G.A.: Non-linear error correction, asymmetric adjustment and cointegration. Econ. Model. **15**(2), 197–216 (1998)
10. Chan, F., Lim, C., McAleer, M.: Modelling multivariate international tourism demand and volatility. Tour. Manage. **26**(3), 459–471 (2005)
11. Alavifard, F.: Modelling default dependence in automotive supply networks using vine-copula. Int. J. Prod. Res. **57**(2), 433–451 (2019)
12. Zhi, B., Wang, X., Xu, F.: Impawn rate optimisation in inventory financing: a canonical vine copula-based approach. Int. J. Prod. Econ. **227**, 107659 (2020)

13. Stasinopoulos, D., Rigby, R.: Generalized additive models for location, scale and shape (GAMLSS). J. Stat. Softw. **23**, 1–46 (2007)
14. Rohmer, J., Gehl, P.: Sensitivity analysis of Bayesian networks to parameters of the conditional probability model using a Beta regression approach. Expert Syst. Appl. **145**, 113130 (2020)
15. Stasinopoulos, M., Rigby, R., Akantziliotou, C.: Instructions on how to use the gamlss package in R (2008). www.gamlss.org/wp-content/uploads/2013/01/gamlssmanual.pdf
16. Stasinopoulos, M., Rigby, B., Voudouris, V., Heller, G., De Bastiani, F.: Flexible regression and smoothing: The GAMLSS packages in R. GAMLSS for Statistical Modelling. GAMLSS for Statistical Modeling (2015)
17. Rojas, F., Ibacache-Quiroga, C.: A forecast model for prevention of foodborne outbreaks of non-typhoidal salmonellosis. PeerJ **8**, e10009 (2020)
18. Rojas, F., Wanke, P., Coluccio, G., Vega-Vargas, J., Huerta-Canepa, G.F.: Managing slow-moving item: a zero-inflated truncated normal approach for modeling demand. PeerJ Comput. Sci. **6**, e298 (2020)
19. Liu, Z.: Refinement of an inequality of Grüss type for Riemann-Stieltjes integral. Soochow J. Math **30**(4), 483–489 (2004)
20. Stasinopoulos, M., Rigby, B., Akantziliotou, C., Heller, G., Ospina, R., Stasinopoulos, M.M.: Package 'gamlss. dist' (2020)
21. Infanger, G., Morton, D.P.: Cut sharing for multistage stochastic linear programs with interstage dependency. Math. Program. **75**, 241–56 (1996). https://doi.org/10.1007/BF02592154
22. Rojas, F., Leiva, V., Wanke, P., Marchant, C.: Optimization of contribution margins in food services by modeling independent component demand. Revista Colombiana de Estadística **38**, 1–30 (2015)
23. Wanke, P., Leiva, V.: Exploring the potential use of the Birnbaum-Saunders distribution in inventory management. Math. Probl. Eng. Article ID **827246**, 1–9 (2015)
24. Wanke, P., Ewbank, H., Leiva, V., Rojas, F.: Inventory management for new products with triangularly distributed demand and lead-time. Comput. Oper. Res. **69**, 97–108 (2016)
25. Wanke, P.F.: Consolidation effects: assessing the impact of tail dependence on inventory pooling using copulas. Inte. J. Inventory Res. **2**, 174–188 (2014)
26. Wanke, P.: Consolidation effects and inventory portfolios. Transp. Res. Part E Logist. Transp. Rev. **45**, 107–124 (2009)
27. Wanke, P.F., Saliby, E.: Consolidation effects: Whether and how inventories should be pooled. Transp. Res. Part E Logist. Transp. Rev. **45**, 678–692 (2009)

Quality in Use Evaluation of a GraphQL Implementation

Antonio Quiña-Mera[1,2(✉)], Pablo Fernández-Montes[2], José María García[2], Edwin Bastidas[1], and Antonio Ruiz-Cortés[2]

[1] Faculty of Engineering in Application Science, eCIER Research Group, Universidad Técnica del Norte, 100105 Ibarra, Ecuador
aquina@utn.edu.ec
[2] SCORE Lab, I3US Institute, Universidad de Sevilla, 41012 Sevilla, Spain

Abstract. The software development trend uses service-oriented software architecture (SOA), which provides efficiency, agility, and ease of growth. The architectural design most commonly used in SOA application development is REST (Representational State Transfer); however, some data management problems have been identified in its Application Programming Interface called API-REST. Several technological options have emerged to appease these problems, such as SPARQL, Cypher, Gremlin, and the most popular GraphQL. GraphQL was developed by Facebook in 2012 and released in 2015 to the community as an open-source project, used by companies such as GitHub, Airbnb, Amazon, Apollo, IBM, and Facebook. The goal of this research is to demonstrate whether GraphQL implementations work. Therefore, we based the research design on Design Science Research (DSR) to evaluate the quality-in-use of a GraphQL implementation that automated the systematic mapping studies (SMS) process for technology researchers at Universidad Técnica del Norte - Ecuador. We used the ISO/IEC 25000 series of standards to evaluate the quality in use; the results showed that the implementation met 84.11% of the established quality model's expected value. The detailed evaluation by quality characteristics was: Effectiveness 96.62%, Efficiency 78.90%, and Satisfaction 70.26%.

Keywords: GraphQL · SOA · Quality in use · ISO/IEC 25000

1 Introduction

A technological trend is the consumption of services through the Internet; this practice is called Cloud Computing; its benefits are cost savings, efficiency, agility, growth opportunities, and innovation [1]. Cloud Computing offers three areas of services: software, platform, and infrastructure [2]. We focus on the software as a service (SaaS) model, which distributes computer applications hosted by cloud service providers and made available to users through a network without the need to download or install them [3]. The most popular software architectures

M. Botto-Tobar et al. (Eds.): CIT 2021, LNNS 405, pp. 15–27, 2022.
https://doi.org/10.1007/978-3-030-96043-8_2

for the development of applications that support SaaS are the Service-Oriented Architectures (SOA) and Microservice Oriented Architectures (MSA), the latter being a modification of SOA [4]. REST (REpresentational State Transfer) is the most widely used SOA and MSA architectural design to develop Application Programming Interfaces called API-REST or API-RESTFUL [5]. Despite the acceptance of REST in the scientific and technological community, it has presented some problems in data handling, such as over-fetching (occurs when the data provider delivers more information than the client requires in a request). And under-fetching (occurs when the data provider does not offer in a query all the information that the client needs, therefore it must make more requests to obtain the complete information) [6]. Several technological options have emerged to improve the REST problems, such as SPARQL, Cypher, Gremlin, and the most popular of these GraphQL [7]. GraphQL is a query language and execution engine for data in client-server applications that has been accepted in the technology community because it was developed and used in the products of the company Facebook [8]. Created in 2012 and released in 2015 to the community as an open-source project and used by technology and application development companies such as 8base, Airbnb, Amazon Web Service, Apollo, Dgraph Labs, Elementl, Facebook, Fauna, Gatsby, Hasura, HomeAway, IBM, Intuit, Neo4j, PayPal, Prisma, Salsify, Shopify, Solo.io, Twitter [9].

Accordingly, we base this study on the Design Science Research (DSR) approach and pose the research question Does the GraphQL implementation work?. The research aims to answer the research question by evaluating the quality-in-use of a GraphQL component's implementation. We evaluated the quality-in-use using the ISO/IEC 25000 series of quality standards known as SQuaRE (System and Software Quality Requirements and Evaluation). The GraphQL implementation consisted of automating Systematic Mapping Studies (SMS) management proposed by technology researchers. We evaluated the quality in use using the ISO/IEC 25000 series of quality standards known as SQuaRE (System and Software Quality Requirements and Evaluation) at the Software Department of the Universidad Técnica del Norte - Ecuador. The rest of paper is structured as follows: Sect. 2) Research Design: we establish the research activities based on DSR, theoretical foundation, and artifact design and build (software implementation). Section 3) Results: results of the evaluation of the artifact quality in use. Section 4) Discussion: discussion of the research. Section 5) Conclusions.

2 Research Design

We designed the research based on the guidelines of the Design Science Research approach [10,11], see Table 1.

2.1 Population and Sample

The population of professors of the Software Department of the Universidad Técnica del Norte is twenty-two. The sample calculation with a margin of error of 5%, a heterogeneity of 99%, and a confidence level of 99% is thirteen subjects.

Table 1. Research design

Activity	Components	Paper section
Problem Diagnosis	Problem; Objective	Introduction
	Population and sample; Acceptance survey	Research design
Theoretical foundation	Software Architecture; SCRUM Framework; GraphQL; Systematic Mapping Study; ISO/IEC 25000 Standards: Quality in Use	Research design
Artefact Design: GraphQL implementation	Requirements; Design (Process and Architecture); Development and Deployment	Research design
Artifact evaluation	Quality in use evaluation of a GraphQL implementation	Results

However, in the quality-in-use model's measurement instruments (practical workshop and satisfaction survey), the non-probabilistic convenience sampling technique was applied. The sample consisted of 40 subjects (15 professors and 25 students) who had adequate knowledge to perform the practical workshop.

Acceptance Survey. To establish the automation topic's acceptance (SMS Management) for GraphQL implementation, we surveyed the study population. The survey consisted of 6 questions validated by three expert researchers in Software Engineering before the survey. The survey results indicated that 45.45% of the respondents knew about literature review methods (SLR[1]/SMS), but no one knew a tool to carry them out; furthermore, 100% indicated that they would like training about SMS management.

2.2 Theoretical Foundation

Software Architecture. Software architecture has evolved in different architectural approaches over the years. In the '80s, structures were vertical and isolated; in the '90s, horizontal models focused on business processes appeared, and nowadays, software architecture is focused on continuous delivery of different services. This evolution leads companies to migrate their monolithic architectures (combining user interface, business logic, and data in the same application towards Service Oriented Architecture (SOA) and Microservices (small services that operate independently) [4,13]. SOA and microservices present an approach to building distributed systems that deliver web services (self-describing, modular business applications that export business logic as a service via the Internet [14]; its implementation has also had an evolution ranging from building web

[1] A systematic literature review (SLR) is a means of identifying, analyzing, and interpreting reported evidence related to a set of specific research questions [12].

services based on WSDL and SOAP protocol to building REST and GraphQL Web APIs [13].

GraphQL. Created in 2012 by Facebook as an internal project and then released in 2015 to the community as an open-source project [15]. GraphQL is a runtime and API data query language [8]. It arises with the need to reduce the overhead of transferred data and the number of API query requests; reduce the number of invalid client query errors; support the evolution of the data model without an API version [16]. GraphQL service is composed of a type system, executable operations, or by extensions of external type systems. The type system defines the schema of the data capability of the GraphQL API service. The schema is defined in terms of types (objects, scalars, Enums, input, interfaces, and unions), types of operations (queries, mutations, and subscriptions), and directives (describes alternate runtime execution) [8].

SCRUM Framework. Establishes processes to manage product development with an interactive and incremental life cycle that promotes adaptive planning, development, and evolutionary delivery of the software product that drives rapid response to change. Scrum provides a usage guide comprised of phases, roles, events, artifacts, and best practices [17].

Systematic Mapping Study (SMS). Provides an overview of a research topic through a classification [18]. The SMS focuses on conducting a literature search that answers research questions usually are what, when, and where have been published about a research topic [19]. The main phases of SMS are review planning, study identification, and data extraction and classification [20].

ISO/IEC 25000 Standards. Called SQuaRE (System and Software Quality Requirements and Evaluation), focuses on the specification, measurement, and evaluation of software product quality requirements. SQuaRE establishes a model for measuring the internal, external, and in-use quality of software. The quality-in-use requirements specify the required level of quality from the end user's point of view; it is a validation of the software product from the needs of a context of use [21]. For quality-in-use evaluation, we define a quality model, measure the model, and evaluate it according to its measurement. We determined the quality-in-use model based on the ISO/IEC 25010 standard, establishing the quality characteristics and sub-characteristics. In the measure of the model, we use the metrics defined in ISO/IEC 25022. For model evaluation, we use the recommendations of ISO/IEC 25040.

2.3 Software Implementation (Artefact Design)

Requirements. The requirements backlog for the Systematic Mapping Study (SMS) management automation we established in 15 user stories.

Design (Process and Architecture). We used the Petersen et al. guide to establish the three-phase process for conducting SMS: Phase 1: Review planning, Phase 2: Study identification, Phase 3: Data extraction and classification. We designed a client-server application based on SOA service-oriented architecture using the provider, an API-GraphQL. The client, a web application integrated with the API-Mendeley for bibliographic reference management, see Fig. 1.

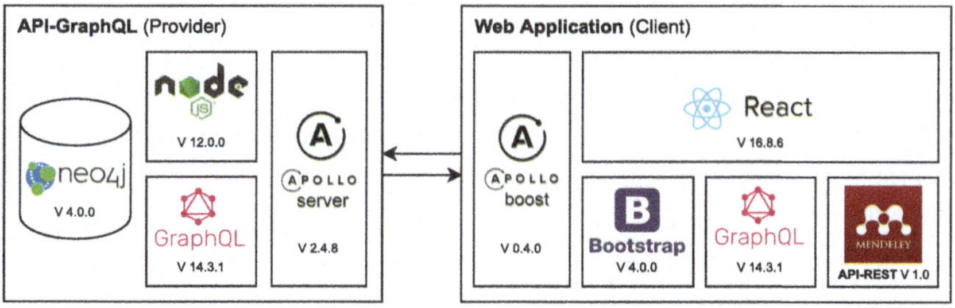

Fig. 1. Software product architecture design.

Development and Deployment. The application development was performed from 03/06/2019 to 06/12/2019 (220 h) using the SCRUM framework. The application (named "SMS-Online") automates the Systematic Mapping Study process. The work team consisted of two stakeholders, one product owner, one scrum master, and one developer. Below is a summary of the phases and artifacts used in the software product implementation, see Table 2.

Table 2. Development and implementation summary

Phase	Sprint (duration)	Deliverables
Pre-Game	Sprint 0 (20 h)	Requirements: - 15 user stories - Product Backlog Design (Architecture): - Process Diagram - Software Architecture - Initial Database Diagram
Game	Sprint 1 (42 h) Sprint 2 (42 h) Sprint 3 (42 h) Sprint 4 (42 h)	- Spring Backlog - Software Product Increment
Post-Game	Sprint 5 (32 h)	- 15 Acceptance Tests - Deployment of the software product - Delivery-Receipt Act

In the development, we verified the GraphQL components implemented in SMS-Online. Besides, we noticed that using the neo4j-graphql.js package (A JavaScript package to make it easier to use GraphQL and Neo4j together) translates GraphQL queries to a single Cypher[2] query, eliminating the need to write queries in GraphQL resolvers and to perform batch queries [23]. Table 3 shows the components implemented manually and generated by neo4j-graphql.js.

Table 3. GraphQL components implemented

Component clasification	Component	Implemented	Provider		Client
			Generated	Developed	Components
Executable definition	Query	✓	12	0	27
	Mutation	✓	62	1	40
	Subscription	X	0	0	0
Type system definition	Schema	✓	0	1	0
	Scalars	✓	0	132	132
	Objects	✓	0	12	12
	Interfaces	✓	0	1	1
	Unions	✓	0	1	1
	Enums	✓	0	2	2
	Input object	✓	63	0	0
	Directives	✓	0	14	0

Access to the SMS-Online application: https://appsms.utn.edu.ec:3000.

3 Results (Artifact Evaluation)

As a result, we show the quality in use evaluation of the GraphQL implementation developed in the previous section. Next, we offer the evaluation process.

3.1 Definition of the Quality in Use Model

The Product Owner and the Scrum Master of the project defined the quality in use model according to the ISO/IEC 25010 guidelines and the implementation context's quality needs. The model was structured with the expected valuation (percentage of the total) of the quality characteristics and sub-characteristics established in the model, up to complete 10% [24], see Table 4.

[2] Cypher is Neo4j's graph query language [22].

Table 4. Quality in use model

Characteristics	Sub characteristics	Percentage sub-characteristic	Percentage characteristic
Effectiveness	Tasks completed	16%	42%
	Objectives achieved	16%	
	Errors in a task	16%	
Efficiency	Task time	16%	32%
	Time efficiency	16%	
Satisfaction	Usefulness	8%	26%
	Trust	8%	
	User experience	8%	

3.2 Measurement of the Model

We identified a set of metrics described in ISO/IEC 25022 [25] associated with measuring the sub-characteristics of the quality-in-use model defined in the previous section. The metrics describe the measurement functions and their elements, see Table 5.

Table 5. Quality in use measures

Characteristic	Sub characteristic	Measurement function	Measurement
Effectiveness	Tasks completed	X = A/B	A = Number of unique tasks completed. B = Total number of unique tasks attempted
	Objectives achieved	$X = 1 - \sum Ai$ $\|X \geq 0$	Ai = Proportional value of each missing or incorrect objective in the task output (maximum value = 1)
	Errors in a task	X = A	A = Number of errors made by the user during a task
Efficiency	Task time	X = T	T = Task time
	Time efficiency	X = A/T	A = Number of objectives achieved T = Time
Satisfaction	Usefulness	X = N/T	N = Number of satisfied users. T = Number of users surveyed
	User trust	C = A/T, X = 1 – C	X = % complaints. C = % Trust. A = Number of complaints filed. T = Total number of survey respondents
	User experience	X = A	A = Psychometric scale value from a pleasure questionnaire (Likert Scale)

We used two instruments to obtain the measurement elements shown in Table 5: (i) Practical workshop on the use of the SMS-Online application, for data on effectiveness and efficiency of quality in use; and (ii) A satisfaction survey (based on SUS[3] [26]) on the use of the SMS-Online application after the practical workshop. We applied the measurement instruments to the sample of subjects (40 subjects) established in Sect. 2.1. After applying the instruments, we assessed reliability using statistical methods.

Statistical Evaluation of the Measurement Instruments. We evaluated the instruments' reliability according to their types of variables and the results obtained in the measurement applied to the study subjects, using the RStudio[4] tool for the calculations.

Practical Workshop. We used the statistical method Cronbach's Alpha coefficient; the analysis matrix structure contains the following measurement elements: tasks attempted, tasks completed, objectives completed, errors in a task, time tasks, objectives achieved, number of complaints, and confidence. The result obtained is 0.89, indicating that the workshop's reliability is acceptable according [28,29], see Fig. 2.

```
Reliability analysis
Call: psych::alpha(x = setValidacion)

 raw_alpha std.alpha G6(smc) average_r S/N   ase mean   sd median_r
      0.89      0.89    0.81      0.81 8.5 0.033 0.45 0.16     0.81
```

Fig. 2. Cronbach's alpha of the practical workshop.

Satisfaction Survey. We used the Confirmatory Factor Analysis (CFA) technique; the factor structure contains the factors: usefulness (ML1) and comfort (ML2). We analyzed the correlation between the initial survey design and the CFA calculation of the survey execution results (see Fig. 3). The initial design of ML1 contains the questions: Q1, Q6, and Q9, of which we discarded Q6 for correlating less than 0.3 (low level to maintain a viable survey). The initial design of ML2 contains the questions: P2, P3, P4, P5, P7, P8, and P9, of which we discarded questions: P3, P5, P7, P9 for having a low correlation. After analyzing and choosing the correlated questions, we calculated the CFI (Comparative Fit Index) and the TLI (Tucker and Lewis) index with the same AFC algorithm. We obtained CFI = 0.94 and TLI = 0.90, considered acceptable [30], and confirm that the survey is valid and reliable.

[3] The System Usability Scale (SUS) provides a "quick and dirty", reliable tool for measuring the usability [26].

[4] RStudio is an integrated development environment (IDE) for R, a programming language for statistical computing and graphics [27].

```
     ML1    ML2
P1   0.73 -0.08
P2  -0.05  0.46
P3   0.77 -0.07
P4   0.02  0.70
P5   0.74  0.02
P6   0.11  0.52
P7   0.87  0.07
P8   0.39  0.47
P9   0.73 -0.03
P10 -0.11  0.75
```

Fig. 3. Correlation between survey questions (CFA calculation).

3.3 Evaluation of the Quality in Use Model

We evaluated based on the expected values established in the quality model and the model's measurement results, see Table 6.

Table 6. Evaluation of the quality in use model

Characteristics	Sub characteristics	Measurement	Expected	Achieved	Completed
Effectiveness	Tasks completed	0.97	16%	15.50%	96.62%
	Objectives achieved	0.97	16%	15.50%	
	Errors in a task	0.96	16%	9.58%	
Efficiency	Task time	0.78	16%	12.51%	78.90%
	Time efficiency	0.80	16%	12.74%	
Satisfaction	Usefulness	0.71	8%	7.05%	70.26%
	Trust	0.70	8%	5.60%	
	User experience	0.70	8%	5.62%	
			Total	84.11%	

The result obtained from evaluating the quality model in use is ***84.11%***, which is considered an ***"opportunity"*** to improve economic, health, or environmental outcomes, according to the risks and opportunities associated with the quality level established in ISO/IEC 25022 [25].

4 Discussion

Before proposing the automation of the Systematic Mapping Study (SMS) process as an implementation of GraphQL called SMS-Online, we checked in the bibliographic databases Scopus, DBLP, and Springer for similar proposals. We found the studies of Knutas et al. [31] and Kohl et al. [32] that expose systematic literature review (SLR) and SMS automation tools. Knutas proposes the application called NAILS, which connects to the bibliographic database "web of science" to choose a data collection of studies and export in files imported into web-based analysis server HAMMER to display a statistical analysis of the entered data. They also offer an option to download an R language-based application for the study of the data collection. The difference with SMS-Online is the systematization of specific data in the SMS data extraction, the documentation of the entire SMS process, and the integration of studies from various bibliographic databases; this helps researchers organize and document their studies in a maintainable way and other researchers to reproduce.

Kohl reviews Online tools supporting the conduct and reporting of systematic reviews and systematic maps, showing 22 applications, of which two applications (Colandr and CADIMA) support SMS systematization. We reviewed the functioning of Colandr (free access), where we were able to perform Planning, Citation Screening, Full-text Screening, Data Extraction, and import of data collections through flat files. The similarities between Conlandr and SMS-Online are that they can conduct the primary SMS phases; the differences, SMS-Online offers help documentation at each step of the process, import primary-studies API-Mendeley and connects them directly to Mendeley Web. We also reviewed the CADIMA tool (free access), which presented detailed documentation at each step of the process; however, it was difficult to use it, which suggested having previous knowledge of the application to use it correctly. The advantage over SMS-Online is the option of collaborative work of several users in the same SMS study. The functionality of finding duplicate studies in SMS-Online is managed with the functionalities of Mendeley.

SMS-Online complements its functionalities with Mendeley's SMS tool functionalities; it displays and exports SMS results for use in documents such as scientific articles, scientific posters, or papers. The limitations found in SMS-Online are that it doesn't collaborate with several users in the same SMS and can not perform a check for duplicates in selecting studies. We propose these limitations as future work in updating the implementation of the SMS-Online application.

5 Conclusions

In this paper, we use the DSR approach to formulate the research question, Does the GraphQL implementation work?. We answer by evaluating the quality-in-use of implementation of GraphQL components. The implementation consisted of automating the management of Systematic Mapping Studies (SMS) named SMS-Online, directed to technology researchers of the Software Department of the Universidad Técnica del Norte - Ecuador in the academic period 2019, those who accepted (survey) with 100% the proposal of the automation topic.

GraphQL implementation was performed in a client-server web application using SOA (Service Oriented Architecture) and integrated with the Mendeley API-REST to complement the bibliographic management functionality. We developed using the SCRUM framework and the following technologies: Provider API-GraphQL (Apollo Express, GraphQL, JavaScript, Node.js, Database: Neo4J), Client Web Application (Apollo Scient, GraphQL, React,

Bootstrap, and Mendeley API-REST). We implemented 301 GraphQL components in the SMS-Online application: 12 queries, 63 mutations, 1 schema, 12 object, 132 scalars, 1 enums, 63 input objects, 1 interface, 1 union, and 14 directives. We evaluated the implementation using the quality-in-use model based on the ISO/IEC 25000 series of standards. The evaluation had three phases: 1) Definition of the quality-in-use model based on ISO/IEC 25010; 2) Measurement of the quality model based on ISO/IEC 25022; and 3) Evaluation of the quality model based on ISO/IEC 25040. The evaluation result showed that the application's implementation complied with 84.11%, which is considered an "opportunity" to improve economic, health, or environmental outcomes; where, the evaluation by characteristics was: Effectiveness 96.62%, Efficiency 78.90%, and Satisfaction 70.26%. According to these results, on the one hand, we proved that the implementation of GraphQL components is acceptable and work in SOA applications, thus answering the research question. On the other hand, we note that it is useful and convenient to use ISO/IEC standards to evaluate the software product's quality and identify improvement features in the applications.

As future work, we note the need to evaluate the implementation of all GraphQL components; improve the limitations found in the SMS-Online application; evaluate the internal and external quality and replicate the quality-in-use of GraphQL components.

References

1. Tsai, W.T., Bai, X.Y., Huang, Y.: Software-as-a-service (SaaS): perspectives and challenges. Sci. China Inf. Sci. **57**(5), 1–15 (2014)
2. Mell, P.M., Grance, T.: The NIST Definition of Cloud Computing. Technical report, National Institute of Standards and Technology, Gaithersburg (2011)
3. Marston, S., Li, Z., Bandyopadhyay, S., Zhang, J., Ghalsasi, A.: Cloud computing - the business perspective. Decis. Support Syst. **51**(1), 176–189 (2011)
4. Bogner, J., Zimmermann, A.: Towards integrating microservices with adaptable enterprise architecture. In: Proceedings - IEEE International Enterprise Distributed Object Computing Workshop, EDOCW, vol. 2016, September, pp. 158–163. Springer, Toronto (2016)
5. Han, H., et al.: A RESTful approach to the management of cloud infrastructure. In: CLOUD 2009 - 2009 IEEE International Conference on Cloud Computing, pp. 139–142 (2009)
6. Vogel, M., Weber, S., Zirpins, C.: Experiences on migrating RESTful web serVICES to GraphQL. In: Braubach, L., et al. (eds.) ICSOC 2017. LNCS, vol. 10797, pp. 283–295. Springer, Cham (2018). https://doi.org/10.1007/978-3-319-91764-1_23
7. Seifer, P., Härtel, J., Leinberger, M., Lämmel, R., Staab, S.: Empirical study on the usage of graph query languages in open source Java projects. In: SLE 2019 - Proceedings of the 12th ACM SIGPLAN International Conference on Software Language Engineering, co-located with SPLASH 2019, pp. 152–166, Athens, October 2019. Association for Computing Machinery, Inc
8. The GraphQL Foundation. GraphQL (2018)
9. GraphQL Foundation. GraphQL Foundation (2019)

10. Maedche, A., Hevner, A., Hutchison, D.: Designing the digital transformation. International Conference on Design Science Research in Information System and Technology, vol. 10243, pp. 231–246. Springer, Kristiansand (2017). https://doi.org/10.1007/978-3-319-59144-5
11. Hevner, A., Chatterjee, S.: Design science research in information systems. In: Design Research in Information Systems. Integrated Series in Information Systems, 39 edn., vol. 22, pp. 9–22. Springer, Boston (2010). https://doi.org/10.1007/978-1-4419-5653-8_2
12. Kitchenham, B., Brereton, P.: A systematic review of systematic review process research in software engineering. Inf. Softw. Technol. **55**(12), 2049–2075 (2013)
13. Seda, P., Masek, P., Sedova, J., Seda, M., Krejci, J., Hosek, J.: Efficient architecture design for software as a service in cloud environments. In: 10th International Congress on Ultra Modern Telecommunications and Control Systems and Workshops, ICUMT 2018, pp. 317–322, Moscow, November 2018. IEEE Computer Society
14. Klein, U., Namjoshi, K.S.: Formalization and automated verification of RESTful behavior. In: Gopalakrishnan, G., Qadeer, S. (eds.) CAV 2011. LNCS, vol. 6806, pp. 541–556 (2011). Springer, Heidelberg. https://doi.org/10.1007/978-3-642-22110-1_43
15. Byron, L.: GraphQL: A data query language - Facebook Code (2015)
16. Bryant, M.: GraphQL for archival metadata: an overview of the EHRI GraphQL API. In: Proceedings - 2017 IEEE International Conference on Big Data, Big Data 2017, vol. 2018, January, pp. 2225–2230, Boston, December 2017. Institute of Electrical and Electronics Engineers Inc
17. Schwaber, K., Sutherland, J.: The Definitive Guide to Scrum: The Rules of the Game. Scrum.org (2020)
18. Petersen, K., Vakkalanka, S., Kuzniarz, L.: Guidelines for conducting systematic mapping studies in software engineering: an update. Inf. Softw. Technol. **64**, 1–18 (2015)
19. Petersen, K., Feldt, R., Mujtaba, S., Mattsson, M.: Systematic mapping studies in software engineering. In: 12th International Conference on Evaluation and Assessment in Software Engineering, EASE 2008, p. 11, Bari. BCS Learning and Development Ltd. (2008)
20. Galindo, J.A., Benavides, D., Trinidad, P., Gutiérrez-Fernández, A.-M., Ruiz-Cortés, A.: Automated analysis of feature models: Quo vadis? Computing **101**(5), 387–433 (2018). https://doi.org/10.1007/s00607-018-0646-1
21. ISO/IEC. NTE INEN-ISO/IEC 25000 ISO/IEC 25000: 2014 Systems and software engineering - Systems and software Quality Requirements and Evaluation (SQuaRE) - Guide to SQuaRE. International Organization for Standardization, 2 edn. (2014)
22. Neo4j. Cypher Query Language - Developer Guides (2020)
23. GRANDstack. neo4j-graphql.js. User Guide—GRANDstack (2021)
24. ISO/IEC. NTE INEN-ISO/IEC 25010. International Organization for Standardization, 1 edn. (2015)
25. ISO/IEC. Systems and software engineering - Systems and software quality requirements and evaluation (SQuaRE) - Measurement of quality in use, vol. 1. International Organization for Standardization, 1 edn., Geneva (2016)
26. Brooke, J.: SUS: a retrospective. J. Usability Stud. **8**(2), 29–40 (2013)
27. Mhairi McNeill. About RStudio - RStudio (2020)
28. Sijtsma, K.: On the use, the misuse, and the very limited usefulness of cronbach's alpha. Psychometrika **74**(1), 107–120 (2009)

29. Juan Mendoza. RPubs - Alfa de Cronbach - Psicometría con R (2018)
30. Tóth-Király, I., Orosz, G., Dombi, E., Jagodics, B., Farkas, D., Amoura, C.: Cross-cultural comparative examination of the academic motivation Scale using exploratory structural equation modeling. Pers. Individ. Differ. **106**, 130–135 (2017)
31. Knutas, A., Hajikhani, A., Salminen, J., Ikonen, J., Porras, J.: Cloud-based bibliometric analysis service for systematic mapping studies. In: ACM International Conference Proceeding Series, vol. 1008, pp. 184–191 (2015)
32. Kohl, C., et al.: Online tools supporting the conduct and reporting of systematic reviews and systematic maps: a case study on CADIMA and review of existing tools. Environ. Evid. **7**(1), 1–17 (2018)

Augmented Reality in Determining the Cognitive Load of Blind Users During Navigation Process

Nancy E. Guerrón Paredes[1,2](✉), Antonio Cobo[2,3], Rodolfo Maestre[2,4], and José J. Serrano[2,3]

[1] Universidad de las Fuerzas Armadas ESPE, Sangolquí 11703, Ecuador
neguerron@espe.edu.ec
[2] Centre for Biomedical Technology, Universidad Politécnica de Madrid, 28223 Madrid, Spain
[3] CIBER-Bioengineering, Biomaterials and Nanomedicine, 28029 Madrid, Spain
[4] Universidad Politécnica de Sinaloa, 82199 Sinaloa, Mexico

Abstract. Augmented Reality Mobile Applications were developed to help blind users navigate unfamiliar spaces and to determine their cognitive load during that task. The present study compared two navigation configurations: guided and unguided, to assess the differences in cognitive load in blind adults to perform a secondary auditory task during navigation.

A 4.3% increase in cognitive load was found for participants using the unguided configuration. The evidence supports improvements in efficiency for guided configuration, with a 53.07% reduction in navigation time compared to unguided configuration (p-value = 0.028). Effectiveness also achieved 21.4% more successes with the guided configuration. Deferred memories were 38.3% higher among blind participants who used the guided configuration (p value = 0.006).

Keywords: Cognitive load · Augmented reality · Applications for blind people · Navigation through unfamiliar spaces

1 Introduction

1.1 Background

Cognitive load is the mental effort that an individual makes to temporarily retain the information that his brain processes to perform a certain activity; its determination is relevant at a psychological, physical and social level, because it is related to the amount of memory needed to perform a particular task, the difficulty associated with it and its adequate completion [1]. A lower mental effort to navigate unknown spaces would bring an increase in the independence and confidence of blind individuals to carry out activities in new places, consequently facilitating their integration into society [2].

Navigation plays a fundamental role, as it allows people to move independently and safely in a heterogeneous environment [3]. For people with visual disabilities, everything is discovered and assessed with their hands while their ears capture sounds, this replace

M. Botto-Tobar et al. (Eds.): CIT 2021, LNNS 405, pp. 28–43, 2022.
https://doi.org/10.1007/978-3-030-96043-8_3

"the vision" of the world around them. Blind people mainly use the ear and touch to obtain information about the environment and to explore their surroundings [4]. There is evidence that in the absence of vision, the other senses act as functional substitutes and are often improved [5]. Technological aids for the blind usually emit auditory and tactile signals to provide protection or information [6].

A navigation aid system consists of a series of directed instructions to facilitate the movement in an oriented, autonomous and safe way within environments of diverse complexity; this requires establishing a route and a method of orientation [7]. The spatial image of the target point can be experienced external to the user's head and body, same who receives the distance and direction information through visual, auditory and haptic stimulation, as well as spatial language [8].

Virtual spaces are a set of software tools used to develop knowledge of a place [9]. Augmented reality applications (ARA) are generated by adding information from the environment to virtual spaces, which could be used by people with blindness to freely navigate an unknown environment. When designing mobile applications for people who are blind, proper interface design appears to reduce the mental effort of users [10, 11]. This would allow a better performance of the individual during navigation and a reduction of the stress generated by the mental overload that this activity generates [12].

The working memory of an individual can be benefited if several sensory channels are used, avoiding overloading only one [13], hence the need to use several interfaces in mobile applications.

Memory, attention, concentration and other spatiality parameters can be evaluated using the Montreal test [14]; this is used in clinical practice for the detection of mild cognitive disorders and early dementia. A part of the Montreal test was applied to the participants of this study, prior to the navigation process assessment, in order to know some executive functions of the participants, such as memory retention capacity and deferred memories, which are directly associated with the measurement of cognitive load.

The e-Glance project [15] developed at the Polytechnic University of Madrid, has investigated the use of sensitive and cognitive interfaces in virtual and augmented reality applications. These applications aim to support the mobility of blind people in unfamiliar environments. The use of these applications for locomotion represents a challenge for people with visual impairment because most of them are designed for people with normal vision [16]; hence the physical interfaces must adapt to their other functional senses.

In this research augmented reality applications were developed with spatial sounds, also called three-dimensional sounds or 3D sounds to direct the user towards the target. A 3D sound consists of sound waves generated by the mobile application to indicate the direction and distance to the target point with respect to the user's position [17]. 3D sound is the convolution in the spatial domain of sound waves that uses transfer functions related to head movement. Unity is a software used to develop video games; it processes 3D sounds using Audio Spatializer SDK with a generic transfer function (HRTF). The sound waves emanate from a point in a three-dimensional space and deceive the brain using the ears, pretending to place different sounds in different places, delivering full reception of spatiality.

In this study, our team developed four ARA, so that blind people could navigate through a maze, using three-dimensional sounds. The objective of the research was to compare two navigation configurations and determine which of them requires less mental effort and also which one is the most efficient and effective for the navigation process.

1.2 Measuring Cognitive Load

Cognitive load cannot be measured directly; Sánchez [18] shows the difficulties around its determination and the different dimensions and aspects it presents [19]. According to Vásquez [12] the cognitive load must be approached from an integral perspective considering: cognitive demands, complexity of the task, characteristic of the task, temporal organization, work rhythm and health consequences. According to [20] the techniques for measuring mental load are: performance measures, subjective measures and physiological measures; each technique presents advantages and limitations.

Mental effort as an indicator of cognitive load can be measured when participants are performing a task or immediately after completing it and the performance will be the quantification of the achievement of the activity developed. Walking without seeing implies a cognitive load that increases when the environment becomes complex [10]. In this study we measured the cognitive load during navigation through a test maze, using a secondary task of listening, which is an activity that could become complex for people without vision. According to [21] the difficulty of the cognitive task has a direct effect on the second auditive task when the user walks through the virtual labyrinth.

A maze was designed between two points located on the first floor of the UPM Biomedical Technology Center so that participants could walk through its corridors with the help of a mobile application. As a second task we used an auditory task, in which the participants while walking were listening to five words spoken by the researcher. According to [22] memory has a limited capacity in size and complexity. The number of words used in this study was five different words from the common vocabulary. We established that this number of elements is less than the maximum contemplated in other studies [23] and would allow users to remember all the words after a navigation session. A greater number of correct remembered words implied less mental effort for the participant during the execution of the activity.

2 Materials and Methods

2.1 Participants

Fourteen blind participants (9 men, 5 women, average age 49.14, SD = 14.66) participated in this research. All the participants had become blind more than eight years ago and they had no other physical impairment. They previously had collaborated with the e-Glance project developed in the University Polytechnic of Madrid, testing several virtual reality applications using different physical and cognitive interfaces. The participants were affiliated with the ONCE Foundation (National Organization for the Blind Spanish people).

2.2 Cognitive and Sensitive Interfaces

The physical interface is the means used to communicate between the user and the mobile device. The cognitive interface indicates how the user will acquire the information to navigate. In previous studies carried out within the e-Glance project [24, 25], it was determined that there were no significant differences between using 3D sounds and using the female human voice to guide users through unknown environments. Based on that, it was decided to use 3D sounds as one of the physical interfaces to inform the user of the direction and distance to the target during the tour. The 3D sound plays on the side of the user's ear that matched the target's position, if the target happened to be in front of the user the sound was reproduced in both ears. A total of three physical interfaces were established in the present study: a) 3D sounds with linear response to provide information related to the position and distance to which the participant should go; b) A 440 Hz tone called a beep was used to warn of nearby obstacles and c) The female voice was used to generate complementary information, linked to the user's progress while he/she walked through the maze.

As a cognitive interface, two mechanisms were established to instruct the user how to navigate the test maze.

Interface 1: Participants freely walked through a test maze using their white cane and dodging obstacles. The app generated a 3D sound, indicating the location of the end of the route. The app generated a 400 Hz warning tone about obstacles that appeared in front of the user, so that they could modify their trajectory and avoid collisions.

Interface 2: Participants walked with their white cane through a maze, by obstacle-free areas, guided by 3D sounds generated by the application. There were five secure areas that were established within the maze. Every time the user reached a safe area, the application generated a female-voice verbal warning that told their had completed that part of the path.

2.3 Variables of Study

Cognitive load was computed as the Mahalanobis distance of two 3-dimensional vectors, where the smallest deviation between the vectors represents the lowest cognitive load. The origin vector (1,1,1) represents that the user has successfully completed the navigation by the unknown space, that he/she has used the least amount of time and that he has remembered all the words spoken by the researcher during the journey through the maze, as shown in Table 1.

Table 1. Variables of the study

Cognitive load		
Efficiency	Effectiveness	Deferred memories

Efficiency was defined as the time spent by the user to navigate the maze, from the beginning to the end of the tour (1). The researcher recorded the time it took for each participant to leave the starting point and reach the target point. There were no time restrictions to go from one point to another. If the user had gone in another direction or could not find the way, the researcher had to decide if the test needed to be suspended and to register this event as a not completed test. The lower the time taken to travel the maze translates to a greater efficiency of the tool as a navigation instrument.

$$Efficiency = \frac{\sum_{n=1}^{6} time_{zone\ n}}{time_{max\_tour}} \quad (1)$$

Effectiveness was defined as a dichotomous variable and corresponds to success to reach the target point. The effectiveness of the tool increases proportionally with the number of users who finished successfully the maze.

Deferred memories are linked to the participants' working memory and were defined as the percentage of correct words remembered by the users at the end of route, regardless of the order in which they were said. One point was assigned for each of the correct words remembered (Wn). The incorrect or omitted words were considered as an information processing *Error* and were attributed to the user's misinterpretation or because there was excess in the mental effort required to complete the task (See (2)). A greater number of correct words remembered expressed a lower mental effort in the execution of the activity.

$$deferred\_memory = \frac{\sum_{k=1}^{n} Wn - Error}{5} \quad (2)$$

2.4 System

2.4.1 Test Scenario

Since large spaces are difficult to control and the participants required sufficient time to complete the route, a test maze was designed in a 16.75 m × 5.63 m × 2.5 m room on the first floor of the Biomedical Technology Center of the Polytechnic University of Madrid. This space had six windows, two columns and two access doors. This space was divided into five sections using chairs and divisions made with cardboard and wood. In each section there was an obstacle free zone. The maze could be traveled in both directions. Four virtual rooms were implemented representing a different test maze. Each of these virtual places has six corridors 2.5 m to 3.5 m apart, as shown in Fig. 1.

2.4.2 Implementation

The system was implemented as a video game for iOS using Unity3D. It was deployed on an iPhone 6S plus running iOS 10.2.1. The app required a smartphone equipped with a gyroscope and a touch screen. The application showed the virtual representation of labyrinth as a first person view of the avatar's point of view. In spite of the fact that visual information is irrelevant for blind people, it is useful for the development process and debugging purposes.

A structure sensor [26] composed of a laser and an infrared was used in conjunction with the mobile device to map the maze and track the position of the user. Since the user requires a free hand to carry the white cane, the navigation device was placed on the participant's chest.

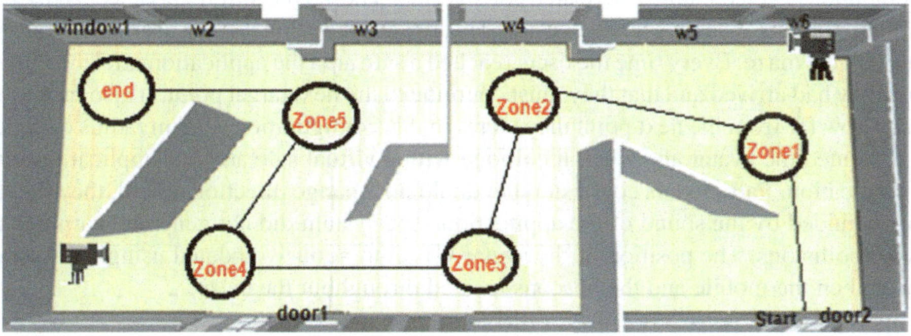

Fig. 1. Zenital plane of the maze

The system tracked the user's progress through the maze and dynamically updated the scene in the virtual world. The sensor set the distance to the objects and structures that appear during the tour, however it accumulated a drift error, which could alter the user's actual position in the application. A correction of the error produced by the structure sensor was performed using Estimation Proximity Beacons [27] conveniently located inside the maze; these were mostly near the windows, which was where the tracking sensor laser started to fail. Auditory notifications generated by the app were played via Bluetooth 4.0 on the bone conduction headphones worn by the users.

The center of the safe areas in the maze was marked on the floor with white tape, so that the researcher could record arrival times and also pronounce the words that users should remember later. Two cameras located at the ends of the room recorded the data on the route.

2.4.3 Configuration of Navigation

Two navigation configurations were developed, one called "guided configuration" and the other called "unguided configuration", which were responsible of guiding participants through the maze.

The unguided navigation configuration provided the user information within reach of the white cane, generating warning tones about obstacles that appear 1 m in front or 0.5 m to the sides of their route. The closer the obstacle was in their frontal area, the greater the intensity of the beep that the mobile device played. In this configuration, a 3D sound is played, which tells the participant the location of the end point of the route. The spatial sound is a continuous tone whose intensity varies linearly with distance and was reproduced in the bone headphones that the participants wore during the tests. In the design of the application, two lateral rectangular prisms of 0.5 m were placed to the left and right of the avatar and another rectangular prism of 1 m in front of the avatar to inform the user about nearby obstacles. These prisms moved with the avatar and informed the user about objects or structures that entered these areas. The blind person could stop or change direction at will, causing the avatar to stop or walk in the application; his position was continuously updated in the app using the tracker and beacons.

The guided navigation configuration used 3D sounds to guide users through the maze along unobstructed paths to safe areas of 1.75 m^2. These zones were modeled as 0.75 m radius cylinders. The program determined in advance the route that the user would follow through the maze. Every time the users reached a safe area the application informed them that they had arrived and that they must continue to the next target point. It also informed them how far from the next point they were. In this configuration a 0.5 m radius cylinder represented the avatar and when it collided with a virtual safe area the application sent a notification. Participants could stop but could not change direction at will, they had to move guided by the sound of the application. The system did not generate warnings to avoid collisions. The position of the avatar was continuously updated using the tracker located on the mobile and the beacons located throughout the maze.

2.5 Study Design

A within-subjects cross-sectional study was conducted. Eligible subjects were those who had been blinded for at least two years prior to the start of the study, those who were totally blind, have had no other physical impairment, and were over eighteen years of age.

Taking into account the availability of physical space and the size of the research team, we decided to limit the duration of the experiment based on previous studies conducted with these participants. Four ARA were developed. Two applications were used to traverse the maze forward (outward journey) and two applications were used to traverse the maze backward (return journey). Two navigation configurations (guided and unguided) were used. In the unguided configuration, 3D sounds, 400 Hz beeps, and the female voice were used as physical interface and cognitive *interface 1*. In the guided configuration, 3D sounds and female voice were used as physical interface and cognitive *interface 2*. The participants were involved in two experiences. During the first experience, they walked through the maze in one direction and in the second experience they walked in the opposite direction. During the round trip they used a different configuration. To minimize the impact of possible carryover effects, both the direction and the configuration were randomized according to the Latin square in Table 2.

Table 2. Latin square of study design.

Participant ID	First experience	Second experience
1	Outward-unguided	Return-guided
2	Return-guided	Outward-unguided
3	Outward-guided	Return-unguided
4	Return-unguided	Outward-guided

2.5.1 Procedure

All participants signed an informed consent prior to the beginning of the study. The activities were recorded by two cameras and through observation records. At the beginning of the session, each participant received a warm welcome and a member of the research team summarized the objectives of the study.

Each experience consisted of four stages:

1. Memory, Concentration and Deferred Memories assessment: A segment of the Montreal test was used to assess the memory, concentration, and deferred memories (See Table 3). The required administration time was approximately five minutes. The maximum score of this test was 10 points. Obtaining a score less than 8 points was considered a user with mild cognitive impairment.

To assess memory, the researcher read a list of 5 words at a normal rate with a spacing of approximately one second between words after having given the following instructions: "This is a memory test. I'm going to read a list of five words for you to remember. Listen carefully and when I finish reading I would like you to repeat all the words you can remember". The researcher read the list of words in Table 3 and marked all the words that the participant repeated and the order in which they were said. At this stage, one point is assigned for each word remembered in the correct order, with a maximum of 5 points.

Table 3. Montreal test for memory, concentration and deferred memories.

Memory	Face	Silk	Church	Carnation	Red	
	1	1	1	1	1	____/5
Concentration: Read the series of letters: The patient should tap every time the letter A is said FBACMNAAJKLBAFAKDEAAAJAMOFAAB						No points are assigned
Delayed recall	Face	Silk	Church	Carnation	Red	
	1	1	1	1	1	____/5

The researcher reads the same words again and informed the participant that he/she would be asked to repeat these words later.

To assess concentration, the researcher read a series of letters at a rate of one per second, after having given the following instructions: "I am going to read a lists of letters. Every time I say the letter A, tap your hand once. At this stage no point was assigned.

To assess deferred memories, the researcher gave the following instructions: "Previously I read a series of words and asked you to remember them. Now tell me all the words you remember". The examiner marked the words that the patient remembered without the need for clues. A point was assigned for each of the words remembered spontaneously.

2. Training: A member of the research team placed the bone headphones on the user's head; then he opened a practice application on the mobile for the user to follow the sound by rotating on its own axis. Next, the navigation device was placed on the participant's chest. The user was asked to move around the training space, following the sound until the application indicated him that he had arrived. At this stage there were no time restrictions. The device was removed once the training was finished.
3. Navigation Stage: A team member selected the application according to Table 2, and then he placed the user at the beginning of the maze and planted the navigation device on the participant's chest. The participant was asked to walk through the maze until the application indicated that he had reached the destination. In the outward-journey-application the user was at entrance door 2, and in the return-journey-application he was near window 1 of the maze (See Fig. 1). As the user progressed through the maze, the researcher was waiting for the participant to reach the area corresponding to zone 1 to pronounce the first word that appears in Table 4; this action was repeated in each of the five zones of the journey. In this stage, (a) the navigation time by zones, (b) the number of zones reached and (c) the success in reaching the destination were recorded.

Table 4. Words spoken by the researcher during the tour.

Zones	1	2	3	4	5
Outward	School	Garden	Exit	Green	Rain
Return	Theater	Museum	Entrance	Blue	Film

4. Post navigation: Each participant was asked to remember the name of the words spoken by the researcher during the tour. The responses were recorded manually. In this stage, the following were recorded: (d) number of words remembered and (e) number of correct words remembered. Finally, a five-minute break was made to continue with the second experience.

3 Results

3.1 Cognitive Evaluation

The Montreal cognitive assessment test was used to classify blind users into two categories: users with and without mild cognitive impairment. The concentration test was successful for the total of the participants; however, the section corresponding to memory and deferred memories showed that 50% of the participants present a mild cognitive impairment, given that the percentage of correct answers in the tests was lower than the 8/10 points of the cognitive evaluation test, as seen in Fig. 2.

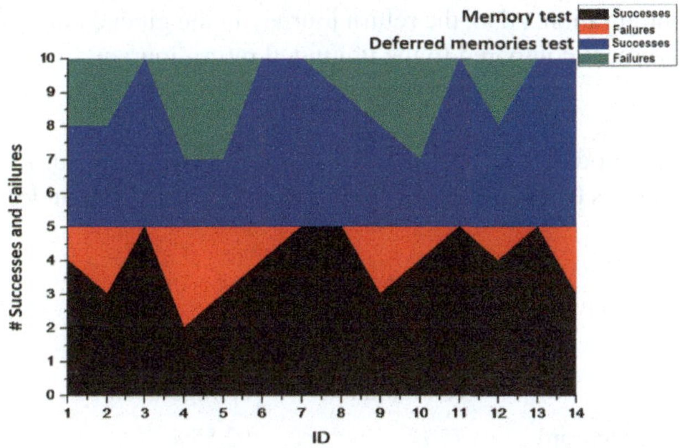

Fig. 2. Result of the cognitive assessment test applied to participants

3.2 Navigation

Efficiency

The efficiency of the navigation process was defined as the total time used to navigate the maze. Figure 3 summarizes the values of efficiency for the two configurations under study.

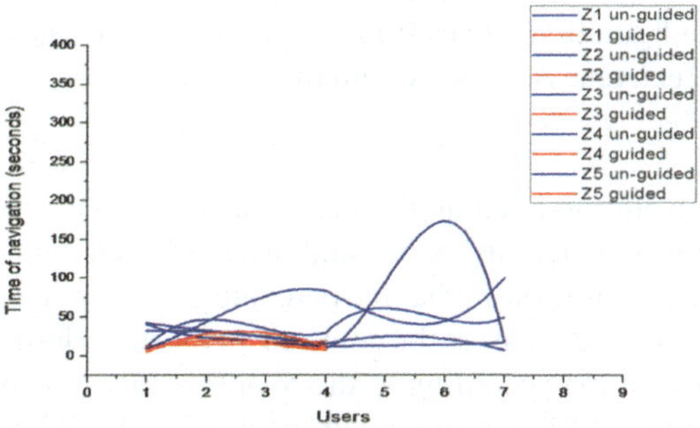

Fig. 3. Time used by participants to complete the tour

The dependence of efficiency with the type of configuration was assessed with a Friedman Test, after checking assumptions for this kind of test. Data from the two groups showed a normal distribution according to the Shapiro-Wilk test (P1 = 0.057; P2 = 0.9). The result was statistically significant (F = 6.3858, DF = 1, p-value = 0.02812). There was a 53.07% improvement in the efficiency of the guided configuration compared to unguided configuration. The outward journey using the application in the guide configuration reached a 57.65% improvement compared to the return journey in the unguided configuration. Also, the return journey in the guided configuration reached a 44.45% improvement compared to the unguided return journey.

Effectiveness
Effectiveness was modeled as the success at completing the journey. The Table 5 summarizes the values of effectiveness for the two configurations under study.

Table 5. Effectiveness.

	Unguided configuration	Guided configuration
Outward	75%	92.85%
Return	66.66%	87.5%

Success was modeled as a dichotomous variable. The dependence of success with the type of configuration was assessed with a Mantel-Haenszel test of general association, which resulted not significant ($M^2 = 21$, p-valor = 0.08). However, a dependency bond was observed, this should be studied in greater detail and with a large number of participants. The guided configuration achieved 21.4% more success than the unguided configuration. In addition, it was observed that for the participants of both groups, the return journey using the guided configuration was more complex than the outward journey with the same configuration. Other studies that implement a new orientation method could improve the effectiveness of the system to reach the target point, in any new space.

Deferred Memories
Deferred Memories were modeled as the percentage of words correctly remembered regardless of the order in which they were said, using (2). We found that the higher the number of remembered words, the higher the percentage of deferred memories. Repeated measures ANOVA were used to assess the deferred memories. The data of the four groups showed a normal distribution according to the Shapiro-Wilk test (p-guided_outward = 0.775; p-guided_return = 0.1218; p-unguided_outward = 0.3137; p-unguided_return = 0.1912). Sphericity was evaluated with the Mauchly test (W = 0.15207, p = 0.0559). Homoscedasticity was evaluated with the Bartlett test (Bartlett statistic = 1.3, DF = 3, p-value = 0.7). ANOVA was not statistically significant (F = 2.27, DF = 3, p = 0.109).

The deferred memories did not show statistical significance related to the type of configuration. However, we observed that the percentage of deferred memories was higher in the guided configuration, so we thought that there was probably some type of dependency bond. Given this, it was decided to separate the analysis between participants with mild cognitive impairment and those without it.

Among participants without cognitive impairment, deferred memories were assessed with a Friedman test (F = 0.41, DF = 3; p-value = 0.608), on data with a normal distribution verified with the Shapiro-Wilk test (P1 = 0.89; P2 = 0.885). Statistical significance was not found.

Friedman test resulted statistically significant for participants with mild cognitive impairment (F = 3.54, DF = 3, p-valor = 0.006). Pairwise comparison between journey types was conducted by applying a Tukey test. It was found that the outward journey using the guided configuration reached an increase of 47.5% in the number of words remembered with p = 0.00151, with respect to the outward journey in the unguided configuration. In addition, the return journey using the guided configuration reduced the mental effort of the participants with mild cognitive impairment, with an increase of 29.1% with p = 0.00182 in the number of words remembered compared to the return journey using the unguided configuration. The Fig. 4a presents the deferred memories of all the participants during the four trajectories performed and Fig. 4b presents the deferred memories of the participants with mild cognitive impairment.

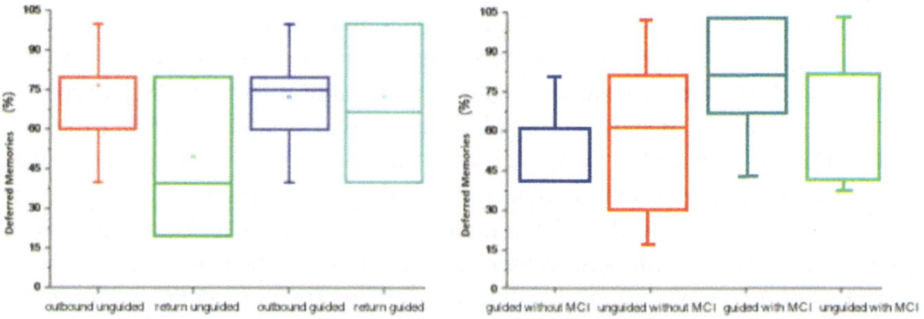

Fig. 4. Percentage of deferred memories of a) all participants, b) participants with and without mild cognitive impairment (MCI)

Cognitive Load

The deviations from the point with the lowest cognitive load are calculated as the Mahalanobis distance of two 3-dimensional vectors, where the lowest cognitive load is represented by the coordinates (1,1,1). The efficiency and deferred memory dimensions vary between zero and one, and the effectiveness can be zero or one. Therefore, the maximum deviation can be calculated as the Mahalanobis distance between (1,1,1) and (0,0,0), which corresponds to the value of 6.67.

Table 6 summarizes the average overall cognitive load scores for the two configurations under study. It should be taken into account that the score represents the value of the deviation from the coordinates that locate the lowest cognitive load. Therefore, the lower the value of the distance score, the lower the cognitive load.

Table 6. Deviation of cognitive load

Unguided configuration	Guided configuration
3.0711	3.954
40.26%	44.71%

The dependency of the cognitive load with the configuration was assessed with a Friedman test that was not statistically significant (F = 0.571, DF = 1, p-valor = 0,449).

4 Discussion

Navigation by unfamiliar places can be strenuous for blind people, given their limitation in recognizing and avoiding obstacles. The extra mental effort to use an ARA to navigate a new site can cause more stress and insecurity for the user. Assisted navigation systems for the blind often use voice to convey instructions to their users. However, this is problematic for short-range navigation systems that need to provide fine but diligent guidance in order to avoid obstacles. This study developed four ARA that could be used to navigate an obstacle-based maze. The application used a laser and infrared sensor to detect nearby objects and the user's progress in the real world.

To test the cognitive load of users while navigating unfamiliar places, an auditory event was added that the user had to complete at the end of the tour. We compared two configurations, one guided and one unguided, that used 3D sounds to indicate the location of the target point to the users. Both systems helped users reach the target point through a test maze. Blind users with mild cognitive impairment who used the application in the guided configuration achieved a 47.5% increase in the fulfillment of the auditory task, with a p-value = 0.0015; this achievement was not found in the rest of users. Therefore, we argue that blind people with mild cognitive impairment who visit new places can use the application in the guided configuration efficiently and effectively.

A group of fourteen blind participants were studied while navigating a test maze. 93.75% of the blind participants reached the goal using the guided configuration, and 71% arrived using the unguided configuration. It was observed during the tests that the participants, who were somewhat nervous during the participation, felt more comfortable and safe using the application in the guided configuration. We believe that the fact of experiencing that the first zone was reached without difficulty, gave them greater confidence to advance to the following spaces.

The time used for navigation was reduced by 53.07% using the guided configuration (t_navigation_unguided = 209 s; t_navigation_guided = 140 s), with an appropriate effect size and level of significance (Cohen's d = 0.006884, p-value = 0.028). While other research [13] has shown the efficiency of using virtual reality apps for navigation, using smaller test mazes for training (t_navigation = 105 s), they also found that time decreases significantly between two test sessions and that participants require an additional 50 s to complete each maze when there is a lower level of training. In our study an average of 69 additional seconds was used to complete the test maze when users used the unguided configuration.

The effectiveness to reach the goal was evident when the participants used the guided configuration, with an improvement of 22.75% compared to the unguided configuration and although there was no statistical significance (p_value = 0.067) regarding the use of the application between all the participants, a subsequent analysis carried out with the users who showed mild cognitive impairment, determined that the applications in the guided configuration were statistically significant. It was found that the participants with mild cognitive impairment presented a statistically significant difference (M^2 =

25.4, p-value = 0.024) using the guided configuration, with 4% more effectiveness compared to the unguided configuration. In addition, it was observed that the return journey using the guided configuration was more difficult for some of the participants than the outward journey, surely due to the orientation and initial turn they had to make to advance through the maze. On the contrary, the outward journey using unguided configuration was more difficult for them to do than the return journey; this was due to the fact that the sound generated by the application pointed towards the final target point which curiously coincided with the direction of the first obstacle; this made them lose several seconds of progress given that the participants walk directly towards to the obstacle and did not changed direction until they heard the obstacle warning beep. In another study [28] a significant difference in performance was found between two training sessions with different proportions of completed mazes, the study indicates that this result is likely due to additional time and more difficult starting orientations of each session, which could possibly make it difficult to complete the mazes.

It was observed a 4.3% reduction in the cognitive load of all the participants who used the application in the guided configuration compared to the unguided one, without statistically significant differences (p-value = 0.449). We presume that the two applications guided the users during the navigation process and helped them to reach the target point, but the task is possibly very simple and neither of the two shows a significant reduction in the value of the cognitive load used by the participants. This could be studied with a greater number of users in the future and with a more demanding task.

Since there are no previous studies that measure cognitive load during navigation on people who are blind, we compared our results with a study [13] where three labyrinths were used to measure alternation and cognitive load in people with normal vision during double tasks. The study found no statistically significant differences; as indicated by the results the Stroop hearing task was not difficult enough to sufficiently impair participants' attention.

In other research looking at dual-task interference during navigation and cognitive performance in a space with and without obstacles [28] they found significant differences in cognitive domains such as executive function, inhibition of habitual response and processing speed. According to this research, that used people with intact cognition, adding a task during walking increased the cognitive processing of the participants and therefore there was a reduction in the speed of walking, especially in obstacles.

References

1. Hossain, G., Yeasin, M.: Cognitive Load and Usability Analysis of R-MAP for the People Who are B lind or Visual Impaired (2011). https://doi.org/10.1145/2038476.2038503
2. Botto-tobar, M.: Advances in intelligent systems and computing artificial intelligence. Comput. Softw. Eng. Adv. **1327**, 326–343 (2020)
3. Cecflio, J., Duarte, K., Furtado, P.: BlindeDroid: an information tracking system for real-time guiding of blind people. Procedia Comput. Sci. **52**, 113–120 (2015). https://doi.org/10.1016/j.procs.2015.05.039
4. Lahav, O., Schloerb, D.W., Srinivasan, M.A.: Rehabilitation program integrating virtual environment to improve orientation and mobility skills for people who are blind. Elsiever Comput. **80**, 1–14 (2015). https://doi.org/10.1016/j.compedu.2014.08.003

5. Cattaneo, A., Vecchi, T.: Blind Vision: The Neuroscience of Visual Impairment. Massachusetts (2011)
6. Sánchez, J., Espinoza, M.: Audio haptic videogaming for navigation skills in learners who are blind. In: Proceedings of the 13th International ACM SIGACCESS Conference Computer Access - ASSETS 2011, 2014, vol. 227 (2011). https://doi.org/10.1145/2049536.2049580
7. Bowman, D.A., Kruijff, E., LaViola, J.J., Poupyrev, I.: An introduction to 3-D user interface design. Presence Teleoper. Virt. Environ. **10**, 96–108 (2001). https://doi.org/10.1162/105474601750182342
8. Loomis, J.M., Klatzky, R.L., Giudice, N.A.: Representing 3D space in working memory: spatial images from vision, hearing, touch, and language. In: Multisensory Imagery, pp. 131–155. Springer, New York (2014). https://doi.org/10.1007/978-1-4614-5879-1_8
9. Esteve-Gonzalez, V., Cela-Ranilla, J.M., Gisbert-Cervera, M.: Using simulation games to improve learning skills. In: 4th International Conference of Education, Research and Innovation, pp. 506–515 (2011)
10. Pigeon, C., Li, T., Moreau, F., et al.: Cognitive load of walking in people who are blind: subjective and objective measures for assessment. Gait Posture **67**, 43–49 (2019). https://doi.org/10.1016/j.gaitpost.2018.09.018
11. Guerreiro, J., Ahmetovic, D., Kitani, K.M., Asakawa, C.: Virtual navigation for blind people: building sequential representations of the real-world. Assets 2017, pp. 280–289 (2017). https://doi.org/10.1145/3132525.3132545
12. Vásquez, P.C., Klijn, T.P., Moreno, M.B., Barriga, O.: Validation of subjective mental workload scale in academics staff. Cienc. Enferm. **20**, 73–82 (2014). https://doi.org/10.4067/s0717-95532014000200008
13. Armougum, A., Orriols, E., Gaston-Bellegarde, A., et al.: Virtual reality: a new method to investigate cognitive load during navigation. J. Environ. Psychol. **65**, 101338 (2019). https://doi.org/10.1016/j.jenvp.2019.101338
14. Kletzel, S.L., Manuel, J., Miskiel, E.F., et al.: Evaluating the performance of the Montreal cognitive assessment in early stage Parkinson's disease. Park. Relat. Disord. **37**, 58–64 (2017). https://doi.org/10.1016/j.parkreldis.2017.01.012
15. CTB. eGlance (2016). http://eglance.ctb.upm.es/es/eglance/
16. Lahav, O., Mioduser, D.: Construction of cognitive maps of unknown spaces using a multi-sensory virtual environment for people who are blind. Comput. Human Behav. **24**, 1139–1155 (2008). https://doi.org/10.1016/j.chb.2007.04.003
17. Härmä, A., Jakka, J., Tikander, M., et al.: Augmented reality audio for mobile and wearable appliances. AES J. Audio Eng. Soc. **52**, 618–639 (2004)
18. Sánchez Colodrón, M., Montorio, I., Fernández de Trocóniz, M., López López, A.: La entrevista de Carga del Cidador: utilidad y validez del concepto de carga. An. Psicol. **14**, 229–247 (1998)
19. Stull, D.E.: The multidimensional caregiver strain index (MCSI): its measurement and structure - PsycNET. Am. Psychol. Assoc. **2**, 175–196 (1996)
20. Cegarra, J., Chevalier Aline, A.: The use of Tholos software for combining measures of mental workload: toward theoretical and methodological improvements. Behav. Res. Methods **40**, 988–1000 (2008). https://doi.org/10.3758/BRM.40.4.988
21. Plummer-D'Amato, P., Brancato, B., Dantowitz, M., et al.: Effects of gait and cognitive task difficulty on cognitive-motor interference in aging. J. Aging Res. (2012). https://doi.org/10.1155/2012/583894
22. Clark, B.R.E., Kirschner, P.A., Sweller, J.: Putting Students on the Path to Learning the Case for Fully Guided Instruction, pp. 6–11 (2012)
23. Robison, M.K., Unsworth, N.: Working memory capacity, strategic allocation of study time, and value-directed remembering. J. Mem. Lang. **93**, 231–244 (2017). https://doi.org/10.1016/j.jml.2016.10.007

24. Cobo, A., Guerrón, N.E., Martín, C., et al.: Computers in human behavior differences between blind people's cognitive maps after proximity and distant exploration of virtual environments. Comput. Human Behav. **77**, 294–308 (2017). https://doi.org/10.1016/j.chb.2017.09.007
25. Guerrón, N.E., Cobo, A., Serrano Olmedo, J.J., et al.: Sensitive interfaces for blind people in virtual visits inside unknown spaces. Int. J. Human Comput. Stud. **133**, 13–25 (2020). https://doi.org/10.1016/j.ijhcs.2019.08.004
26. Occipital. Structure (2017). https://structure.io/
27. Estimote Inc. Estimote Products. In: Estimote (2018). https://estimote.com/products/?gclid=EAIaIQobChMIyvW-otbJ5wIVzJ6zCh1wAQ1sEAAYASAAEgK8OfD_BwE. Accessed 11 Feb 2020
28. Rothacher, Y., Nguyen, A., Lenggenhager, B., et al.: Walking through virtual mazes: spontaneous alternation behaviour in human adults. Cortex (2020). https://doi.org/10.1016/j.cortex.2020.01.018

Machine Algorithm-Based Web Prototype for Crop Pest Detection

Alexander Columba-Guanoluisa, Jefferson Aimacaña-Chuquimarca, Mauro Rosas-Lara, and Julio C. Mendoza-Tello(✉)

Faculty of Engineering and Applied Sciences, Central University of Ecuador, Quito, Ecuador
{arcolumba,jpaimacana,mrosas,jcmendoza}@uce.edu.ec

Abstract. Agriculture is an essential activity because it provides food, raw materials, and employment. This activity is affected by the emergence of pests at any stage of the crop life cycle. In turn, it causes a phytosanitary problem that causes losses in the agricultural production and affects the quality of the final product. Faced with these challenges, this research develops a machine algorithm-based web prototype for crop pest detection. This study focuses on pests of four crops of Ecuadorian highlands, namely: potato, corn, tomato, and apple. The proposed solution is based on machine learning algorithm that best suits our case study. For this, data mining phases, image processing techniques and model evaluation metrics are used. With these requirements, a web prototype is designed and development. Thus, this research provides a computer tool to receive and validate information about the causes and pest treatment. Conclusions and future research are described at the final section of the document.

Keywords: Machine learning · Artificial intelligence · Pest · Data mining

1 Introduction

In Ecuador, agriculture is the sixth productive sector, and represents 9% of gross domestic product [1]. This activity provides sustainability to the economy and improves the quality of life of citizens. However, agricultural production is harmed by the emergence of pests, which reduces the quality and quantity of crops. In addition, these drawbacks affect the healthy diet plan of citizens and reduce the financial performance of companies. Thus, pests affect to both social and economic sphere.

Currently, pest management is carried out through biological, natural, and chemical controls. With these considerations, it is necessary to detect crop pests before those consequences are serious for the harvest. In this line, data mining, artificial intelligence and crop pest image processing can support this task in three ways. First, data mining provides guidelines and procedures to identify patterns in big data. Second, artificial intelligence proposes machine learning algorithms for image classification tasks. Third, digital processing proposes techniques for image segmentation, understanding, extraction, and recognition.

M. Botto-Tobar et al. (Eds.): CIT 2021, LNNS 405, pp. 44–55, 2022.
https://doi.org/10.1007/978-3-030-96043-8_4

This study focuses on pests of four crops, namely: tomato, potato, corn, and apple. Faced with this challenge, this research proposes a machine algorithm-based application that best suits our case study.

The sections of this document are structured as follows. In Sect. 2, the methodology provides a research phase overview. Section 3 explains the theoretical concepts on which the research is based. In Sect. 4, data mining phase is applied to our study case. With this considerations, Sect. 5 describes prototype development. Finally, conclusions and future works are at the end document.

2 Methodology

This research is focuses on two consecutives phases: data mining and software prototype development. Data Mining supports decision making through identification of patterns within big data. For this, it uses techniques to implement predictive and descriptive models. According to Cross Industry Standard Process for Data Mining (CRISP-DM), five sequential activities were realized, namely: business understanding, data understanding, data preparation, modeling, evaluation, and deployment [2].

As the identification of pests is based on crop leaves photographs, digital image processing algorithms are used. Then, supervised machine learning models were created and evaluated to select the best model suits our research. With this consideration, a web prototype is developed using waterfall model. Figure 1 shows an overview of all these phases, including a brief conceptualization.

3 Materials and Background

Briefly, this section describes materials and technical concepts used, as follow:

- Hardware and software. A computer with the following characteristics is used: 8 GB RAM, 1 TB HDD, 2 GB graphic card, 8th generation Intel Core i7 processor, and Windows 10 OS. In this line, software used are the following: Visual Studio Code, Python 3.7, and Google Colab for running code in the cloud. It should be noted that Google Colab provides a 12 GB NVIDIA Tesla K80 GPU for free.
- Image processing techniques. It is a set of activities necessary to extract data from an image. It provides the following versatility, namely: (a) optimal storage and encoding to facilitate image transmission, (b) image enhancement, and (c) pattern recognition and image classification. These techniques support artificial intelligence applications and supervised learning. In order to take advantage of the benefits described, this research uses smoothing methods and edge detection [3].
- Supervised learning algorithms. Machine learning is an artificial intelligence application that enables systems to learn and improve on their own. This process identifies patterns from the supplied data; with these patterns, systems can acquire knowledge to support decision making [4]. A training algorithm is called a supervised learning algorithm. Supervised term refers to the fact that the training of an algorithm is like having a supervising teacher of tasks. For this, the training data must be labeled; that is, the data must match the outputs. After training, new unobserved data entries are

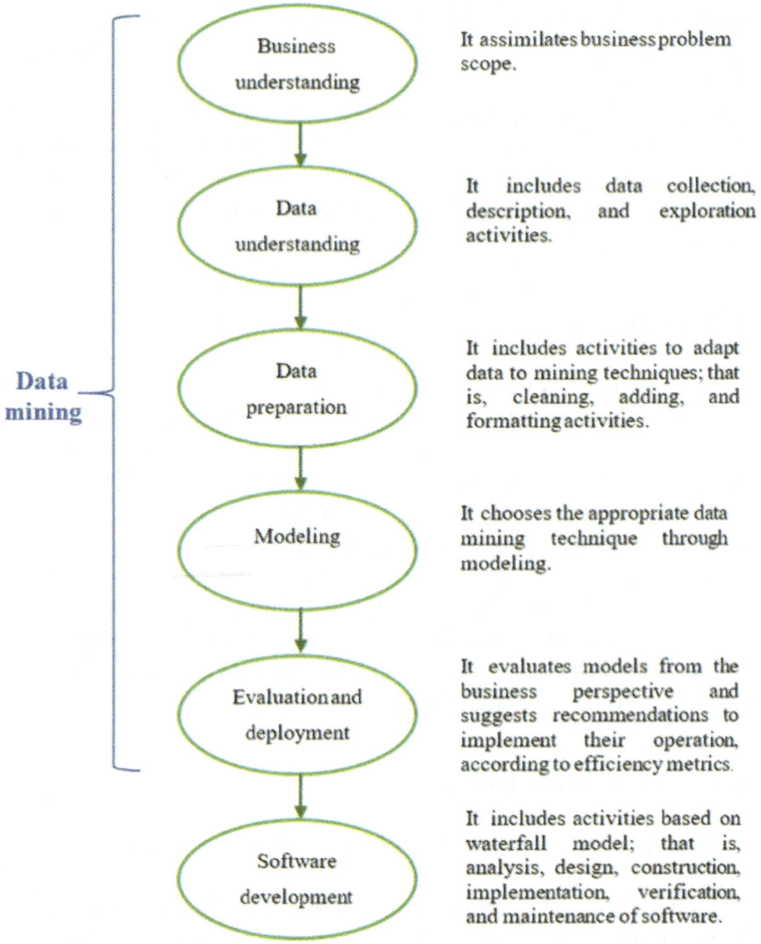

Fig. 1. Research phases overview

obtained, and the algorithm decides the classification label. This model predicts the correct labels for new input data. Two types of algorithms stand out in supervised learning, namely: regression and classification. A regression algorithm predicts output labels whose numerical values are continuous. In contrast to this, a classification algorithm predicts outputs labels or classes, whose values are categorical. For example, the prediction: if an email is spam or not, if a tumor is benign or malignant, if a plant has a plague or not. In supervised learning, the following algorithms stand out: logistic regression, support vector machine (SVM), random forest [5], and neural networks [6].

- Model evaluation metrics based on confusion matrix. This matrix labels the analysis results as positive or negative [7]. These results can be categorized, as follow: true positive (TP: data correctly labeled as positive), false positive (FP: negative data mislabeled as positive), true negative (TN: negative data correctly labeled as negative), and false negative (FN: positive data mislabeled as negative. With these indications, four metrics are used, namely: precision (data rate classified as positive that are effectively positive), recall (positive data rate correctly labeled), F1 (represents the harmonic mean between precision and recall), and accuracy (data rate that the model got right).

4 Data Mining: Experimental Phase and Results

In this section, CRISP-DM phases are applied to our study.

4.1 Business Understanding

This stage considers that the data mining approach must be aligned with the business approach. In this line, the business objective is to know the status of crops through images of leaves. In this way, it is possible to identify whether the crops are infected by parasites, viruses, mineral deficiencies, and climatic conditions. To guarantee the success of the project, it is necessary that users (such as farmers) can mitigate the pest using the information provided by the application. For this, it is necessary that the user interacts appropriately with the application. Thus, the margin of loss of harvest is reduced, the production is increased, and the quality of the product is improved. About data mining, it is intended to identify the different crop pests using supervised learning technique, and digital image processing algorithms.

4.2 Data Understanding

Initially, 20,722 tree leaves images were collected from PlantVillage using Mendeley Data [8]. The initial samples were organized within 4 datasets, as follow: tomato (11580), corn (3852), potato (2201), and apple (3089). Each dataset is divided into categories. Each category or class corresponds to crop pest and healthy leaves. Tables 1, 2, 3, and 4 describe the initial sample for each dataset. Each image has JPG format with a 256 $\times$ 256 RGB pixels size. A minimal amount of image noise was identified.

4.3 Data Preparation

For best results, it is necessary to obtain images without background noise and generate new images with format. In this context, five sequential activities are carried out, as follows.

First, data selection. Duplicate photographs are removed from the study. In addition, the number of images was balanced according to each dataset. Consequently, the number of images for each dataset is as follows: 2924 (tomato), 2189 (corn), 1444 (potato), and 1167 (apple).

Second, data cleansing. All background noises in the images were removed (Fig. 2). For this, an algorithm was designed, developed (using JavaScript), and executed (using Photoshop).

Table 1. Tomato initial dataset.

Category and description	Image
Pest: Late blight Description: Moist blackish - brown spots. Initial sample: 1909 (16.48%)	
Pest: Leaf mold. Description: Yellowish-green spots. Initial sample: 952 (8.22%)	
Pest: Septoria leaf spot. Description: Small circular brown and blackish spots. Initial sample: 1771(15.29%)	
Pest: Yellow leaf curl virus. Description: Small yellowish leaves with curved edges. Initial sample: 5357 (46.26%)	
Healthy leaf. Description: Dark green pinnate leaves. Initial sample: 1591 (13.73%)	
Total: 11580 (100%)	

Table 2. Corn initial dataset

Category and description	Image
Pest: Common rust. Description: Circular brown spots Initial sample: 1192 (30.94%)	
Pest: Leaf blight Description: Foliar and elongated lesions with an elliptical grayish brown Initial sample: 985 (25.57%)	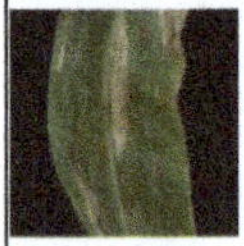
Pest: Cercospora Description: small and elongated grayish-brown necrotic spots. Initial sample: 513 (13.31%)	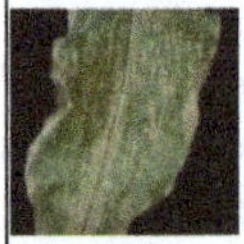
Healthy leaf. Description: Without lesions. Initial sample: 1162 (30.16%)	
Total: 3852 (100%)	

Third, data construction. New images were generated through the addition of new attributes or records. For this, each image was rotated 90, 180, and 270 degrees with respect to the original image. Images were kept with 256 × 256 size using JPG format. An algorithm was designed and developed (using Python). Consequently, the number of images for each dataset is as follows: 8754 (corn), 5776 (potato), 11694 (tomato), and 6668 (apple).

Fourth, integration and data format. During all phases, the datasets were kept within the same local machine. JPG images were converted to pixel arrays using Numpy and OpenCV libraries (based on Python). The images were resized to 128 × 128 pixels (from an original 256 × 256 size). In addition, matrix data were normalized between 0 and 1.

Fifth, dataset preparation for training, validation, and testing. Each dataset is fragmented according to two requirements, namely: 80% for training, and 20% for validation. In addition, a reduced set of images was used for final model test. With these considerations, Tables 5, 6, 7, and 8 show in detail the number of images used in each dataset for training, validation, and final tests.

Table 3. Apple initial dataset

Category and Description	Image
Pest: Black rot Description: Leaf edge with light and dark brown spots Initial sample: 582 (18.8%)	
Pest: Scab Description: Small circular blackish spots. Initial sample: 600 (19.42%)	
Pest: Rust Description: Small yellowish spots. Initial sample: 262 (8.48%)	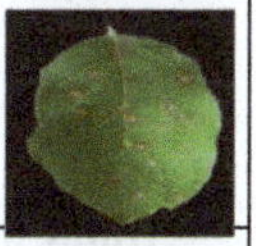
Healthy leaf Description: Oval green leaf with sawn edge without lesions Initial sample: 1645 (53.25%)	
Total: 3089 (100%)	

Table 4. Potato initial dataset

Category and Description	Image
Pest: Early blight Description: Small irregular brown spots. Initial sample: 1000 (45.43%)	
Pest: Late blight Description: Large irregular light and dark-brown spots Initial sample: 1000 (45.43%)	
Healthy leaf Description: Without lesions. Initial sample: 201 (9.13%)	
Total: 2201 (100 %)	

(a) (b)

Fig. 2. (a) Original image, and (b) image without background noise

4.4 Creation of Machine Learning Models

Preliminarily, four supervised learning models or algorithms were selected for the following reasons. First, the logistic regression algorithm does not require high computational resources and is easy to interpret. In addition, it produces predefined probabilities, supports large datasets, and is easy to implement. Second, random forest algorithm reduces the correlation between trees.

Third, support vector machines (SVM) have high performance in pattern recognition for both large and small volumes. In addition, they are highly effective in hyperdimensional space.

Fourth, neural networks have high performance in data processing. Furthermore, they provide two robust architectures [9]: VGGNet-19, and InceptionV3. With these considerations, the algorithms were run using four datasets, namely: tomato, corn, potato, and apple.

Table 5. Tomato final dataset

Class	Training	Validation	Test	Total
Late blight	1855	464	45	2364
Leaf mold	1834	459	45	2338
Septoria leaf spot	1724	431	45	2200
Yellow leaf curl virus	1926	481	45	2452
Healthy leaf	1836	459	45	2340
Total	**9175**	**2294**	**225**	**11694**

Table 6. Corn final dataset

Class	Training	Validation	Test	Total
Common rust	1754	438	40	2281
Leaf blight	1760	440	40	2240
Cercospora	1525	381	40	1946
Healthy leaf	1837	459	40	2336
Total	**6876**	**1718**	**160**	**8754**

Table 7. Potato final dataset

Class	Training	Validation	Test	Total
Late blight	2056	514	30	2600
Early blight	1874	468	30	2372
Healthy leaf	619	155	30	804
Total	**4549**	**1137**	**90**	**5776**

Table 8. Manzana final dataset

Class	Training	Validation	Test	Total
Black rot	1442	360	30	1832
Scab	1496	374	30	1900
Rust	814	204	30	1048
Healthy leaf	1486	372	30	1888
Total	**5238**	**1310**	**120**	**6668**

4.5 Evaluation and Deployment of Machine Learning Models

This section evaluates the models from three approaches: training time, validation, and test. First, the training time is directly proportional to the number of images used in each dataset. For each dataset, individual times were obtained. With these measurements, an average time for each model (or supervised learning algorithm) is obtained (in minutes). In this line, InceptionV3 is the algorithm with the longest delay time (53 min) and Random Forest is the most agile training model (2 min). Table 9 describes these values by model and dataset.

Second, with respect to validation. In practice, four metrics based on the confusion matrix were used, namely: (a) accuracy measures the percentage of successes of the model, (b) precision measures the fidelity percentage to classify an image, (c) recall measures the percentage of leaves with pests that the model can identify, and (d) F1 measures the percentage of optimization of the model. Based on this description, the evaluation of the models in each dataset is as follows:

Table 9. Algorithm training times (in minutes)

Dataset	Logistic regression	VCG-19	InceptionV3	SVM	Random Forest
Potato	26	30	31	11	2
Corn	56	42	51	32	2
Tomato	66	56	68	45	3
Apple	18	43	61	22	2
Approximate average	42	43	53	28	2

Table 10. Precision metric (%).

Dataset	Logistic regression	VGG-19	InceptionV3	Random Forest	SVM
Tomato	74	97	97	80	74
Corn	80	97	97	85	80
Potato	74	97	97	80	74
Apple	76	97	99	83	75

Table 11. Accuracy metric (%)

Dataset	Logistic regression	VGG-19	InceptionV3	Random Forest	SVM
Tomato	75	97	97	80	73
Corn	80	96	97	85	80
Potato	75	97	97	80	73
Apple	76	97	98	81	75

- Tomato dataset. With average values between 96% and 97%, neural network-based models obtained the best scores. In contrast to this, with average values between 72% and 74%, SVM-based model obtained the worst scores.
- Corn dataset. With average values between 96% and 97%, neural network-based models obtained the best scores. In contrast to this, with average values of 80%, the logistic regression model together with SVM model obtained the worst scores.
- Potato dataset. With average values between 96% and 97%, neural network-based models obtained the best scores. In contrast to this, with average values between 72% and 74%, SVM-based model obtained the worst score.
- Apple dataset. With average values between 97% and 99%, neural network-based models obtained the best scores; in this point, InceptionV3 stands out with values of 99%. In contrast to this, with average values between 72% and 74%, SVM-based model obtained the worst results. Tables 10, 11, 12, and 13 show the measurements obtained by model and dataset.

Table 12. Recall metric (%).

Dataset	Logistic regression	VGG-19	InceptionV3	Random Forest	SVM
Tomato	75	96	97	80	72
Corn	80	96	97	84	80
Potato	75	96	97	80	72
Apple	77	96	99	81	76

Table 13. F1 metric (%)

Dataset	Logistic regression	VGG-19	InceptionV3	Random Forest	SVM
Tomato	74	97	97	80	72
Corn	80	96	97	84	80
Potato	74	97	97	80	72
Apple	76	96	99	82	75

Third, with respect to tests. According to the validation results, neural network-based models performed better than others. However, it is necessary to test whether these are suitable for implementation in the business area; that is, models capable of predicting the status of crops. In this context, InceptionV3 and VGG-19 were tested as follows:

- Tomato dataset. 225 test data were used. InceptionV3 hit 220 predictions; it failed: 3 times when predicting leaf mold pest and 2 times for late blight pest. VCG-19 hit 197 right predictions; it failed: 9 times when predicting leaf mold pest, 3 times for late blight pest, and 16 times for yellow virus pest.
- Corn dataset. 160 test data were used. InceptionV3 hit 158 predictions and failed 2 times when predicting cercospora pest. VCG-19 hit 152 right predictions; it failed: 3 predictions for leaf blight pest and 5 for cercospora pest.
- Papa dataset. 90 test data were used. InceptionV3 hit all predictions right; that is 90. VCG-19 hit 82 right predictions; it failed: 5 times when predicting healthy leaves and 3 times for late blight pest.
- Apple dataset. 120 test data were used. InceptionV3 hit 117 predictions and failed 3 times when predicting healthy leaves. VCG-19 hit 97 right predictions; it failed: 3 times predicting the scab pest, 6 for black rot, and 14 for rust pest.

According to test results, InceptionV3 is the supervised learning model that best fits our case study.

5 Prototype Development and Results

Software development is based on a waterfall model. In this line, five phases were executed, as follows:

- Requirements. Three functionalities were defined, as follows: user management, crop information management, and crop pest prediction. Information management includes CRUD operations; that is, create, read, update, and delete. Crop pest prediction is based on machine learning model (InceptionV3). Due that the users are Ecuadorian farmers, the web page language is Spanish.

- Design. A database was defined and built using relationship-entity model (Fig. 3) and SQlite3, respectively. Additionality, web application was built using Django framework, which modular components pattern contain model, view, and template. Interface user can be deployed using web browser and app (Fig. 4).
- Construction. Phyton was used as a development language. Django framework for back-end application was used. Additionality, CSS and JavaScript were used as languages for HTML design using Bootstrap framework for front-end application.
- Verification. Proofs were conducted to validate functionality. In this way, Figs. 5 and 6 show a representative crops information management functionality.
- Maintenance. Currently, the prototype has been tested and has good results. We hope that new requirements arise to add them to the project. In this way, the farmer will have a versatile application with greater functionalities.

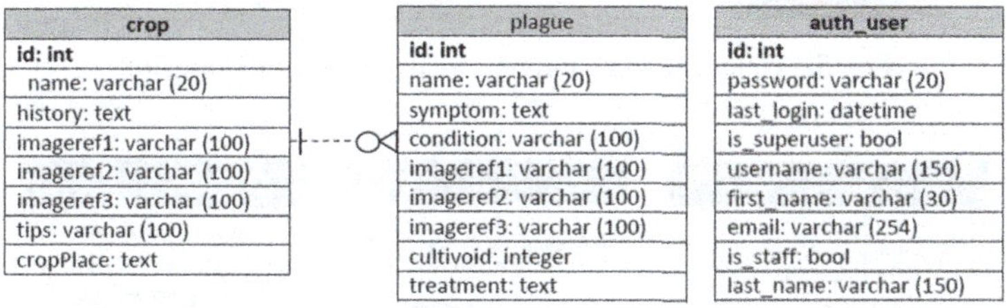

Fig. 3. Relationship-entity model

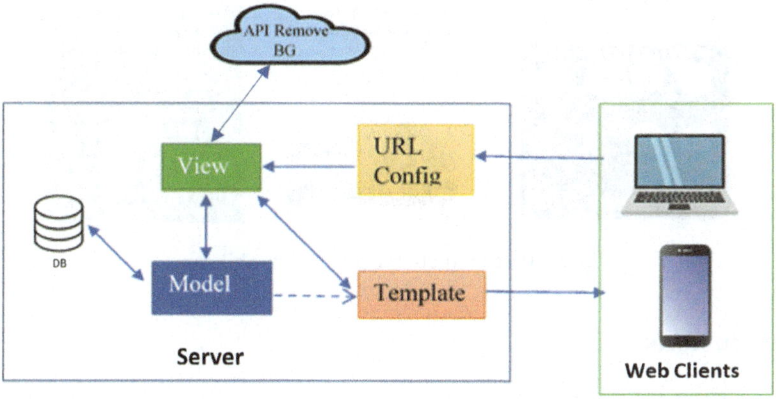

Fig. 4. Web Architecture using Django.

Fig. 5. Welcome page and pest analysis using mobile app.

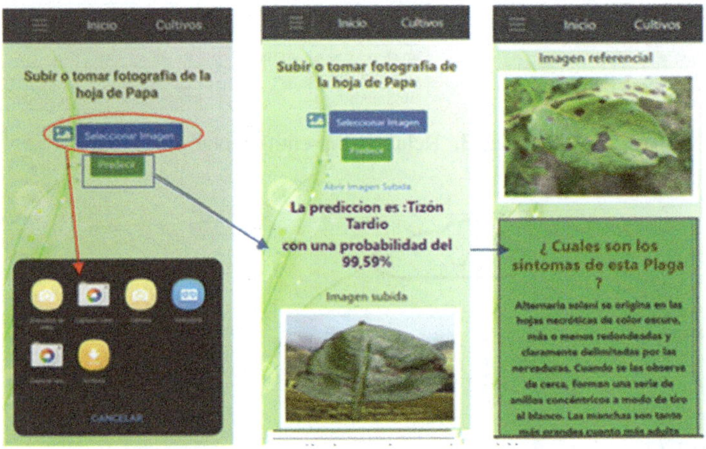

Fig. 6. Pest Prediction using web mobile app.

6 Conclusions

The research study case is the pest detection of four crops in Ecuadorian highlands, namely: tomato, corn, potato, and apple. This document highlights how the mining process can support this phytosanitary problem. In this process, five phases were executed to select the supervised learning model that best suits the study. In addition, image processing techniques were used to prepare data, prior to the construction of models. Next, measures of precision, accuracy, and F1 were used to evaluate models within the business process. All models performed very well. In this line, Inception-V3 stands out because its evaluation exceeds 97%; with this selection, a mobile app prototype based on Inception-v3 model was developed.

The main contribution of this research is to provide a computer tool for the timely detection of pests. In this way, farmers will be able to reduce crop damage and use of toxic pesticides.

As future work, this research can be extended to other crops. This requires that agricultural institutions provide access to more datasets. Thus, collaborative efforts between companies and research groups can maximize common benefits.

References

1. The World Bank Group. Agriculture, forestry, and fishing, value added (% of GDP). https://data.worldbank.org/indicator/NV.AGR.TOTL.ZS
2. Schröer, C., Kruse, F., Marx, J., Kruse, F., Marx, J.: A systematic literature review on applying process model on applying CRISP-DM process model. Procedia Comput. Sci. **181**, 526–534 (2021). https://doi.org/10.1016/j.procs.2021.01.199
3. Moorthy, G., Karthikumar, S., Thuy, N., Chi, L.: Versatile image processing technique for fuel science: a review. Sci. Total Environ (2021). https://doi.org/10.1016/j.scitotenv.2021.146469
4. Westreich, D., Lessler, J., Funk, M.J.: Propensity score estimation: neural networks, support vector machines, decision trees (CART), and meta-classifiers as alternatives to logistic regression. J. Clin. Epidemiol. **63**, 826–833 (2010). https://doi.org/10.1016/j.jclinepi.2009.11.020
5. Antoniadis, A., Lambert-Lacroix, S., Poggi, J.M.: Random forests for global sensitivity analysis: a selective review. Reliab. Eng. Syst. Saf. **206**, 107312 (2021). https://doi.org/10.1016/j.ress.2020.107312
6. Fávero, L.P., Belfiore, P.: Binary and multinomial logistic regression models. In: Fávero, L.P., Belfiore, P. (eds.) Data Science for Business and Decision Making, Chapter 14, pp. 539–615. Academic Press, New York (2019)
7. Luque, A., Carrasco, A., Martín, A., de las Heras, A.: The impact of class imbalance in classification performance metrics based on the binary confusion matrix. Pattern Recogn. **91**, 216–231 (2019). https://doi.org/10.1016/j.patcog.2019.02.023
8. Arun Pandian, J., Geetharamani, G.: Data for: identification of plant leaf diseases using a 9-layer deep convolutional neural network. Mendeley Data. **1** (2019). https://doi.org/10.17632/tywbtsjrjv.1
9. Abbassi, N., Helaly, R., Hajjaji, M.A., Mtibaa, A.: A deep learning facial emotion classification system: a VGGNet-19 based approach. In: Proceedings - STA 2020: 2020 20th International Conference on Sciences and Techniques of Automatic Control and Computer Engineering, pp. 271–276. Monastir, Tunisia (2020)

MQTT Based Event Detection System for Structural Health Monitoring of Buildings

Iván Palacios[1], José Placencia[1], Milton Muñoz[2], Víctor Samaniego[3], Santiago González[1(✉)], and Juan Jiménez[2,3]

[1] Department of Electric, Electronic and Telecommunication Engineering, University of Cuenca, Av.12 de Abril, 010203 Cuenca, Ecuador
{ivan.palacios,jose.placencial, santiago.gonzalezm}@ucuenca.edu.ec

[2] Red Sísmica del Austro, University of Cuenca, Av.12 de Abril, 010203 Cuenca, Ecuador
{milton.munozc,juan.jimenez}@ucuenca.edu.ec

[3] Department of Civil Engineering, University of Cuenca, Av.12 de Abril, 010203 Cuenca, Ecuador
victor.samaniegog@ucuenca.edu.ec

Abstract. Structural Health Monitoring (SHM) consists in a fundamental research field which aim to evaluate the current status of an infrastructure with the main purpose to identify damages and prevent catastrophic events. This paper presents an SHM solution that implements an automatic system based on the MQTT protocol and IoT devices for detecting seismic events. In particular, the architecture consists of a set of accelerometer sensors which communicate by means of a decentralized network topology (i.e., an Ad hoc Network configuration). Moreover, the system has the capacity to transmit the information about the events detected in real-time using cloud services. In order to verify the proper operation, the system was deployed on an actual building and the information acquired by the sensors was registered along four months. In this context, a relevant event detected was selected for analyzing the dynamic response of the building during a seism. Results show that the acceleration values increase as a function of the building height. Regarding the seismic event analyzed, the RMS values of acceleration identified on the basement were 0.26, 0.22, and 0.22 cm/s^2 and in the case of the eighth floor were 1.18, 1.33, and 0.59 cm/s^2 for the longitudinal, transverse, and vertical axes, respectively. Additionally, a first assessment regarding the structural health status of the building was performed through the OMA methodology (Operational Modal Analysis). Specifically, the FDD (Frequency Domain Decomposition) mechanism was used to determine the first four frequencies and its respective vibration modes.

Keywords: MQTT · IoT · Event detection · WSN · Structural Health · Operational Modal Analysis

1 Introduction

Structural Health Monitoring (SHM) consists in a crucial research field which aim to evaluate the current status of an infrastructure (e.g., bridges, heritage buildings, and

M. Botto-Tobar et al. (Eds.): CIT 2021, LNNS 405, pp. 56–70, 2022.
https://doi.org/10.1007/978-3-030-96043-8_5

strategic facilities, among others). This analysis is performed with the main purpose to identify damages and prevent catastrophic events. In particular, there are two methodologies developed in order to characterize the dynamic response of structural elements which are known as Experimental Modal Analysis (EMA) and Operational Modal Analysis (OMA), [1]. The EMA strategy is based in the application of artificial sources (e.g., controlled explosions or the use of instrumentation such as a vibrating table) therefore this methodology is usually applied in a laboratory environment. On the other hand, OMA is carried out under the natural conditions which structures are exposed daily (i.e., seismic events or the load conditions), [2].

Furthermore, the OMA strategy involves deploying a set of sensors (e.g., accelerometers) along the structure under analysis. Consequently, the number and the proper location of the sensors are the main factors to take into account, as is discussed in [3]. In this context, in [4] and [5] are presented studies where is proposed the use of triaxial accelerometers in order to evaluate the modal parameters of buildings. In particular, the authors point out that through the vertical component it is feasible to register the phenomena of rocking. In addition, the authors propose the installation of sensors located parallel to each other with the aim to identify torsional modes.

Following with this methodology, with respect to the data captured by the sensors, it is necessary to ensure the continuous register of the information. Currently, this challenge can be faced through emerging technologies such as Internet of Things (IoT) and Wireless Sensor Networks (WSN). For example, in [6] is described an SHM architecture that incorporates IoT technologies with the purpose to determine structural damages as well as in order to generate warning messages. A similar study is described in [7], in this case, the proposal includes three functional layers, data acquisition, information management, and SHM services. Moreover, in [8], the authors propose a system based on IoT for structural health monitoring in real-time. Finally, in [9] is presented a study that incorporates both IoT and WSN technologies in order to deploy an early warning system.

Regarding WSN it is worth mentioning that technology consists in a specialized application of the communication paradigm known as Ad hoc Networks. That sort of network has the capacity to self-configuring without the need of a central station or a preexisting infrastructure, that represents a valuable feature, particularly during critical conditions where the conventional communication networks can be affected. For example, in [10] a resilient communications system based on Ad hoc Networks is proposed as alternative for emergency conditions after natural disasters. Additionally, in [11] the authors propose the application of Mobile Ad hoc Networks (MANET) on different emergency scenarios. In [12] a similar work is presented, but in this case, the study focuses on disaster scenarios caused by seismic events.

On the other hand, regarding the management of sensor devices used in SHM systems, the MQTT protocol (Message Queue Telemetry Transport) represents an efficient and current mechanism characterized by a simple design, high level of security, low bandwidth requirement, among others advantages [13, 14]. The MQTT operation is based on a publish-subscribe schema where the information is organized in a hierarchy of topics.

In such a context, in [15] is described a proposal for detecting natural disasters, particularly, the MQTT client functionality was implemented on a Raspberry Pi platform while the server operation was configured on a computer. Additionally, in [16] the authors propose a solution based on the MQTT that takes advantage of the sensors available on smartphones for deploying an early warning system focused on seismic events. Also, in [17] is described a study for monitoring the status of a building by means of an accelerometer that publishes the information captured through an MQTT topic.

In particular, seismic events are the main scenario to consider during the implementation of an SHM system. In that sense, there are various mechanisms in order to detect a seism, which differ on complexity and computational cost. For example, the most basic technique consists in analyzing if the amplitude of seismic signals exceeds a certain threshold value. Another algorithm widely used is the STA/LTA mechanism (Short Term Average to Long Term Average), which computes the moving average of the signal amplitude using a short and a long-time window. The short time window (STA) allows for detecting seismic events while the long window (LTA) provides information about the seismic noise. Consequently, when the STA/LTA relation exceeds a threshold value, this behavior is considered as the start of a seismic event [18]. In such a context, in [19] and [20] the authors propose algorithms based on STA/LTA with the purpose to identify specifically the P-wave (primary wave). It is worth indicating that there are other variants of this algorithm, for example, the mechanisms Carl STA/LTA, Delayed STA/LTA and Recursive STA/LTA, where this later variant presents higher computational efficiency and smoother response, as is discussed in [21] and [22].

Additional techniques for detecting seismic events are described in the literature. For example, in [23] is proposed a solution based on Convolutional Neural Networks. On the other hand, in [24], the authors present a mechanism using deep-learning techniques. Finally, in [25] is detailed an adaptive mechanism for detecting seismic events considering the conditions of environmental noise.

In such a context, with the aim of evaluating the structural status of buildings that can be affected by environmental conditions (i.e., earthquakes), this paper proposes a solution for deploying a remote SHM architecture that incorporates emerging technologies. Particularly, WSN, IoT as well as an automatic mechanism for detecting seismic events that is based on the MQTT protocol.

The main contributions of this work compared with related works, are the capacity for analyzing events in real-time, the extraction and transmission only of data related to the events that result in a reduction of the bandwidth demand and consequently an improvement with respect to the delay in the communication system.

Moreover, with the purpose of deploying a resilient system, the accelerograph stations communicate through a multi-hop Ad hoc wireless network configured along a real scenario. Results show that the proposed system operates successfully and allows for characterizing the dynamic response of a building to seismic events.

The rest of the paper is organized as follows. In Sect. 2, the methodology and components are detailed. The experiments and results are discussed in Sect. 3. Finally, in Sect. 4, the main conclusions of this proposal are presented.

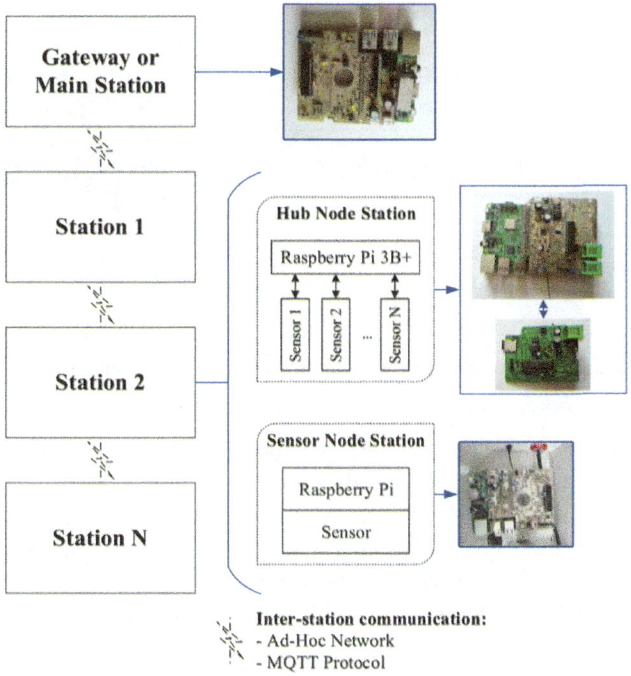

Fig. 1. Architecture of the SHM system.

2 Methodology and Components

This section presents the architecture designed for the SHM system, particularly details the functionality of the main station or gateway and the communication system among the accelerograph stations that was stablished through an Ad hoc network. Additionally, the solution based on the MQTT protocol for detecting seismic events is described. Regarding the scenario under analysis, the architecture was deployed on the headquarter building of The Electricity Company of the city of Cuenca, Ecuador.

2.1 Architecture of the SHM System

The architecture designed for the SHM system is based in a client-server model, where the server consists in a main accelerograph station while the clients correspond to a set of accelerograph stations located in different places along a structure under analysis. In Fig. 1, a general diagram of the proposed architecture is depicted. Particularly, the system allows for deploying a set of stations which can be of two types. The devices named as Sensor Node Station incorporates on the same equipment a Single Board Computer (SBC), specifically a Raspberry Pi, as well as a micro-controlled system which has three main functions: reading samples from an accelerometer sensor, generating a data frame including the sample number and timestamp and finally transmitting data to the SBC platform. The development and evaluation of this instrumentation were previously analyzed in [26].

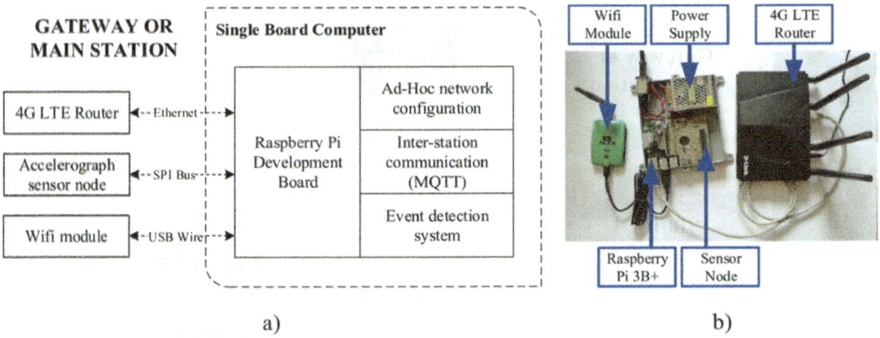

Fig. 2. Main station. a) Architecture and functionality. b) Implementation and components.

Concerning the stations named as Hub Node, the design presents a variant, in this case, the micro-controlled system and the SBC are separated and the communication was established through the RS-485 standard following and architecture master-slave. This modification allows for transmitting data from more than one sensor (located in different places) to the SBC. Regarding the Main Station, it presents a similar design to the Sensor Node Stations and includes additional capacities as is presented following.

2.2 Main Station

The main station or gateway was implemented on a Raspberry Pi 3B+, its architecture is depicted in the diagram of Fig. 2a. As can be appreciated, the station supports the Ad hoc mode for network communication, this functionality was enabled by means of an external module detailed in [27] that also allows for extending the distance of the link.

With regard to the application layer, the communication is managed by means of the MQTT protocol. It is worth highlighting that the main station not only performs management tasks but also incorporates the capacity of collecting acceleration data through a sensor node attached to the SPI interface of the Raspberry Pi. This last functionality allows for the implementation of a referential system for detecting events.

In addition, with the purpose of providing the system with Internet access, the main station includes an external Router 4G LTE [28] attached to the ethernet interface. Figure 2b presents a picture of the implementation performed, also the main components are highlighted.

On the other hand, it is important to indicate that the continuous register of information in a sensor node generates a file size of around 200 MB per day. Consequently, in order to reduce the volume of information to be transmitted, it was necessary to design an automatic system for detecting and extracting only the events (i.e., vibrations caused by earthquakes). For this purpose, the Recursive STA/LTA algorithm was implemented using the C programming language, Fig. 3, presents the flow chart for the solution developed.

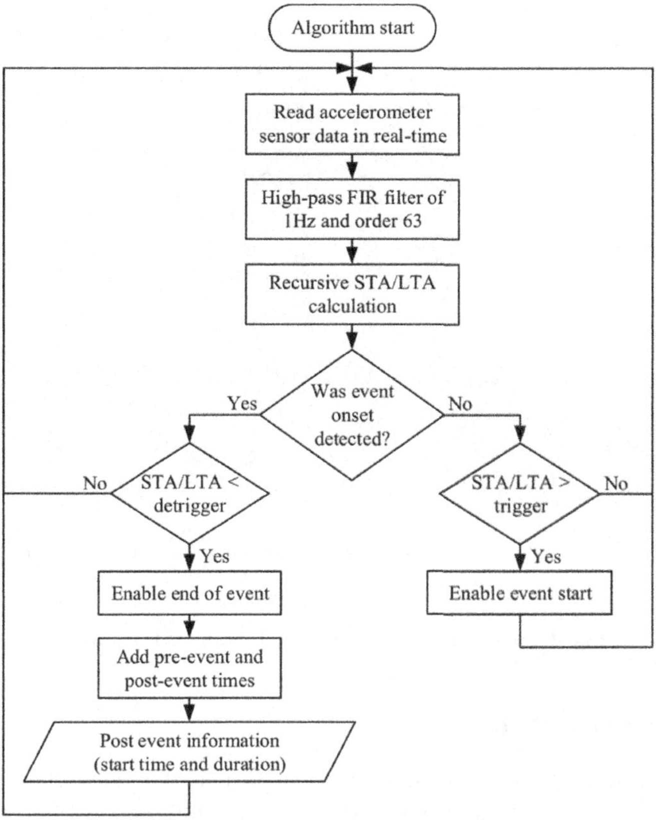

Fig. 3. Flow chart for the automatic event detection system.

The process starts by reading the data from the accelerometer sensor, following an FIR filter (Finite Impulse Response) is applied to the signal in order to eliminate the DC components (Direct Current). Then, the STA/LTA relation is computed and when this value exceeds a trigger threshold (previously defined) the system interprets this behavior as the start of a seismic event. Moreover, the end of an event is defined when the STA/LTA relation is less than the detrigger threshold level. Also, with the purpose of evaluating the seismic noise (i.e., environmental noise) an additional pre- and post-event interval are considered.

It is worth clarifying that the parameters used for processing the signal, i.e., the trigger and detrigger values, the time intervals of pre- and post-event as well as the parameters of the FIR filter, have been determined experimentally following a trial-and-error methodology.

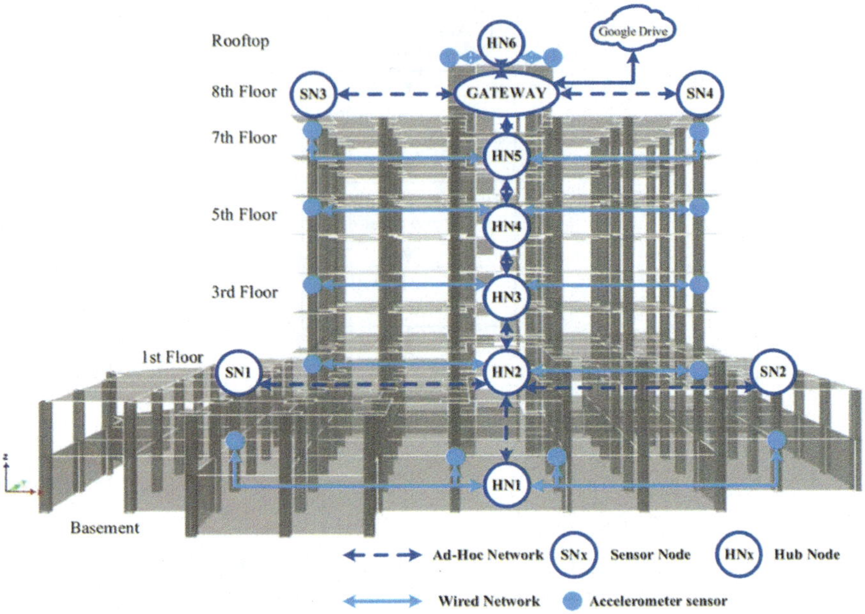

Fig. 4. Network topology designed for the SHM system.

2.3 Communication System

The communication among the main station and the rest of accelerograph stations (i.e., sensor and hub stations) was established through a multi-hop Ad hoc network. The architecture of the system is illustrated in Fig. 4. The location and the number of stations were defined following the methodology described in [4] and [5], from this analysis a set of 18 triaxial accelerometers were distributed along seven floors of the building under study. Specifically, four sensors were placed on the basement, two on opposite columns and two on the concrete walls. On the first floor four additional sensors were installed, in this case, two on indoor columns and two on external concrete slabs. Regarding the floors number 3, 5, 7, it was defined the same sensor configuration consisting in two nodes installed on indoor columns and in the case of the eighth floor two sensors were placed on the concrete slab. Finally, at the rooftop were installed two additional sensors on the core of the concrete walls.

Concerning the main station (gateway) it was placed on the eighth floor as is highlighted on the diagram of Fig. 4. In addition, an application based on the MQTT protocol was developed to manage the exchange of information among the nodes and the gateway. Specifically, it was defined a topic named as EventDetected, as is presented in the functional diagram of Fig. 5.

The application was implemented using the Python programming language, Fig. 6, presents the flow chart of the solution. As can be appreciated, in first place, both the main station and the accelerograph stations are configured. Subsequently, when a seismic event takes place and it is detected by the referential system, the main station publishes the respective notification through the MQTT topic. Following, the accelerograph stations

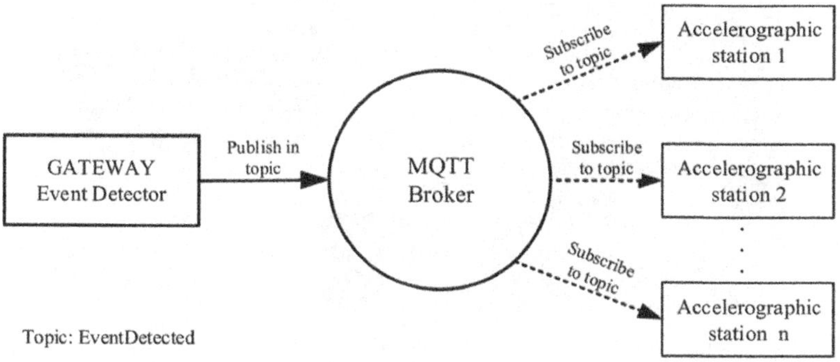

Fig. 5. Schema of the application based on the MQTT protocol for detecting events.

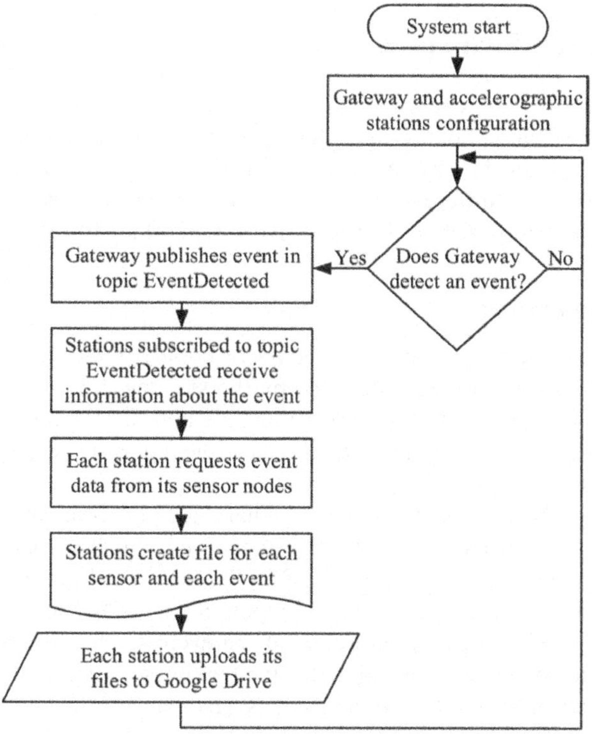

Fig. 6. Flow chart for exchanging information of events between the accelerograph stations.

subscribed to the topic receive the message and generate a request for the data captured in each one of its sensors. Finally, the stations upload the files with the information of the events to the Google Drive platform.

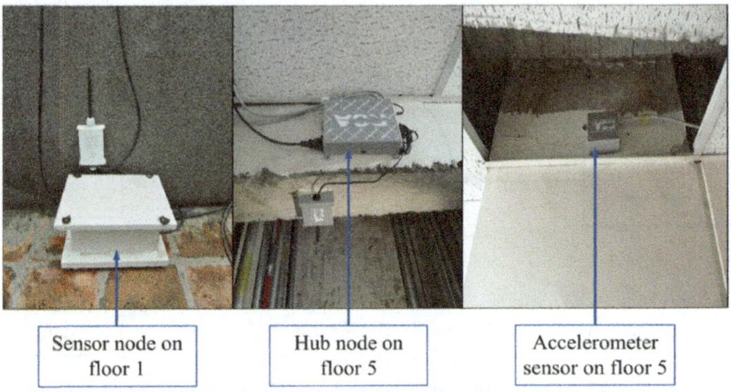

Fig. 7. System set up on the building under analysis.

3 Experimental Evaluation

This section describes the experimental evaluation of the SHM system. Particularly, the set of accelerograph stations were deployed on the headquarter building of The Electricity Company of the city of Cuenca, Ecuador. Figure 7 presents some pictures of the stations set up on different floors. In addition, results regarding a relevant event detected by the system as well as the respective structural analysis are discussed.

3.1 Seismic Monitoring

The operation of the SHM system was tested over a period of four months, during which a total of ten seismic events were detected by the stations. It is worth indicating that the occurrence of these events was verified against both the reports generated by the USGS Earthquake Hazard Program [29], as well as the reports provided by the regionals networks of seismology [30, 31].

In this context, the event detected on 21 January 2021 deserves special attention due to the magnitude and distance with the site under study (approximately 87 km of the building). The epicenter was located at 2.267° S y 79.458° W (6 km of Coronel Marcelino Maridueña, Ecuador), the event magnitude reported was of 4.9 Md and occurred at a depth of 90.2 km [32]. The accelerograms of the event for the different floors of the building are depicted in Fig. 8. Additionally, the maximum, minimum, and RMS (Root Mean Square) values for the longitudinal, transverse, and vertical axes are detailed in Table 1.

Results show that the acceleration values registered on the basement present the lowest amplitude, whereas the highest values of acceleration were detected on the eighth floor. For instance, regarding the longitudinal axis, the acceleration values on the basement range from -2.35 cm/s^2 to 1.93 cm/s^2 and in the case of the eighth floor range from -10.77 cm/s^2 to 9.28 cm/s^2. Similarly, the RMS values on the basement are 0.26, 0.22, and 0.22 cm/s^2 for the longitudinal, transverse, and vertical axes, respectively. Regarding an intermediate area of the building, for example, on the third floor the RMS values are 0.66, 0.72, and 0.41 cm/s^2 and finally, concerning eighth floor the RMS values are 1.18, 1.33, and 0.59 cm/s^2. Consequently, it can be noted that the acceleration values increase on the upper floors of the building. These results are consistent with the method for computing the acceleration detailed in [33], particularly, the acceleration value increases linearly as a function of the height of a building.

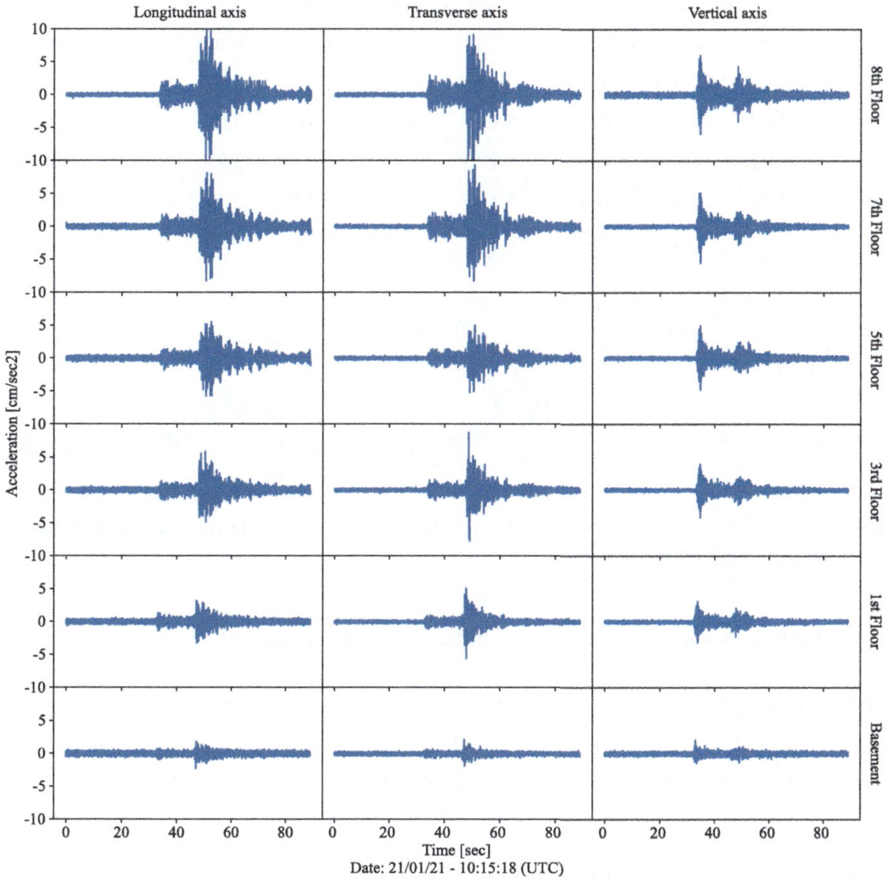

Fig. 8. Accelerograms for the different floors of the building under study.

Furthermore, results show that in all the cases for the different floor of the building the RMS values of the vertical axis are lower than the acceleration values registered for the other two axes, i.e., a seismic event has the greatest effect on the magnitude of the movements along the longitudinal and transverse axes of the building.

3.2 Structural Health Assessment

In order to perform an initial assessment regarding the dynamic response of the building, the technique of Operational Modal Analysis was applied, particularly, the Frequency Domain Decomposition (FDD) method which is a mechanism widely used for Structural

Table 1. Acceleration values registered in the building during the seismic event.

Floor	Longitudinal axis			Transverse axis			Vertical axis		
	Max (cm/s^2)	Min (cm/s^2)	RMS (cm/s^2)	Max (cm/s^2)	Min (cm/s^2)	RMS (cm/s^2)	Max (cm/s^2)	Min (cm/s^2)	RMS (cm/s^2)
8	9.28	−10.77	1.18	10.53	−10.52	1.33	6.08	−5.98	0.59
7	8.20	−8.32	1.14	9.44	−8.30	1.04	5.22	−5.61	0.50
5	5.62	−5.78	0.88	5.11	−5.26	0.70	5.04	−4.84	0.48
3	5.92	−4.91	0.66	8.77	−7.78	0.72	3.99	−4.25	0.41
1	3.28	−3.29	0.40	5.16	−5.67	0.49	3.14	−3.25	0.34
Basement	1.93	−2.35	0.26	2.21	−1.98	0.22	2.14	−1.48	0.22

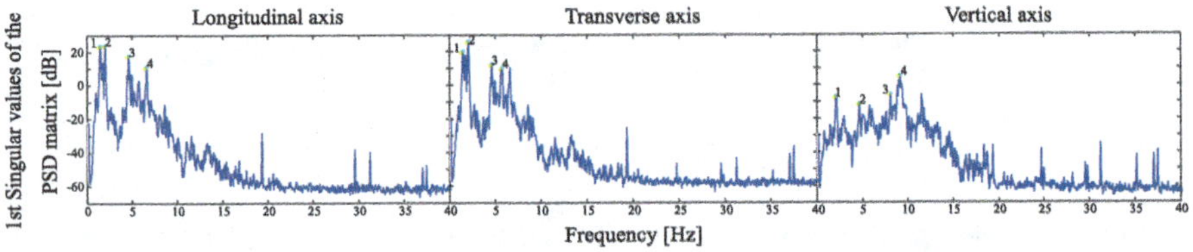

Fig. 9. Peak values selected from the Power Spectral Density Matrix for the longitudinal, transverse, and vertical axis.

Table 2. Values of frequencies identified for each axis using the FDD method.

Axis	Longitudinal				Transverse				Vertical			
Mode	1	2	3	4	1	2	3	4	1	2	3	4
f (Hz)	1.31	1.92	4.55	6.50	1.31	1.92	4.55	5.65	1.93	4.52	7.99	9.06

Health Monitoring as is highlighted in several studies in the literature, for example in [34] and [35].

In this context, the acceleration values of the longitudinal, transverse and vertical axes were analyzed based on the information registered by the sensors during the seismic event occurred on 21 January. Figure 9 presents the behavior of the Power Spectral Density (PSD), where the first four modes of vibration with the highest values of amplitude were selected for the structural assessment.

Regarding the FDD method, it is worth clarifying that the selection of the peak values is performed by inspection, these values are presented in Table 2. As can be seen, for the longitudinal and transverse axes, the first three frequencies (i.e., 1.31, 1.92, and 4.55 Hz) are the same. These results indicate that such modes of vibrations are not separate modes but composite modes and consequently are present in both axes. Additionally, the vibration modes for the longitudinal axis are depicted in Fig. 10, in this case by means of a simplified model of the building, where it can be appreciated the dynamic behavior of the structure for the four modes obtained. As result of this first assessment, it can be indicated that the composite modes reveal an unsuitable response of the building during a seismic event. Particularly, a seismic-resistant building must be characterized by a dynamic response which not includes composite modes as is detailed in the ACI recommendations [36].

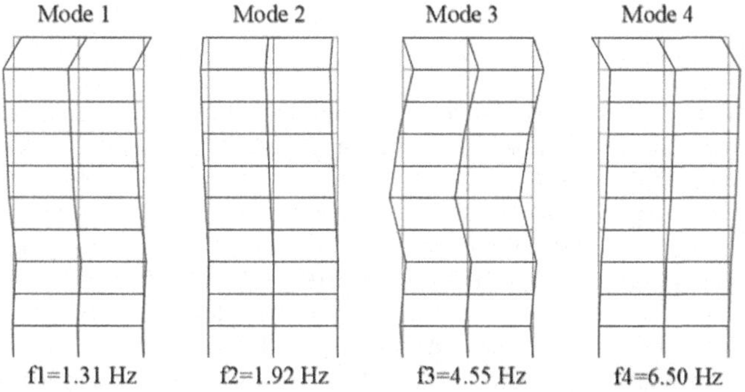

Fig. 10. Vibration modes obtained for the longitudinal axis through the FDD method.

4 Conclusions

In this paper a system for Structural Health Monitoring has been presented, the architecture incorporates an automatic mechanism for detecting seismic events based on the MQTT protocol as well as IoT devices. The main advantages of this solution are scalability, easy adaptation to existing buildings, low bandwidth requirements, and the capacity for monitoring seismic events remotely in real-time.

Regarding the experimental evaluation, it was carried out on an actual building (headquarter building of The Electricity Company of the city of Cuenca, Ecuador), the information captured by the sensors was registered along four months, during which a total of ten seismic events were detected by the stations and the most relevant event was selected for evaluating the dynamic response of the building. Particularly, the maximum, minimum, and RMS values of acceleration were determined for the different floor of the scenario under study. Results show that the magnitude of the acceleration values increases as a function of the building height. Moreover, it was detected that the magnitude of the movements along the longitudinal and transverse axes are higher than the results obtained for the vertical axis.

On the other hand, an initial assessment regarding the structural health status of the building was performed through the OMA methodology. Specifically, the FDD mechanism was used in order to determine the first four frequencies and its respective vibration modes. This first analysis reveals the existence of composite vibration modes where are present in both the longitudinal and the transverse axes which represent a not recommended response of a building during a seismic event. Finally, as future work, evaluations including additional vibration modes will be performed.

Acknowledgment. This work is part of the research project "IoT Technologies and Wireless Sensor Networks for Structural Health Monitoring of Essential Facilities of the City of Cuenca". The authors gratefully acknowledge to the Research Management Department of the University of Cuenca-Ecuador (DIUC), the RSA Department (Red Sísmica del Austro), and The Electricity Company of the city of Cuenca-Ecuador (EERCS) for the support and resources provided during the development of this research work.

References

1. Yun, D.Y., et al.: Field measurements for identification of modal parameters for high-rise buildings under construction or in use. Autom. Constr. **121**, 103446 (2021). https://doi.org/10.1016/j.autcon.2020.103446
2. Pachón, P., Castro, R., García-Macías, E., Compan, V., Puertas, E.: E. Torroja's bridge: tailored experimental setup for SHM of a historical bridge with a reduced number of sensors. Eng. Struct. **162**, 11–21 (2018). https://doi.org/10.1016/j.engstruct.2018.02.035
3. Pachón, P., et al.: Evaluation of optimal sensor placement algorithms for the Structural Health Monitoring of architectural heritage. Application to the Monastery of San Jerónimo de Buenavista (Seville, Spain). Eng. Struct. **202**, 109843 (2020). https://doi.org/10.1016/j.engstruct.2019.109843
4. Boscato, G., Dal Cin, A., Ientile, S., Russo, S.: Optimized procedures and strategies for the dynamic monitoring of historical structures. J. Civ. Struct. Health Monit. **6**(2), 265–289 (2016). https://doi.org/10.1007/s13349-016-0164-9
5. Zhou, Y., Zhou, Y., Yi, W., Chen, T., Tan, D., Mi, S.: Operational modal analysis and rational finite-element model selection for ten high-rise buildings based on on-site ambient vibration measurements. J. Perform. Constr. Facil. **31**(5), 04017043 (2017). https://doi.org/10.1061/(asce)cf.1943-5509.0001019
6. Lamonaca, F., Scuro, C., Sciammarella, P.F., Olivito, R.S., Grimaldi, D., Carnì, D.L.: A layered IoT-based architecture for a distributed structural health monitoring system. Acta IMEKO **8**(2), 45–52 (2019). https://doi.org/10.21014/acta_imeko.v8i2.640
7. Wang, J., Fu, Y., Yang, X.: An integrated system for building structural health monitoring and early warning based on an Internet of things approach. Int. J. Distrib. Sens. Netw. **13**(1) (2017). https://doi.org/10.1177/1550147716689101
8. Abdelgawad, A., Yelamarthi, K.: Structural health monitoring: internet of things application. In: Midwest Symposium on Circuits System, vol. 0, pp. 16–19 (2016). https://doi.org/10.1109/MWSCAS.2016.7870118
9. Alphonsa, A., Ravi, G.: Earthquake early warning system by IOT using Wireless sensor networks. In: Proceedings of the 2016 IEEE International Conference Wireless Communication Signal Processing Networking, WiSPNET 2016, pp. 1201–1205 (2016). https://doi.org/10.1109/WiSPNET.2016.7566327
10. Jagannath, J., Furman, S., Jagannath, A., Ling, L., Burger, A., Drozd, A.: HELPER: Heterogeneous Efficient Low Power Radio for enabling ad hoc emergency public safety networks. Ad Hoc Netw. **89**, 218–235 (2019). https://doi.org/10.1016/j.adhoc.2019.03.010
11. Sikora, A., Niewiadomska-Szynkiewicz, E., Krzysztoń, M.: Simulation of mobile wireless ad hoc networks for emergency situation awareness. In: Proceedings of the 2015 Federated Conference on Computer Science and Information Systems, FedCSIS 2015, vol. 5, pp. 1087–1095 (2015). https://doi.org/10.15439/2015F52
12. Martinez, Z.O.N., Arias, O.M., Lopez, P.A., Ugarte, S.A.: Hybrid wireless ad hoc network design based on WIFI technology for facing seismic catastrophes. In: Canadian Conference on Electrical and Computer Engineering, vol. 2016 (2016). https://doi.org/10.1109/CCECE.2016.7726638

13. MQTT.org: MQTT: The Standard for IoT Messaging (2021). https://mqtt.org/. Accessed 15 Mar 2021
14. Yang, J., Wu, P., Chen, Y., Wei, W.: Development and testing of the earthquake early warning information push platform based on MQTT protocol. In: Science Conference Mechatronics Engineering and Computing Science, SCMC 2019, no. Scmc, pp. 399–405 (2019). https://doi.org/10.25236/scmc.2019.082
15. Greco, L., Ritrovato, P., Tiropanis, T., Xhafa, F.: IoT and semantic web technologies for event detection in natural disasters. Concurr. Comput. **30**(21), 1–9 (2018). https://doi.org/10.1002/cpe.4789
16. Zambrano, A.M., Perez, I., Palau, C., Esteve, M.: Technologies of internet of things applied to an earthquake early warning system. Futur. Gener. Comput. Syst. **75**, 206–215 (2017). https://doi.org/10.1016/j.future.2016.10.009
17. Pierleoni, P., et al.: IoT solution based on MQTT protocol for real-time building monitoring. In: 2019 IEEE 23rd International Symposium on Consumer Technologies, ISCT 2019, pp. 57–62 (2019). https://doi.org/10.1109/ISCE.2019.8901024
18. Trnkoczy, A.: Understanding and parameter setting of STA/LTA trigger algorithm. New Man. Seismol. Obs. Pract. **2**, 1–20 (2012). https://doi.org/10.2312/GFZ.NMSOP-2
19. Li, X., Shang, X., Wang, Z., Dong, L., Weng, L.: Identifying P-phase arrivals with noise: an improved Kurtosis method based on DWT and STA/LTA. J. Appl. Geophys. **133**, 50–61 (2016). https://doi.org/10.1016/j.jappgeo.2016.07.022
20. Zhang, J., Tang, Y., Li, H.: STA/LTA fractal dimension algorithm of detecting the P-wave arrival. Bull. Seismol. Soc. Am. **108**(1), 230–237 (2018). https://doi.org/10.1785/0120170099
21. Sya'bani, Y.A., Novianty, A., Prasasti, A.L.: Implementation of automatic first arrival picking on P-wave seismic signal using logistic regression method. In: 2020 8th International Conference on Information and Communication Technology, ICoICT 2020, pp. 1–5 (2020). https://doi.org/10.1109/ICoICT49345.2020.9166345
22. Indi, M.W.P., Novianty, A., Prasasti, A.L.: Automatic first arrival picking on P-wave seismic signal using support vector machine method. In: 2020 8th International Conference on Information and Communication Technology, ICoICT 2020, pp. 1–6 (2020). https://doi.org/10.1109/ICoICT49345.2020.9166267
23. Dokht, R.M.H., Kao, H., Visser, R., Smith, B.: Seismic event and phase detection using time-frequency representation and convolutional neural networks. Seismol. Res. Lett. **90**(2A), 481–490 (2019). https://doi.org/10.1785/0220180308
24. Wu, Y., Lin, Y., Zhou, Z., Bolton, D.C., Liu, J., Johnson, P.: DeepDetect: a cascaded region-based densely connected network for seismic event detection. IEEE Trans. Geosci. Remote Sens. **57**(1), 62–75 (2019). https://doi.org/10.1109/TGRS.2018.2852302
25. Anchal, S., Mukhopadhyay, B., Kar, S.: UREDT: unsupervised learning based real-time footfall event detection technique in seismic signal. IEEE Sensors Lett. **2**(1), 1–4 (2017). https://doi.org/10.1109/lsens.2017.2787611
26. Muñoz, M., Guevara, R., González, S., Jiménez, J.C.: Reliable data acquisition system for a low-cost accelerograph applied to structural health monitoring. J. Appl. Sci. Eng. Technol. Educ. **3**(2), 181–194 (2020). https://doi.org/10.35877/454ri.asci159
27. A. N. Inc.: ALFA AWUS036NH (2021). https://www.alfa.com.tw/products/awus036nh?variant=36481029374024. Accessed 15 Mar 2021
28. D.-L. Corporation: DWR-M921 D-Link (2021). https://la.dlink.com/la/4g-lte/dwr-m921/. Accessed 15 Mar 2021
29. U. E. H. Program: Search Earthquake Catalog (2021). https://earthquake.usgs.gov/earthquakes/search/. Accessed 15 Mar 2021
30. I. G. EPN: Informes de los Últimos Sismos (2021). https://www.igepn.edu.ec/ultimos-sismos. Accessed 15 Mar 2021

31. R. S. U. Cuenca: Red Sísmica U Cuenca | LinkedIn (2021). https://www.linkedin.com/in/red-sísmica-u-cuenca-a223821a6/detail/recent-activity/. Accessed 15 Mar 2021
32. U. E. H. Program: M 4.9 - 6 km SSW of Coronel Marcelino Maridueña, Ecuador (2021). https://earthquake.usgs.gov/earthquakes/eventpage/us7000d1zv/executive. Accessed 15 Mar 2021
33. American Society of Civil Engineers: Minimum Design Loads and Associated Criteria for Buildings and Other Structures, no. 7–98 (2016)
34. Lacanna, G., Ripepe, M., Marchetti, E., Coli, M., Garzonio, C.A.: Dynamic response of the Baptistery of San Giovanni in Florence, Italy, based on ambient vibration test. J. Cult. Herit. **20**, 632–640 (2016). https://doi.org/10.1016/j.culher.2016.02.007
35. Neu, E., Janser, F., Khatibi, A.A., Braun, C., Orifici, A.C.: Operational Modal Analysis of a wing excited by transonic flow. Aerosp. Sci. Technol. **49**, 73–79 (2016). https://doi.org/10.1016/j.ast.2015.11.032
36. ACI Committee 318: Building Code Requirements for Structural Concrete (ACI 318-14) (2014)

Elliptic Curves Cryptography for Lightweight Devices in IoT Systems

Ana Simon Francia[1], Javier Solis-Lastra[1], and Erik Alex Papa Quiroz[1,2(✉)]

[1] Universidad Nacional Mayor de San Marcos, Lima, Peru
{ana.simon1,jsolis1,epapaq}@unmsm.edu.pe
[2] Universidad Privada del Norte, Lima, Peru
erick.papa@upn.edu.pe

Abstract. To protect data confidentiality, cryptographic methods are widely used, however, traditional methods require a higher level of computational resources to achieve a higher level of security. For this reason, ECC (Elliptic Curves Cryptography) is considered as a promising solution. This paper shows a systematic review about ECC methods for IoT systems. Our research questions were related to find out the principal applications of ECC, criteria for implementing ECC and the appropriate elliptic curves for lightweight devices. The papers were collected from Scopus, Science Direct and SpringerLink and after doing the study, we conclude that IoT is the principal application of ECC, the processing time is the most common implementation criteria and the Koblitz elliptic curve over prime fields are accurate for IoT.

Keywords: Cryptography · IoT · Elliptic curve cryptography

1 Introduction

In the recent era, rapid development of IoT technology enables to connect and manage heterogeneous devices [1], bringing new opportunities for industrial market [2], biomedical applications [3], smart cities [4], transportation [5], etc. As the number of connected devices experiences an exponential growth, new cyber-attacks are IoT enabled. An IoT vulnerability can be exploited as a first step to compromise a critical system if security issues are not properly considered in the design [6].

User safety could be compromised just by accessing to a simple lightweight device on an IoT network. If the device directly transmits or receives sensitive information, the use of inefficient cryptographic algorithms could cause data leakage [7]. In the scenarios where lightweight devices work with non-relevant data for user security, vulnerabilities still exist since a bad management of keys could cause the device to be accessed remotely by an attacker [8], opening the possibility of privilege escalation in the network until reaching a critical device. Furthermore, it is also possible to use the peripherals devices to compromise the privacy of users, whether in a domestic, industrial, or medical environment.

M. Botto-Tobar et al. (Eds.): CIT 2021, LNNS 405, pp. 71–82, 2022.
https://doi.org/10.1007/978-3-030-96043-8_6

The tendency of having smaller devices in IoT leads to more limited battery capacity. Therefore, reducing the amount of energy required is extremely important to ensure stable service and reduce operating costs [9]. If a device is required to perform tasks with a higher degree of intelligence, the complexity of the operations performed also increases, this implies more memory resources consumed [10].

Due to the limited capacity and resources of IoT devices, traditional cryptography may not be a viable option [11]. Traditional cryptography was designed for devices with a constant power source, increased memory resources, and processing power. In contrast, lightweight devices lack these features, that is why any process required in IoT devices must consume the least number of resources, while maintaining the expected level of efficiency.

As described above, it is necessary to determine what cryptography authentication methods based in ECC are suitable for IoT devices to maintain equilibrium between security and computational resources. To provide this information, which is the objective of this paper, we present a systematic review on the use of ECC for lightweight devices in IoT. We use the methodology of Kitchenham and Charters [12], to response the following three questions: What are the principal applications of ECC? What are the principal criteria for implementing ECC? and What elliptic curves are applicable to lightweight devices? After performing the analysis of the research, we obtain that IoT is the principal application of ECC, the execution time and computational cost are the most common implementation criteria and the Koblitz elliptic curve over prime fields are accurate for IoT respectively.

The results of this paper are important to know the actual state of the art of the applications of ECC in IoT and the elliptic curve algorithms used in IoT. Observe that this paper is one of the first works on the state of the art in this line of research.

The organization of the paper is the following: in Sect. 2 we present the research methodology, in Sect. 3 we analyze the obtained information, in Sect. 4 we analyze the results and finally in Sect. 5 we present the conclusions.

2 Research Methodology

This systematic review considers the methodology proposed by Kitchenham and Charters [12], which involves three relevant phases: planning the review, conduct the review, and report the review. Some recent works related on the above methodology has been presented by Cano and Papa Quiroz [13], Vargas and Mauricio [14] and Wong et al. [15]. We describe briefly these three phases:

Planning the Review: In this phase the objective of the review is identified, the research questions and the review protocol are defined. The answers of research questions permit achieve the objective of the study.

Conducting the Review: In this phase, the planning review is executed, and the main studies are selected, according to the inclusion and exclusion criteria. Furthermore, data extraction, and data synthesis are carrying out.

Reporting the Review: In this phase, the statistics, analysis of the selected studies, and discussion over the research question are presented.

2.1 Planning the Review

As the objective of this paper is to determine which cryptography authentication methods based in ECC are suitable for IoT devices, we define the following research questions:

Q1: Which are the principal applications of ECC?
Q2: Which are the principal criteria for implementing ECC?
Q3: Which elliptic curves are applicable to lightweight devices?

The considered databases for this research were ScienceDirect, Scopus and Springer Link. The stream search was applied on the tittle, abstract, and keywords following: (TITLE-ABS-KEY ("Elliptic curve application")) OR (TITLE-ABS-KEY ("Elliptic curve criteria")) OR (TITLE-ABS-KEY("Elliptic curve in constrained devices")) OR (TITLE-ABS-KEY("Elliptic curve in lightweight devices")).

2.2 Conducting the Review

We show inclusion and exclusion criteria considered in Table 1 and Table 2 respectively. In inclusion criteria, we selected papers focused on ECC and published in indexed journals. Because the main interest of this paper is IoT, it was considered as our important factor.

Table 1. Inclusion criteria

Criterion	Justification
Research focus	Papers with keyword "Elliptic curve"
Quantitative information	Articles must be published in indexed journals
Importance factor	ECC applied in IoT
Language	English

In exclusion criteria, we discarded papers published before 2015, the ones that focus on a topic different from ECC and the ones that do not show a logical and comprehensible development.

Table 2. Exclusion criteria

Criterion	Justification
Type of publication	Papers published before 2015
Analysis unit	Papers that do not consider ECC as the interest focus
Importance factor	Researches that does not show a clear development

Figure 1 shows the process followed in the research, applying the inclusion and exclusion criteria to the eligible papers found in the databases indicated.

2.3 Reporting the Review

59 554 research articles related to the classification of ECC were found in the three considered databases. Table 3 depicts the number of articles found in the three selected databases.

Table 3. Selected papers

Source	Potentially eligible studies	Eligible studies
Scopus	3 435	1 584
SpringerLink	55 739	15 268
Science Direct	380	162
TOTAL	59,554	17,014

3 Study Analysis

From 16 selected investigations, three main questions (Q1, Q2 and Q3) are answered following the frame shown in Fig. 1.

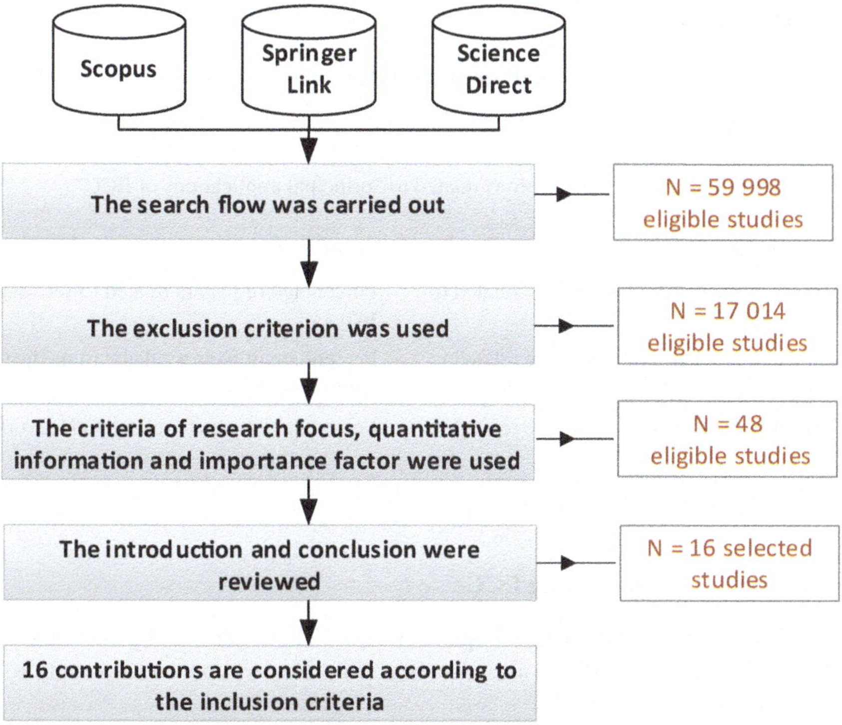

Fig. 1. Inclusion and exclusion criteria flow

As shown in Fig. 1, we collected all the papers related to ECC following the stream search described in Subsect. 2.1. From the 59 554 eligible studies, we applied exclusion criterion shown in Table 2, filtering papers by date, focus and clarity. After this step, there were 17014 eligible studies, were we applied inclusion criterion described in Table 1, giving as a result 48 eligible studies. Finally, we reviewed every introduction and conclusion, resulting in 16 eligible studies that were the most relevant papers for our research.

The following Fig. 2, shows the selected topics that we have chosen to ask the research questions Q1–Q3.

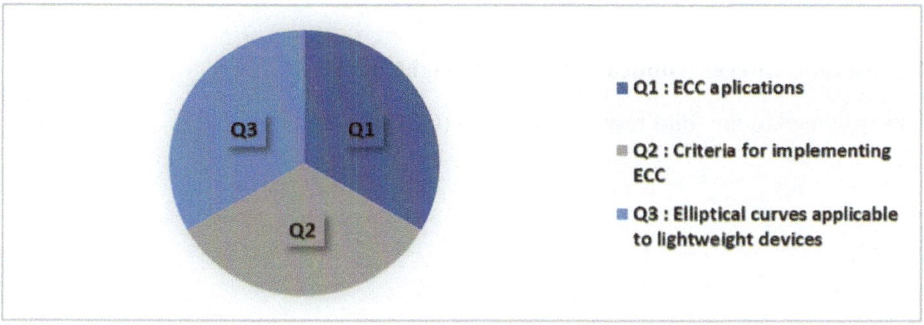

Fig. 2. Proposed framework for the revision of literature

3.1 ECC Applications

This is related to the first research question (Q1), whose sub-questions are shown in Table 4.

Table 4. Research questions related to "principal applications of ECC"

Research questions	Contributions
Q1.1 What percentages of works are related to security in IoT?	Percentage of papers related to security in IoT
Q1.2 What percentages of works are related to medical environment?	Percentage of papers related to medical environment
Q1.3 What percentages of works are related to Networking	Percentage of papers related to Networking

3.2 Criteria for Implementing ECC

This is related to the second research question (Q2). The criteria that present the greatest recurrence in investigations are execution time (encryption and decryption) and computational cost (Table 5).

Table 5. Research questions related to "criteria for implementing ECC"

Research questions	Contributions
Q2.1 Which are the principal criteria for implementing ECC in IoT security?	Execution time, computational cost and memory
Q2.2 Which are the principal criteria for implementing ECC medical environment?	Execution time and computational cost
Q2.3 Which are the principal criteria for implementing ECC in Networking?	Execution time, computational cost, communication cost and memory

3.3 Elliptic Curves Applicable to Lightweight Devices

This is related to the third research question (Q3) (Table 6).

Table 6. Research questions related to Elliptic curves applicable to lightweight devices

Research questions	Contributions
Q3.1 What elliptic curves are used for implementing ECC in IoT security?	Elliptic curves used for implementing ECC in IoT security
Q3.2 What elliptic curves are used for implementing ECC in medical environment?	Elliptic curves used for implementing ECC in medical environment
Q3.3 What elliptic curves are used for implementing ECC in Networking?	Elliptic curves used for implementing ECC in Networking

4 Analysis of Results

4.1 ECC Applications (Q1)

Figure 3 depicts the percentages of the 16 selected studies. We may observe that the principal applications of ECC are in IoT (48%).

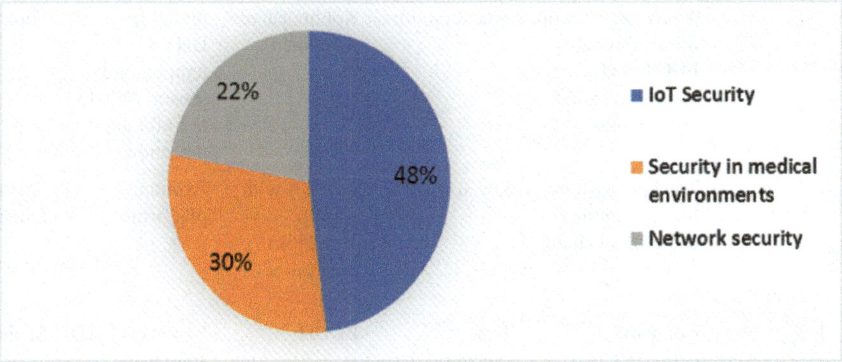

Fig. 3. Research areas

4.2 Criteria for Implementing ECC (Q2)

The most used criteria are execution time and computational cost.

4.3 Elliptic Curves Applicable to Lightweight Devices (Q3)

One of the accurate curves for lightweight devices in IoT are the Koblitz curves, as shown in [16] where a ECDSA scheme over prime field is designed to develop an authentication method with efficient energy consume. In medical environments, where data privacy is essential, ECC can be applied to RFID tags and central server authentication as shown in [3]. Also, in Networking, [17] shows how a mobile payment system can be deployed by utilizing ECC with Koblitz curves over binary field by employing a central router to encrypt and decrypted information can only be accessed by the constrained device with the authorized key. Table 7 summarizes some relevant articles reviewed for this analysis.

Table 7. Overview of analyzed articles.

Ref.	Area	Contribution	Criteria	ECC algorithm	Tools	Database
[18]	IoT	Smart parking ECC application which runs on different host operating systems	Execution time, computational cost, communication cost, and memory	Elliptic Curve Diffie - Hellman with 160-bit key size	Wiselib algorithm library, C++, ATMEGA1281, 500 MHz linux-based system, microcontroller Jennic JN5139, microcontroller MSP430	Science Direct
[19]	IoT	Security scheme for IoT	Execution time, computational cost, and memory	Elliptic Curve Diffie Hellman with curve m-221	Wireless sensor Tmote Sky (8 MHz)	Science Direct
[20]	IoT	Authentication of device-server with novel signature ECC based algorithm called cBLAKE2b	Key size and execution time	ECDSA with Koblitz curves over binary field	Raspberry Pi 3 model B, DHT-11 temperature sensor, MQ-135 harmful gas detection	Science Direct
[5]	IoT	Tracking system for car actions using Blockchain system	Key size	ECDSA with Koblitz curve secp256k1 over prime field	Exonum platform	Science Direct
[4]	IoT	Comparison between DH, RSA, and ECDH with different key sizes	Time execution, computational cost and memory	ECDH	Synopsys CAD suite using 90 nm UMC Faraday library	Scopus
[21]	IoT	Smart home security centralized system on Wi-Fi network	Key size	ECDH	STM32F4 ARM Cortex-M4F microprocessor, Raspberry Pi Linux and AllJoyn framework	Scopus

(*continued*)

Table 7. (*continued*)

Ref.	Area	Contribution	Criteria	ECC algorithm	Tools	Database
[22]	IoT	Lightweight no-pairing attribute-based encryption based on Elliptic Curve Decisional Diffie-Hellman (ECDDH) assumption	Time execution, computational cost, communication cost and memory	ECDDH with security of 160 bits	Selective-Set security model	Scopus
[23]	IoT	Design and implementation of ECC curves at different security levels for AVR-based an MSP4300-based sensor node	Execution time, key size, communication cost and energy cost	ECDH with P159 curve	ATmega128 8-bit processor, MSP430 processor with 16-bit multiplier, and mathematical computational algebra system Magma	Scopus
[16]	IoT	Security improvement of Message Queue Telemetry Transport (MQTT) and MQTT-SN (Sensor Network) protocol using ECC	Key size and execution time	ECC based on Decisional Bilinear Diffie-Hellman (DBDH) problem with key sizes of 256 and 512 bits	Java Platform, Eclipse Paho client and open source Mosquitto broker	Scopus
[24]	Medical environment	Secure multi-server authentication protocol using biometric-based smart card and ECC	Execution time, computational cost and memory	ECDH over prime field with private key of 2028 bit	AVISPA (Automated Validation of Internet Security Protocols and Applications) tool using HLPSL (High Level Protocol Specification Language) and OFMC backend	Scopus
[17]	Medical environment	Secure sensor association and key management protocol based on ECC and hash chains	Execution time, computational cost, and memory	ECDH over prime field	8 MHz TI-MSP430 microcontroller, 10 KB RAM and 48 KB ROM	Scopus

(*continued*)

Table 7. *(continued)*

Ref.	Area	Contribution	Criteria	ECC algorithm	Tools	Database
[3]	Medical environment	Method based on ECC for authentication between RFID cards, card readers and servers for medical information systems	Execution time and computational cost	Elliptic curves over prime field with 160 bits key length	Server, RFID card and RFID card reader	Springer
[25]	Medical environment	Propose framework based on hyperelliptic curves based public key cryptosystem to address Health Care Systems security flaws	Execution time	Hyper Elliptic Curve Digital Signature Algorithm (HcDSA) over prime field with security of 10024 bit	Automated Validation of Internet Security Protocol and Applications (AVISPA) tool	Springer
[26]	Networking	Security scheme for a mobile payment system	Key size and execution time	ECCC with Koblitz curves over binary field	The International Mobile Equipment Identity (IMEI) number	Springer
[27]	Networking	Mutual authentication using Conditional Privacy-Preserving Authentication (CPPA) scheme for Vehicular Ad-Hoc Network (VANET)	Execution time and computational cost	ECDH over prime field with 80 bits security	Cryptographic library MIRACL and Intel I7-4770 processor with 3.40 GHz clock frequency, 4 gigabytes memory with Windows 7 operating system	Scopus
[11]	Networking	Mechanism for client authentication and re-authentication to avoid WPA2 PSK attacks	Execution time	ECDH over prime field	Access point and client station	Springer

5 Conclusions

This research presents a systematic review on the use of ECC for lightweight devices in IoT using the methodology of Kitchenham and Charters [12]. We found that between IoT, medical environment and Networking, the principal applications of elliptic curves to lightweight devices are in IoT with a 48%. The most used criteria for implementing ECC are execution time and computational cost, and the elliptic curves applicable to lightweight devices are the Koblitz curves over binary field.

The implications of this study will motivate further research regarding different elliptic curves, for example the family of elliptic curves called SafeCurves, [28], now applied to IoT.

References

1. Lara-Nino, C.A., Diaz-Perez, A., Morales-Sandoval, M.: Elliptic curve lightweight cryptography: a survey. IEEE Access **6**, 72514–72550 (2018)
2. Li, R., Song, T., Capurso, N., Yu, J., Couture, J., Cheng, X.: IoT applications on secure smart shopping system. IEEE Internet Things J. **4**(6), 1945–1954 (2017)
3. Noori, D., Shakeri, H., Niazi Torshiz, M.: Scalable, efficient, and secure RFID with elliptic curve cryptosystem for Internet of Things in healthcare environment. Eurasip J. Inf. Secur. **2020**(13), 1–11 (2020)
4. Goyal, T.K., Sahula, V.: Lightweight security algorithm for low power IoT devices. In: 2016 International Conference in Advance Computing, Communications and Informatics, ICACCI 2016, pp. 1725–1729 (2016)
5. Narbayeva, S., Bakibayev, T., Abeshev, K., Makarova, I., Shubenkova, K., Paskevich, A.: Blockchain technology on the way of autonomous vehicles development. Transp. Res. Proc. **44**, 168–175 (2020)
6. Stellios, I., Kotzanikolaou, P., Psarakis, M., Alcaraz, C., Lopez, J.: A survey of IoT-enabled cyberattacks: Assessing attack paths to critical infrastructures and services. IEEE Commun. Surv. Tutor. **20**(4), 3453–3495 (2018)
7. Zou, Y., Zhu, J., Wang, X., Hanzo, L.: A survey on wireless security: technical challenges, recent advances, and future trends. Proc. IEEE **104**(9), 1727–1765 (2016)
8. Tewari, A., Gupta, B.B.: A lightweight mutual authentication protocol based on elliptic curve cryptography for IoT devices. Int. J. Adv. Intell. Paradigm **9**(2–3), 111–121 (2017)
9. Park, J., Kwon, H., Kang, N.: IoT–Cloud collaboration to establish a secure connection for lightweight devices. Wireless Netw. **23**(3), 681–692 (2016). https://doi.org/10.1007/s11276-015-1182-y
10. Mahto, D., Khan, D.A., Yadav, D.K.: Security analysis of elliptic curve cryptography and RSA. In: Proceeding of the Word Congress on Engineering, WCE 2016, London, U.K, vol. 1 (2016)
11. Singh, P.K., Vij, P., Vyas, A., Nandi, S.K., Nandi, S.: Elliptic curve cryptography based mechanism for secure Wi-Fi connectivity. In: Fahrnberger, G., Gopinathan, S., Parida, L. (eds.) ICDCIT 2019. LNCS, vol. 11319, pp. 422–439. Springer, Cham (2019). https://doi.org/10.1007/978-3-030-05366-6_35
12. Kitchenham, B.A., Charters S.: Guidelines for performing systematic literature reviews in software engineering. EBSE 2007-001. Keele University and Durham University Joint report (2007)

13. Cano Lengua, M.A., Papa Quiroz, E.A.: A systematic literature review on support vector machines applied to classification. In: Proceedings of the 2020 IEEE Engineering International Research Conference, EIRCON 2020 (2020)
14. Vargas, P.S.C., Mauricio, D.A.: Review of literature about models and factors of productivity in the software factory. Int. J. Inf. Technol. Syst. Approach **11**(1), 48–71 (2018)
15. Wong, L.R., Mauricio, D., Rodriguez, G.D.: A systematic literature review about software requirements elicitation. J. Eng. Sci. Technol. **12**(2), 296–317 (2017)
16. Singh, M., Rajan, M.A., Shivraj, V.L., Balamuralidhar, P.: Secure MQTT for Internet of Things (IoT). In: 2015 Fifth International Conference on Communication Systems and Network Technologies, pp. 746–751 (2015)
17. Shen, J., Tan, H., Moh, S., Chung, I., Liu, Q., Sun, X.: Enhanced secure sensor association and key management in wireless body area networks. J. Commun. Netw. **17**(5), 453–462 (2015)
18. Chatzigiannakis, I., Vitaletti, A., Pyrgelis, A.: A privacy-preserving smart parking system using an IoT elliptic curve based security platform. Comput. Commun. **89–90**, 165–177 (2016)
19. Wang, K.H., Chen, C.M., Fang, W., Wu, T.Y.: A secure authentication scheme for Internet of Things. Perv. Mobile Comput. **42**, 15–26 (2017)
20. Rao, V., Prema, K.V.: Light-weight hashing method for user authentication in Internet-of-Things. Ad Hoc Netw. **89**, 97–106 (2019)
21. Santoso, F.K., Vun, N.C.H.: Securing IoT for smart home system. In: 2015 International Symposium on Consumer Electronics, ISCE 2015, pp. 5–6 (2015)
22. Yao, X., Chen, Z., Tian, Y.: A lightweight attribute-based encryption scheme for the Internet of Things. Future Gener. Comput. Syst. **49**, 104–112 (2015)
23. Liu, Z., Huang, X., Hu, Z., Khan, M.K., Seo, H., Zhou, L.: On emerging family of elliptic curves to secure internet of things: ECC comes of age. IEEE Trans. Depend. Secure Comput. **14**(3), 237–248 (2017)
24. Odelu, V., Das, A.K., Goswami, A.: A secure biometrics-based multi-server authentication protocol using smart cards. IEEE Trans. Inf. Forensics Secur. **10**(9), 1953–1966 (2015)
25. Kavitha, S., Alphonse, P.J.A., Reddy, Y.V.: An improved authentication and security on efficient generalized group key agreement using hyper elliptic curve based public key cryptography for IoT health care system. J. Med. Syst. **43**(8), 1–6 (2019)
26. Vincent, O.R, Okediran, T., Abayomi-Alli, O.A.: An identity-based elliptic curve cryptography for mobile payment. SN Comput. Sci. **1**, 112 (2020)
27. He, D., Zeadally, S., Xu, B., Huang, X.: An efficient identity-based conditional privacy-preserving authentication scheme for vehicular ad hoc networks. IEEE Trans. Inf. Forensics Secur. **10**(12), 2681–2691 (2015)
28. SafeCurves Homepage. http://safecurves.cr.yp.to/index.html. Accessed 08 Apr 2021

Use of Blockchain in IoT Devices Security: A Systematic Mapping Study

Jhon Arcos(✉), Elizabeth Morejón, Danilo Martínez, Christian Parra, and Joseph Cruz

Universidad de las Fuerzas Armadas "ESPE", 171103 Sangolquí, Ecuador
{jdarcos,memorejon,mdmartinez,cfparra1,jacruz7}@espe.edu.ec

Abstract. Information security in IoT devices is considered a great challenge within the field of computing, which has been growing based on the architecture and the resources that compose it. The implementation of Blockchain has provided improvements in security and reliability for handling data. However, there is no concrete information on techniques, advantages, or results of the use of Blockchain in IoT devices. To identify relevant information on the use of Blockchain in smart devices and present the results as a basis for future research, we promoted a systematic mapping study to classify the selected studies and summarize their content. A limited number of articles was evidenced according to the proposed search chain, only since 2017 an interest in the subject has been observed, the same interest that grows rapidly over time. In addition, disadvantages and shortcomings were identified when applying Blockchain in IoT. Therefore, there is a need to evaluate the performance of Blockchain in different environments and new investigations that provide better results as computer vulnerabilities and their possible effects on devices used in IoT increase.

Keywords: Blockchain · Internet of Things · Security · Technology · Techniques · Information security · Smart devices

1 Introduction

Over the years, IoT has gained space in the social, economic, and technological spheres, being especially used in the latter specifically in the development of engineering applications that aim to meet the needs in terms of process automation. Khan et al. [1] mention that since 2015 the use of devices that are not personal computers was 16% and in 2019 it increased almost to 70%, being 2020 a year in which the use of techno-technology exploded in overdrive, being used up to 95%.

However, as promising as this technology is, it must also face great challenges that hinder the exploitation of its great potential. At a time when information is almost everywhere and it is easier to breach it, especially when security levels are not optimal. Attacks can be generated at any time of the day and in several ways, clear examples are phishing and malware.

Being the attacks on Internet-connected devices, those most likely to be violated, these attacks are focused on violating users privacy, specifically the economic sphere

M. Botto-Tobar et al. (Eds.): CIT 2021, LNNS 405, pp. 83–98, 2022.
https://doi.org/10.1007/978-3-030-96043-8_7

where it is getting to the bank information provided for online purchases or other transactions.

According to Khan et al. [1], both the scientific and industrial communities consider Blockchain as a disruptive technology that can consolidate as an important entity in the management, control and protection of devices, Blockchain can be a key technology to solve IoT security problems.

Blockchain offers the ability to share valuable, tamper-proof data securely. Its distributed architecture solves many of the limitations and problems of the client-server model, such as availability, synchronization times and communication failures between IoT devices. This makes it possible to avoid phishing attacks or man-in-the-middle attacks, which can generate incorrect transactions or modify their details.

To identify evidence on the impact of Blockchain on the Internet of Things, we conducted a systematic mapping study SMS, reviewing of previous studies, papers and publications.

A research question and three sub-questions were determined, which subsequently served to relate the selected studies to each of them.

However, a SMS was established classifying and relating them, obtaining information on how these results and possible research gaps are related, tracing precise routes of inquiry for researchers and readers. Finally, we present our conclusions as a possible background for the development of future research and propose areas for further investigation.

2 Background

According to González [2] the devices used in IoT can range from sensors and mechanical elements to common objects such as household appliances or clothing. Any object can be connected to the internet at any time and place, this implies that any object can be a source of data. This promises to transform the way people live, work and even play, offering applications or services that serve as a bridge between the virtual and the real.

IoT establishes several interactions between people and everyday devices with internet connection, it also has a powerful capacity for data analysis, generating an efficient and adaptable environment for the user's requirements. The IoT also has a powerful data analysis capacity, generating an efficient and adaptable environment to the user's requirements, which implies a daily evolution in the ways of interacting with daily processes or automated processes. Reyes [3] mentions that due to the rapid evolution of IoT, certain aspects such as scalability, security and privacy have been threatened; these shortcomings must be solved in the short term.

Fattori [4] points out that current IoT systems are based on the client-server model, where if communication with the server is lost, data can be lost. The client-server model is fundamental to connect devices to each other. However, it is not certain that it can withstand the challenges posed by the constant technological evolution and attacks that IoT devices may face.

Similarly, Fattori [4] states that for a solution to a certain security gap to be successful, it is necessary to apply Blockchain both on the server side and on the client side, which would be the IoT devices, thus avoiding the disclosure of unauthorized information, applying an authenticity validation mechanism through the verification of positions in a token matrix.

IOT security is important due to its application in everyday life, as its use within the frequency spectrum makes it vulnerable to attacks that can block its services or create a threat to privacy and data confidentiality. IoT devices are organized as interconnected networks of heterogeneous elements, therefore, they are susceptible to the most common security problems of computer networks.

The use of IoT devices connected to a network is becoming more frequent, and with the needs that today's world demands, it is possible to perceive that distances are shorter thanks to greater use of these devices and considerable improvements in connection speeds. This progress has several challenges, ranging from the deployment and use of infrastructure, to problems in the security of user information.

This being the aspect where computer security also known as cybersecurity or information technology security comes into play, it focuses especially on the protection of the integrity and privacy of the information contained in a computer system and transiting through computer networks, minimizing the physical and logical risks to which it is exposed.

According to Baca [5] Blockchain is nothing more than a distributed and decentralized registry that contains blocks of connected transactions; since each Blockchain node contains data of all transactions and in turn these are validated between the same nodes within a chain.

According to Tudela [6], Blockchain was initially intended for the use of cryptocurrencies, however, this technique offers several benefits in other fields, its distributed nature and its security are advantageous for architectures that use IoT applying resource sharing and transaction logging.

Samaniego et al. [7] claim that Blockchain enhances security, offers data transparency and auditability, ensuring tamper-proof storage of approved and executed transactions. Due to its distributed and decentralized operation, Blockchain is used in IoT to manage device configurations, data storage and enable micro payments.

Many IoT systems depend on the security of data stored, transported or manipulated by users or between the devices themselves. Blockchain is probably one of the most secure methods of exchanging and hosting information, as it implements strict security conditions. The application of this technology has caused major changes in information security and thanks to its characteristics, it can be used in the vast majority of devices used in IoT.

3 Research Method

We drive a systematic mapping study considers the guidelines of Petersen et al. [8] and Kitchenham et al. [9] in addition to the research conducted by Fernandez et al. [10]. A SMS allows classifying and summarizing the existing information on a research question in an unbiased way. The process can be seen in Fig. 1.

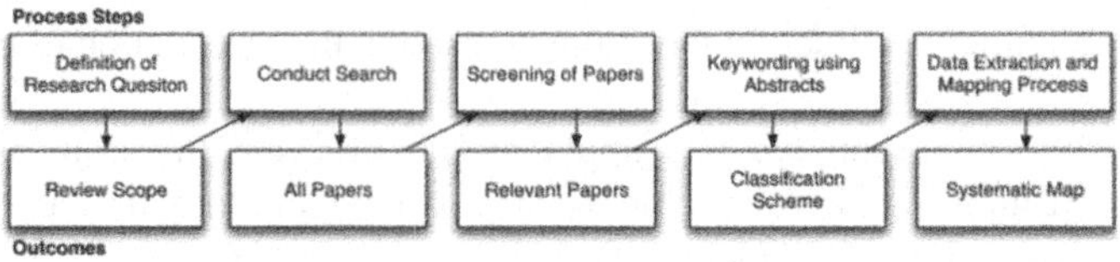

Fig. 1. Systematic mapping process. Taken from Petersen et al. [8].

3.1 Planning

We carried out the following activities to determine appropriate mapping parameters: research question, search strategy, primary study selection, quality assessment, data extraction strategy and synthesis methods.

Research Questions. The aim of our study is to identify evidence about the impact that Blockchain has had on the storage and exchange of information performed by the different devices used in IoT, so we plan the following research question: What is the impact of Blockchain on the computer security of the devices used in IoT? This will allow us to summarize and typify the current knowledge about the security provided by Blockchain when applied to information transactions between IoT devices. Due to the breadth of the research question, it is broken down into sub-questions shown in Table 1:

Table 1. Research sub-questions

Ord	Research sub-question	Motivation
1	*Q1.* What are the Blockchain-based security techniques for smart devices?	To learn how Blockchain works applied to smart devices
2	*Q2.* What advantages does Blockchain provide in IoT information security?	To learn about the benefits of the blockchain technique with respect to information security in the IoT environment
3	*Q3.* What results have been obtained from the application of blockchain in IT security?	To learn about the effectiveness of blockchain in IT security

Search Strategy. To search for primary studies, digital libraries recognized in the computer field were used, such as: IEEEXplore, Springer Link and Science Direct. To search the selected libraries, we use the following search string: (impact OR efect OR repercussion) AND (informatic security OR information security) AND (device OR artifact) AND (IoT OR internet of things) AND (Blockchain). In which a Boolean OR is used to join similar terms or synonyms, as well as a Boolean AND to join different terms (see Table 2).

Selection of Primary Studies. All search results in the different libraries were evaluated by the authors to decide whether or not they should be included in the study, considering the title, abstract and keywords. Any discrepancies were resolved by consensus of all the authors. Studies that met one or more of the following inclusion criteria were included:

- Articles containing information on Blockchain applied to the security of data handled by IoT devices.
- Articles with information about the impact of Blockchain on IoT.

Articles that met at least one of the following exclusion criteria were discarded: Articles that are not focused on computer security.

- Articles that only present definitions, descriptions, or principles of some of the keywords.
- Articles that propose security techniques other than Blockchain.
- Articles not written in English or Spanish.

Table 2. Keywords used for the literature search.

Concept	Alternative terms or synonyms	Connector
Impact	(impact OR effect OR repercussion)	AND
Informatic security	(informatic security OR information security)	AND
Device	(device OR artifact)	AND
IoT	(IoT OR internet of things)	AND
Blockchain	(blockchain)	

Data Extraction Strategy. The data extraction strategy used is based on providing a set of possible answers for each research sub-question, previously posed. This ensures the application of the same data extraction criteria for all selected work, as shown Table 3:

Table 3. Research classification.

Question	Option	Description
Question 1	In development	If the security technique proposes the use of Blockchain, but it has not been applied or tested on smart devices
	Existing	The technique already uses Blockchain as part of its methodology or is entirely Blockchain-based and has been used in smart devices

(*continued*)

Table 3. (*continued*)

Question	Option	Description
Question 2	Important	If the benefits of Blockchain are positively reflected when applied to the security of the information handled and stored in IoT
	Intrascendent	If it is mentioned that Blockchain has irrelevant shortcomings or advantages when applied to the security of the information handled and stored in IoT
Question 3	Positives	The techniques applied by Blockchain provide high information security, so they are in application or use and show positive results
	Uncertain	Possible results or assumptions are mentioned when applying Blockchain techniques for information security, but the effects and evaluation of said techniques are not detailed
	Negative	Negative results are mentioned in the security of the information provided by Blockchain

Synthesis Methods. Within the implemented methodology we apply 2 types of methods based on:

Quantitative data. The number of papers found within the various digital libraries was counted and classified according to their bibliographic source.
Qualitative data. We considered several important studies related to the research sub-questions that helped us to define the results obtained in the previous point regarding quality assessment.

Driving Stage. The application of SMS (systematic mapping) helped us to yield a variety of results which through a thorough analysis allowed us to consider the most relevant data as shown in Table 4.

Table 4. Results of the driving stage.

Source	Potential studies	Selected studies
IEEExplore	11	9
Springer Link	4	1
Science Direct	4	4
ACM DL	6	4
Results	**25**	**18**

4 Results

With the primary studies identified, we have attempted to answer each of the research sub-questions. For which we have proceeded to distribute the primary studies according to the information identified that helps to answer the research sub-questions shown in Table 5.

The most relevant information was extracted from the 18 primary studies, leaving aside redundant content.

In the following subsections we present the results of each research sub-question.

Table 5. Systematic mapping results.

Research sub-questions	Possible answers	Results	
		No. of Studies	Percentage (%)
Q1. Blockchain-based security techniques for smart devices	In development	9	50
	Existing	9	50
Q2. Advantages provided by Blockchain in the information security	Important	16	88.89
	Intranscendent	2	11.11
Q3. Results obtained from the application of Blockchain in the IT security	Positives	2	11.11
	Uncertain	14	77.78
	Negative	2	11.11

4.1 Blockchain Based Security Techniques for Smart Devices

The results obtained revealed that 50% of the studies reviewed propose the use of security techniques based on Blockchain applied in smart devices that are still under development and try to improve existing techniques. The remaining 50% of the studies mention the use of existing security techniques. Blockchain-based security techniques found in the selected studies are detailed in Table 6.

4.2 Advantages that Blockchain Provides in Information Security

88.89% of the studies determine that the advantages provided by Blockchain are positively reflected in information security and only 11.11% indicate that the advantages provided are inconsequential due to irrelevant advantages or possible shortcomings.The important advantages of Blockchain in computer security found in the selected studies are the following:

Provide solutions for malicious cyberattacks and prevent data loss to ensure secure device-to-device (D2D) communication.

- Better security and functionality.
- Low communication and computational costs.

Table 6. Blockchain-based security techniques found.

Technique	Description
BATM	Blockchain Authentication and Trust Module. Moinet et al. [11] proposes to use blockchain as a database to store public keys, digital signatures, and peer information, allowing each component of the network to validate information about all other nodes in the network
BCTrust	Hammi et al. [12] proposes a robust, transparent, flexible and energy efficient Blockchain, specially designed for devices with computational, storage and energy consumption restrictions
LX	This technique takes the name of the authors, Li and Xion (LX) where Li et al. [13] proposes a heterogeneous online and offline signal encryption scheme to secure communication between a sensor node and an Internet host
TTW	This technique is named after the authors, Ting, Tsai and Wu (TTW) don-de Ting et al. [14] proposes an efficient and heterogeneous online/offline encryption scheme. This scheme achieves the security objectives of confidentiality, integrity, authentication, and non-repetition in a single logical step
HCK	This technique is named after the authors, Huh, Cho and Kim (HCK) don-de Huh et al. [15] proposes the use of Blockchain, allowing to control and configure IoT devices. Manage keys using RSA public key cryptosystems where public keys are stored in Ethereum and private keys are stored on individual devices
IoT-Ethereum Framework	Javaid et al. [16] proposes a variant of blockchain, with smart contracts to replace the traditional centralized IoT infrastructure with a decentralized one
BAKMP-IoMT	Garg et al. [17] designed a new blockchain-enabled authentication key agreement protocol for the Internet of Medical things (IoMT) environment, which provides secure key management between implantable medical devices and personal servers and between personal servers and cloud servers

- The security of IoT devices, especially the authentication process, is easier and more efficient.
- Ensures and improves security vulnerabilities and IoT susceptibility to cyber threats.
- Unique features of storing transactions securely and decentralized.
- Useful for processing the huge amount of data generated by IoT.
- Guarantees access control in heterogeneous scenarios.
- Low cost, efficient and flexible for the IoT-based environment.
- Facilitates elimination of the bottleneck problem faced during growing dependency on IoT when working with centralized servers.
- No third party is required to handle confidential data of IoT users.
- Autonomy of IoT devices and communication become easy in a decentralized network.

- Blockchain is the beginning of a series of service monetization opportunities.
- Blockchain-based solutions could be superior to the current IoT ecosystem, which relies primarily on cloud servers centralized through service providers.
- Blockchain certifies the identity of each device with the authentication of each node belonging to this security fence.
- Allows you to mitigate threats that currently security systems do not detect.
- All types of attacks can be avoided thanks to the permanent evaluation of the network devices and the monitoring of their transactions.
- On the other hand, possible disadvantages or shortcomings were found in the techniques that use Blockchain, which are detailed below:
- BCTrust lacks a key management mechanism to generate, store and manage the startup keys on which the entire protocol is based.
- There are great challenges to be solved for Blockchain in IoT, among them scalability, dynamism, mobility, heterogeneity, interoperability, existence, adaptability, and standardization.
- Despite having numerous benefits over traditional techniques, a notable drawback of this technology is the scalability problem. The increasing number of devices could become an obstacle in the next Blockchain applications

4.3 Results Obtained from the Application of Blockchain in Computer Security

Only two selected articles mention positive results. Xu et al. [18] shows an improvement in network security performance, for example, the server attack rate decreases by 66.7% compared to other methods. Golam et al. [19] mentions that the results of a simulation carried out illustrate how effectively the applied method can work to restore security in D2D communication.

Other articles mention uncertain results regarding the application of Blockchain in computer security, since they do not show simulations or real results, these are compiled below:

- Compares with other related existing schemes and also performs better in terms of security features and functionality.
- Less communication and communication costs for the authentication and key management phase compared to the other schemes.
- The results show that the system can protect the device. Provide location information services to users under the premise of location privacy.
- The evaluation of the work presented good results within an environment of emulation of IoT, communication and transactions were established.

On the other hand, Kshetri [20] mentions that Blockchain can contain a security breach in IoT in a specific way after it is discovered.

4.4 Mapping Results

The combination of the three research sub-questions allowed to establish a systematic mapping that provides an overview of the security provided by the Blockchain technique applied in IoT devices. The relationship between the questions provides us with expanded information on the results of our research.

The results obtained from the systematic mapping in the research sub-questions Q1 (Security techniques based on Blockchain) and Q2 (Advantages of Blockchain in information security in IoT), detailed in Fig. 2, may indicate that:

- There are very important computer security advantages, whether in existing or developing techniques that apply Blockchain in an IoT environment.
- The results obtained allow us to identify that security techniques go hand in hand with the advantages that Blockchain offers with respect to device security.
- There are very few disadvantages or irrelevant benefits in the Blockchain technique when applied to IoT devices.

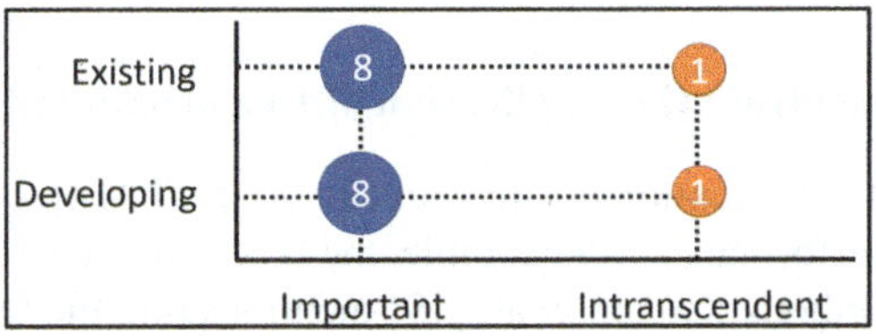

Fig. 2. Sub Mapping of results of sub question 1 and 2.

The results obtained from the systematic mapping in the research sub-questions Q2 (Advantages of blockchain in information security in IoT) and Q3 (Results of blockchain in computer security), detailed in Fig. 3, may indicate that:

- The highest percentage of results are related to very important articles that satisfy the needs of users regarding the advantages that Blockchain provides in the security of IoT information.
- The advantages provided by Blockchain are consistent with the results obtained within the area of computer security, which shows that Blockchain is a very important technique for the provision of security within the IoT environment.
- Very few articles consider the results of blockchain techniques within the IoT area to be irrelevant or negative.
- Few articles show positive results so further testing is required to reflect actual results so far uncertain that are important to the advantages of Blockchain.
- There are no articles that present uncertain results and at the same time consider the advantages that Blockchain provides inconsequential.

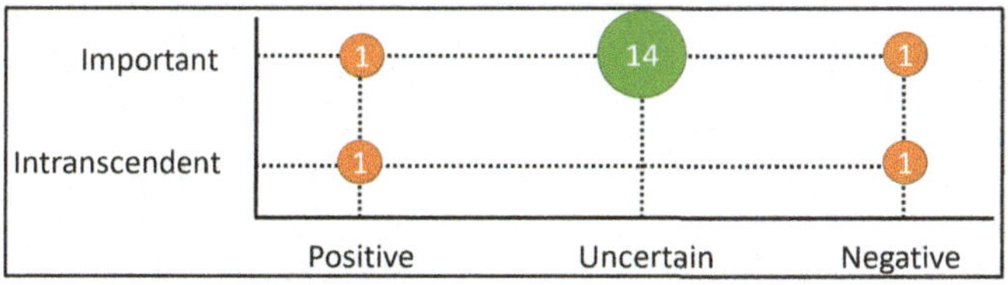

Fig. 3. Mapping of results of sub question 2 and 3.

The results obtained from the systematic mapping in research sub-questions Q1 (Blockchain-based security techniques) and Q3 (Results of blockchain in computer security), detailed in Fig. 4, may indicate that:

- Existing and developing Blockchain-based security techniques provide uncertain results that have not shown real positive effects through the use of this technique.
- The existing techniques showed positive results in contrast to the techniques under development, which showed possible shortcomings or negative results.
- There are no positive results in the techniques under development; on the contrary, the existing techniques show no negative results.

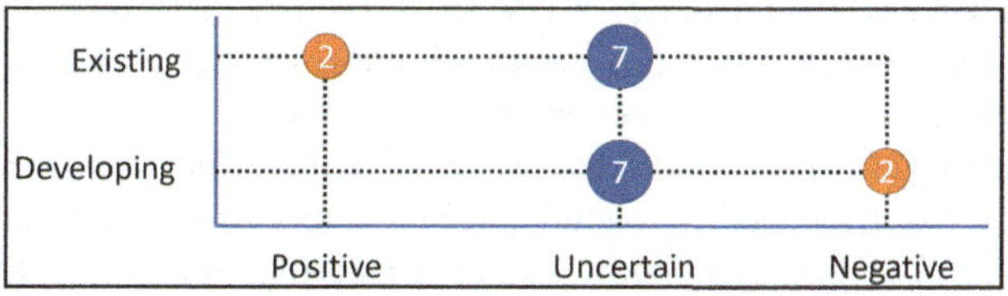

Fig. 4. Mapping of results of sub question 1 and 3.

4.5 Interest of the Topic

According to the studies carried out in the various digital libraries through a timeline, it was possible to determine the publications on Blockchain applied in IoT devices in order to provide security in the management of information, taking into account its year of publication and the bibliographic reference as detailed in Fig. 5.

Figure 5 shows the number of publications on Blockchain and IT security applied to IoT devices by year and source. The analysis of the number of studies investigated showed an interest in this topic initially in 2017 and that there is an increase in interest in this topic of 400% until 2020 (from two studies selected in 2017 to ten studies in 2020).

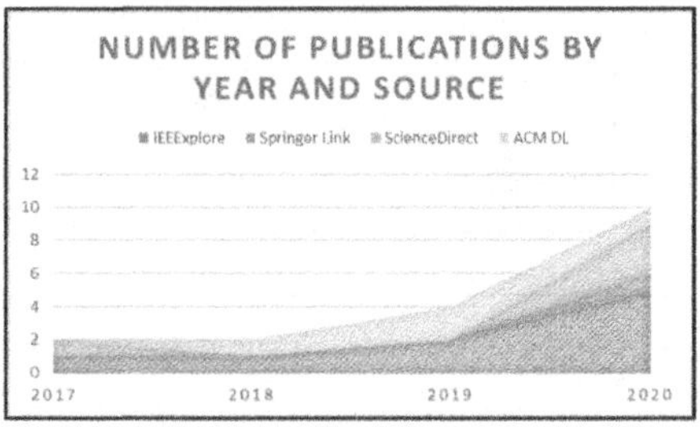

Fig. 5. Number of publications by year and source.

5 Discussion

The main goal of this systematic mapping study was to identify evidence on the impact of Blockchain on the IT security of IoT devices. The main findings of this systematic study are summarized below, the limitations that may threaten its validity are highlighted, and the implications for future research and practice are discussed.

5.1 Main Findings

AlAbdullatif et al. [21] performs an analysis of the main solutions that use Blockchain, among them are BATM, BCTrust, LX, TTW, HCK and IoT - Ethereum Framework. The BCTrust protocol, compared to other solutions, improves the security of IoT devices, especially authentication as it is easier and more efficient.

Mahmoud et al. [22] also propose the use of BCTrust, which is based on the concept of "the friend of my friend is my friend", which refers to the fact that all the nodes participating in an operation must know, and in turn certify, of this form all those that are part of the set are known ensuring the availability and consistency of the information. BCTrust is considered a flexible and efficient Blockchain-based security solution for IoT.

Xu et al. [18] compile various techniques for IoT that use Blockchain technology so that data can be exchanged safely. The technique is designed on a public blockchain, Ethereum, to keep track of tampering and eliminate middlemen.

On the other hand, Golam et al. [19], Garg [17] and Pawar et al. [23] propose the use of Blockchain-based architectures and techniques for IoT environments.

These results contrasted with the temporal dispersion show a trend towards the development of new proposals and the development and research of new methods that apply Blockchain based on techniques already developed and thus provide better information security handled by intelligent IoT devices.

Regarding the advantages that Blockchain provides, 88.89% of the studies positively reflect the security of information in IoT environments and only 11.11% show inconsequential advantages or possible shortcomings.

Kshetri [20] highlights how Blockchain-based techniques could be superior to the current IoT ecosystem, using practical applications and real-world examples, the paper argues that the decentralized feature of the Blockchain is likely to result in a low susceptibility to manipulation. and spoofing by malicious participants.

El Houda et al. [24] mention that Blockchain provides a low cost, efficient and flexible for an environment based on IoT, Golam et al. [19] for them part, states that Blockchain provides a solution for malicious cyberattacks and prevents data loss to ensure security in device-to-device (D2D) communication.

On the contrary, Sharma et al. [25] mentions that there are great challenges to be solved for Blockchain in IoT, among them scalability, dynamism, mobility, heterogeneity, interoperability, existence, adaptability and standardization.

In the reviewed literature, it was found that Blockchain is a solution that provides high performance, and its advantages favor its use in an IoT environment during information management. Despite having a lot of benefits, the rise of devices could be a stumbling block in upcoming Blockchain applications.

The selected studies address several security and privacy device issues, these were identified based on the observations in the operation and interaction of the IoT components. Blockchain technology is considered a solution to meet these challenges.

The selected studies show results obtained from the application of Blockchain in computer security. However, 77.78% of them show uncertain results, based on assumptions and without proving their claims.

Only Xu et al. [18] show a comparison between Blockchain and other methods. While Golam et al. [19] show results of a simulation carried out. These positive results are only 11.11% of the selected studies.

Kshetri [20] mentions a possible security breach in IoT after it is discovered. This represents a negative result since, if true, all the advantages mentioned above by other authors would be disregarded and therefore the positive results that Blockchain has shown so far would be totally affected.

These findings show us that evaluation methods have not been established that allow a detailed analysis of the performance of Blockchain in the security of IoT devices, the results of the articles are based on empirical data as a product of controlled tests or simple assumptions.

In the mapping of the selected studies, it was found that there are very few disadvantages in Blockchain techniques, whether existing or under development, when applied to IoT devices. Few articles show positive or negative results, suggesting that more real tests are required to reflect the results.

Existing techniques show positive results as opposed to developing techniques, which show possible shortcomings or negative results. Existing techniques can be determined to be safe by the results they show. However, with long term research it is possible to find serious flaws with respect to Blockchain.

5.2 Limitations of the Systematic Study

The main limitations of this study have been established based on the bias presented by the articles selected for our mapping, and the possibility of a misclassification during data extraction.

One of the limitations of systematic mapping is that its objective is aimed at categorizing the articles and identifying the most relevant ones instead of including evidence of empirical results, which must be analyzed generating more specific research questions.

The present study presents very recent results, it is only in 2017 when works related to the raised topic appear, due to this the articles considered are of a very small number, a growth in the interest of this topic is observed in the year 2020, so it is expected that in the coming years this trend will continue and new information on the subject of study will appear quickly.

The number of studies selected is very small, which is why the results may not represent in a real way the generalized information existing in virtual libraries, so the conclusions we reach may not be real.

5.3 Research Implications

The findings of our systematic mapping study have implications for researchers who propose new solutions based on Blockchain as well as for those professionals who work in computer security within IoT and want other forms of security than conventional ones.

Our findings showed that some of the papers report that a flaw in the techniques that use Blockchain for information management is the scalability that an IoT environment requires. Therefore, we believe that further field tests should be carried out and its effectiveness verified in environments with many IoT devices.

We have observed that methodologies have not been established to evaluate the performance of the Blockchain in the security of IoT devices. Performance evaluations within the field of safety are very important, as they help to improve the quality of the final product.

The main problem seems to be the barrier of the technology itself, referring to companies with their own systems, this form of closed architecture prevents a standard from being offered for the majority, since each company has its own quality standards.

6 Conclusions and Future Work

In recent years, the use of IoT technology began to occupy more space in daily life, this implies that most users do not consider the security aspect, making it careless and easy to violate; This is when the search for an effective solution to provide security begins.

This article has presented a systematic mapping study that summarizes the existing evidence about the use of Blockchain in the IT security of IoT devices. From a set of 25 articles, a total of 18 were selected and the results obtained have allowed us to draw conclusions about the state of Blockchain.

Interest in Blockchain as a security measure begins in 2017, having a significant increase in 2020. Although the number of studies related to the research topic is scarce due to the recent boom, use and development of IoT, it is expected that these will increase rapidly due to its importance and applicability in various environments.

Blockchain offers permanent security in the transmission and management of information between IoT devices, ensuring that users share data unequivocally, however, the results obtained show a possible flaw in Blockchain regarding scalability. Therefore, there is a need to evaluate the performance of Blockchain in more complex and extended environments.

There is no evidence of articles on the implementation of Blockchain in industrial environments that use IoT, so research is required that includes all technical areas and provides conclusive results that improve existing techniques, identifying errors and correcting them as vulnerabilities increase. We hope that our findings will be useful in the development of research on Blockchain as a security measure applied in devices used in IoT, providing a guide that allows to identify the effectiveness of this technique, as well as its possible failures.

References

1. Ahmad Khan, M., Salahb, K.: IoT security: review, blockchain solutions, and open challenges. Fut. Gener. Comput. Syst. **82**, 395–411 (2018)
2. Rodríguez González, D.: Arquitectura y Gestión de la IoT. Telemática J. **12**(3), 49–60 (2013). ISSN: 1729-3804
3. Reyes Delgado, D.: Aplicación de Blockchain para la Seguridad de los datos del Internet of Things (2018)
4. Fattori de Andrade, A.C.: Análisis del consumo de energía promedio en dispositivos IoT de baja potencia con Blockchain como solución de seguridad (2019)
5. Baca Urbina, G.: Introducción a la Seguridad Informática, México: Grupo Editorial Patria (2016)
6. Tudela Díaz, I.: Arquitectura Blockchain para la securización de Dispositivos IoT Mediante Smart Contracts (2019)
7. Samaniego, M., Jamsrandorj, U., Deters, R.: Blockchain as a service for IoT. In: IEEE/ACM International Conference on & International Conference on Cyber, Physical and Social Computing (CPSCom) Green Computing and Communications (GreenCom) (2016)
8. Petersen, K., Vakkalanka, S., Kuzniarz, L.: Guidelines for conducting systematic mapping studies in software engineering: an update. Inf. Softw. Technol. **64**, 1–18 (2015)
9. Kitchenham, B.A., Budgen, D., Brereton, O.P.: Using mapping studies as the basis for further research – a participant-observer case study. Inf. Softw. Technol. **53**(6), 638–651 (2011)
10. Fernandez, A., Insfran, E., Abrahão, S.: Usability evaluation methods for the web: a systematic mapping study. Inf. Softw. Technol. **53**(8), 789–817 (2011)
11. Moinet, A., Darties, B., Baril, J.-L.: Blockchain based trust & authentication for decentralized sensor networks (2017)
12. Hammi, M.T., Bellot, P., Serhrouchni, A.: BCTrust: A decentralized authentication blockchain-based mechanism. In: IEEE Wireless Communications and Networking Conference (WCNC) (2018)
13. Li, F., Xiong, P.: Practical secure communication for integrating wireless sensor networks into the Internet of Things. IEEE Sensors J. **13**, 3677–3684 (2013)
14. Ting, P.-Y., Tsai, J.-L., Wu, T.-S.: Signcryption method suitable for low-power IoT devices in a wireless sensor network. IEEE Syst. J. **12**, 2385–2394 (2018)
15. Huh, S., Cho, S., Kim, S.: Managing IoT devices using blockchain platform. In: International Conference on Advanced Communication Technology (ICACT) (2017)
16. Javaid, U., Siang, A.K., Aman, M.N., Sikdar, B.: Mitigating loT device based DDoS attacks using blockchain. In: CryBlock 2018: Proceedings of the 1st Workshop on Cryptocurrencies and Blockchains for Distributed Systems, pp. 71–76 (2018)
17. Garg, N., Wazid, M., Das, A.K., Singh, D.P., Rodrigues, J.J.P.C., Park, Y.: BAKMP-IoMT: Design of Blockchain Enabled Authenticated Key Management Protocol for Internet of Medical Things Deployment. IEEE Access **8**, 95956–95977 (2020)

18. Xu, D., Xiao, L., Sun, L., Lei, M.: Game theoretic study on blockchain based secure edge networks. In: IEEE International Conference on Communications in China (ICCC) (2017)
19. Golam, M., Lee, J.-M., Kim, D.-S.: A UAV-assisted blockchain based secure device-to-device communication in Internet of military Things. In: International Conference on Information and Communication Technology Convergence (ICTC) (2020)
20. Kshetri, N.: Blockchain's roles in strengthening cybersecurity and protecting privacy. Telecommun. Policy **41**, 1027–1038 (2017)
21. AlAbdullatif, A., AlAjaji, K., Al-Serhani, N.S. Zagrouba, R., AlDossary, M.: Improving an identity authentication management protocol in IIoT. In: Cyber Security and Digital Forensics (2019)
22. Mahmoud, C., Acuag, S.: Security for Internet of Things: a state of the art on existing protocols and open research issues. In: Proceedings of the 9th International Conference on Information Systems and Technologies, pp. 1–6 (2019)
23. Pawar, R., Kalbande, D.: Use of blockchain technology in wireless body area networks. In: International Conference on Intelligent Sustainable Systems (ICISS) (2020)
24. El Houda, Z.A., Hafid, A., Khou, L.: Co-IoT: a collaborative DDoS mitigation scheme in IoT environment based on blockchain using SDN. In: IEEE Conference and Exhibition on Global Telecommunications (GLOBECOM) (2019)
25. Sharma, S., Chaudhry, R., Bhardwaj, D.: Blockchain for secure Internet of Things. In: Nanda, A., Chaurasia, N. (eds.) High Performance Vision Intelligence. SCI, vol. 913, pp. 33–54. Springer, Singapore (2020). https://doi.org/10.1007/978-981-15-6844-2_3

Proposal of Sensor-Based System to Control the Store Capacity for Prevent Infection by Covid-19

Dayana Anchapaxi(✉), Dennis Ayo, Jefferson Bazantes, Karem Chafla, and Brandon Romero

Universidad De Las Fuerzas Armadas "ESPE", AV. General Rumiñahui S/N, Sangolquí, Ecuador
{dpanchapaxi1,dcayo,jpbazantes,kechafla,baromero6}@espe.edu.ec

Abstract. Nowadays, the world is facing a pandemic of COVID 19, a very contagious disease. However, people do not take it seriously and choose not to respect the necessary biosecurity measures to avoid a possible infection. In this paper, we present a proposal of a sensor-based systems to control the user's capacity in stores of shopping centers that are points with a high risk of mass infections. The prototype consists of a set of sensors and a system of a web systems that allows controlling the capacity of the stores as people enter and leave the place. When exceeding the store capacity an alert indicates that due to security, no more people can enter the store.

Keywords: Control · Capacity · System · Covid-19 · Sensors

1 Introduction

COVID-19 is an infectious disease caused by the SARS-CoV- 2 virus that generates a severe acute respiratory syndrome. The pandemic generated by COVID-19 has influenced the way we carry out daily activities around the world. One of the affected sectors has been the commercial field, where the capacity of people in the different business centers where there is no adequate control method directly affecting the health system due to the excess of infections.

According to Orozco et al. [2], the panorama that we are facing, it is necessary to implement guidelines for the health surveillance of employees of different economic activities establishing social health and safety methods.

Where technology plays an important role, through applications and software based on geo-location, contact registration via Bluetooth, using techniques such as big data, geographic information systems (GIS), and artificial intelligence, to strengthen the national health authority.

Therefore, a technological solution is proposed, through a people counting system, both at the entrance and at the exit of the shopping center, through the use of infrared sensors. The application will identify the total number of people who are inside the

M. Botto-Tobar et al. (Eds.): CIT 2021, LNNS 405, pp. 99–109, 2022.
https://doi.org/10.1007/978-3-030-96043-8_8

premises, and the number of people who can still enter the establishment. Therefore, if it exceeds the store capacity, it will issue an alert.

Our study consists of four main axes. The first is oriented to the health area to collaborate in the prevention of COVID-19 infections by controlling agglomerations. The second axis is to provide statistical information on the management of the pandemic based on the figures collected by other authors or government entities. The third axis is research on existing applications on the market to help control the pandemic. Finally, fourth part, we present our solution and its influence as part of a protocol to mitigate excessive COVID-19 infections.

The main contribution of our study focuses on both the technological area and the health area. In the technological area, we contribute with a prototype to control the maximum capacity of people in shopping centers. Meanwhile, in the health area, a tool is contributed to control the capacity of people to reduce the number of infections.

2 Background

There are some projects developed against COVID-19, the same ones that are undoubted of great contribution to society in search of mitigating the spread of the virus and reducing the number of infected. Some projects already existed, but they have taken a course of development and adaptation according to the social stereotypes of each country, to be used in the midst of the current health crisis.

Lyseen's et al. [4] mentions that the analysis of diseases and health services are research field where the development of technologies, application field for GIS, GPS, remote sensing will help and promote research in GIS and health respectively.

Several systems are being perfected focused on avoiding massive contagions, where Barrientos et al. [5] presents an application that registers nearby users who are carriers of the virus, through the Bluetooth sensor, which works only on smartphones, thus keeping users informed.

Several researches have been carried out on the usefulness of geographic mapping, where Boulos et al. [6] provides information about the measures that were taken in China where people provide information through an application regarding personal decisions about self-quarantine and medical treatment on a spatial scale. They using big data that the government provides on the movement of people and thus avoid crowds.

On the other hand, Big Data models [15] focus on decision-making based on the information and data obtained, through the use of technology that allows monitoring high agglomerations of people. Bluetooth [13], is mainly based on registering and verifying the close contacts detected by the Bluetooth sensor, being that it run only on smartphones. Thus, there is also the design of a scalable electronic prototype [12], which is based on a tool that helps control and monitor agglomerations in times of pandemics. RGB D camera [11], focuses on robustly performing automatic people counting and somehow it would not be feasible due to the high cost of this tool.

3 Methodology

The methodology proposed in this article is based on an evolutionary model of prototypes [16]. The idea is to develop the implementation of a system, which meets the requirements necessary to control the capacity in a store, which must be built quickly and efficiently. In addition, the model can be divided into two stages: the first stage establishes the general structure of the system, where the necessary requirements for its operation are taken into account, for which a simulation of the prototype was carried out, thus being able to observe the operation and thus refine or improve the system for its optimal operation. The other stage was performed through a final simulation test of the prototype where acceptance of the development team could be observed. Users will be tested in a small space room to assess the benefit of such a system.

The methodology focuses on representing the most important aspects of the system, adjusting to the user's requirements, hoping that the simulation is visible to the end-user.

This quick representation shows how it is focused on the prototype design that can be evaluated by the clients and store security assistants. For the feedback on the operation and effectiveness of this system. Therefore, it is possible to refine requirements of the software to be developed and to improve the system as well with the help by users or other personnel.

We apply the data collection method, where various techniques are used to collect information [17, 18]. Since the statistical data collected about COVID-19 infections were analyzed from the databases of the Ecuadorian Ministry of Public Health (MSP) [8], the National Institute for Public Health Research (INSPI), and the Committee of Emergency Operations (COE), who are in charge of the management of COVID-19 at the Ecuador level and who in turn are the ones who are jointly in charge of decreeing the security protocols, which are issued through the Service's website Risk and Emergency Management [19].

For this reason, within the research study framework for the project. There are also data about these security protocols, essentially how they are doing it to channel them with the help of technology of other countries.

The information collected has been implemented in a statistical graph as shown in Fig. 1.

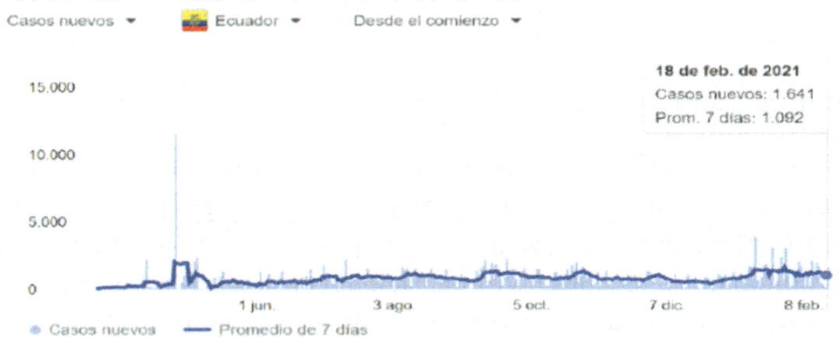

Fig. 1. COVID-19 Contagion Board in Ecuador, with evidence from February 18, 2021, [7].

4 Proposed Solution

Figure 2 show a graphical representation of both hardware and software elements and how they relate to each other to contribute to this research and solve our problem.

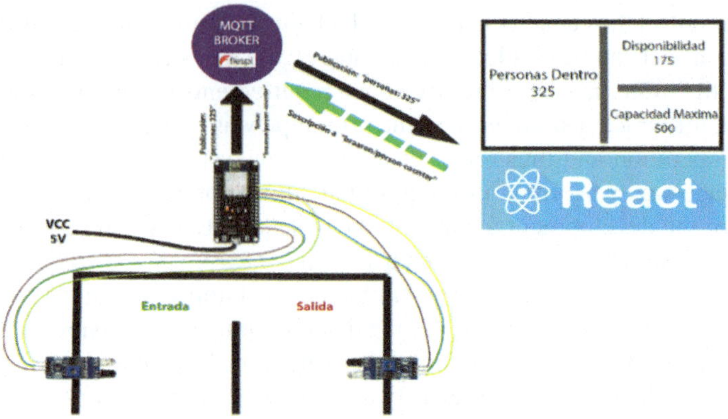

Fig. 2. Graphic representation of the elements involved in the solution prototype

4.1 React

React is a JavaScript-based development framework, which is designed to create user-friendly interfaces and facilitate application development [20].

4.2 MQQT Broker

MQTT (Message Queue Telemetry Transport) is a lightweight messaging protocol, which works by publishing and subscribing messages between different devices. One of the most efficient ways is data-centric communication in publish and subscribe messaging system. Currently, this protocol is used with great frequency in the interconnection of several devices that are linked to the Internet of Things (IoT).

MQTT is designed in such a way that it can run on low-end, battery-operated sensor or actuator devices over bandwidth-restricted WSN (Wireless sensor networks) such as ZigBee-based networks [22].

4.3 Infrared Sensors

Infrared sensors are based on the combination of an emitter and a receiver located close to each other, which will help to send a frequency of the people who are entering the shopping. To implement these sensors, an adequate distance must be had between the emission circuit and the reception circuit. The reception circuit delivers an amplitude variation of the output voltage proportional to the distance from the reflecting surface [23].

With the help of infrared sensors placed on the doors of the premises, they will allow a count simply because this system will be typical of the premises and will not require that people have a smartphone or install an application.

4.4 Arduino IDE

Arduino IDE is a development environment that is normally used on the Arduino board but that also serves to program different electronic modules with the help of special libraries for each module [21].

Our proposal consists of the development of a prototype that integrates a web page in React. With this framework we will connect to the Flespi MQTT server and a graphical interface will be developed, which will show the following information: (1) How many people are inside store, (2) what is the maximum capacity of people within the store and (3) how many people are still allowed access, based on the people inside and the maximum capacity of the store.

To solve this problem, we will use 2 infrared sensors, which are capable of measuring infrared electromagnetic radiation that gets in your field of vision, allowing us to detect if a person enters or leaves a certain place. The person in charge of reading the information sent by the sensors is the NodeMCU module, which is a board that comes integrated with the ESP8266 chip, which is nothing more than a small low-cost Wi-Fi module that uses the TCP/communication protocol. IP, for this it is necessary to first program the NodeMCU with the help of the Arduino IDE. Once the NodeMCU has been programmed, it will connect to an MQTT server provided by the Flespi site for free and will communicate through the MQTT protocol, which can be implemented on any device, and will make it easier to collect data. Sensor data and publication of the obtained values.

With the site already developed, it is advisable to place a screen at the entrance of the premises to show this information, in this way it helps the people in charge of access to have better control of the capacity of the store, in addition to keeping the customers informed, whether they are in front of the information screen or using the website and see that they comply with the biosafety protocols established for COVID-19.

5 Results

To carry out the tests of our prototype, a 9 m × 9 m room was used, with capacity for 3 clients and a vendor. A drill was carried out in which the biosecurity parameters against COVID-19 were maintained, which emphasizes the distance between people with a minimum of two meters as shown in Fig. 3.

Figure 4 represents the system that was configured with initial parameters of a maximum capacity of 3 people. When entering the web page, the general information is displayed: (1) date and (2) clock. Next, the information of the place or store is shown: (1) people inside, (2) availability, (3) the status, which shows if a person entered or left, and (4) total availability.

With the system developed and configured, users can be informed about the state of the store and take the proper biosecurity precautions. As shown in Fig. 5, users waiting to enter the premises can see on the website that the biosafety standards are being met.

Once the user or the person in charge verifies the availability, the user enters the premises passing by the proximity sensor as shown in Fig. 6. This sensor is programmed to increase the number of people entering the store and reduce the availability number on the website.

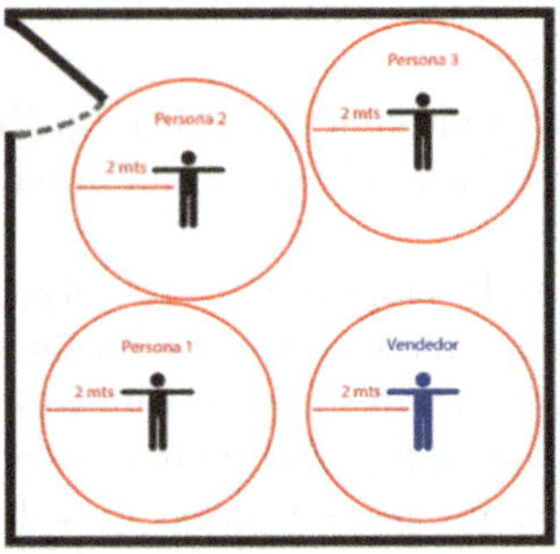

Fig. 3. Test scenario

Fig. 4. Web page made with React.

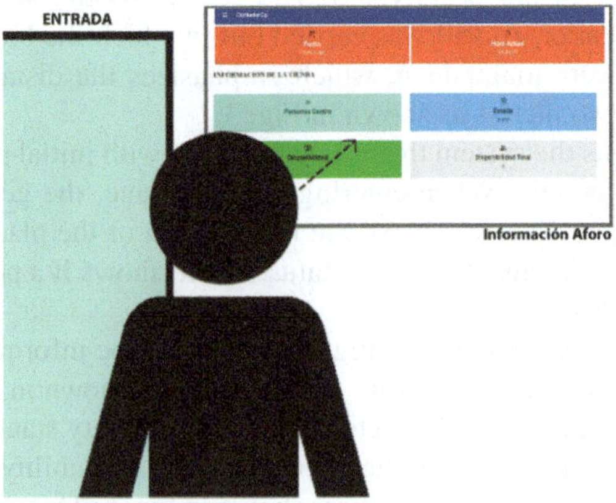

Fig. 5. User verifying the information of the premises

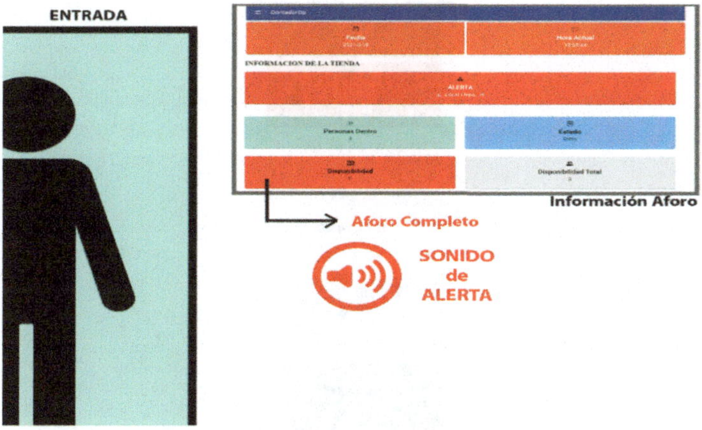

Fig. 6. User passing by the proximity sensor to enter the premises.

Once the number of people inside is the same as the total availability number, it indicates that the place is at its maximum capacity. If a person intends to enter even when it is observed through the website that the place does not admit more people at the moment, a visual and sound alert will be issued indicating that more people are not allowed to enter until there is availability as indicated shown in Fig. 7.

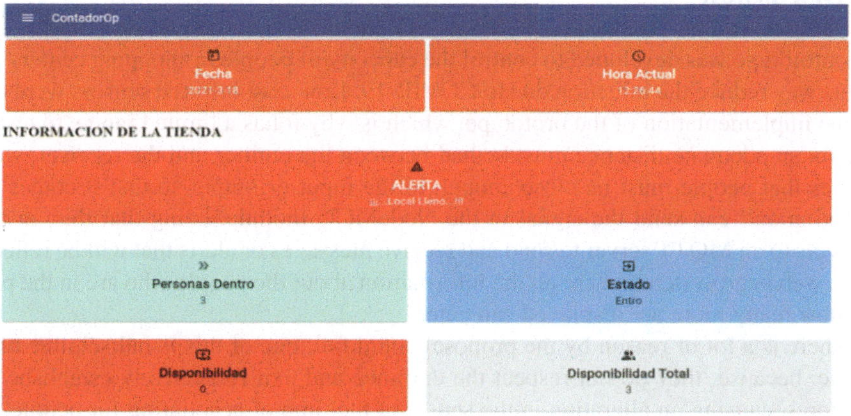

Fig. 7. Web page issuing an alert indicating that the place is at its maximum capacity and does not admit more people.

For this web application to be used in the best way, it is recommended that it be available both at the entrance of the store and in the same way for a mobile device, which will require internet access. Thus, the user can enter the page or application to verify the availability of the capacity of the premises as shown in Fig. 8, and be informed and verify that the capacity of the premises does not exceed the limits, putting their health at risk.

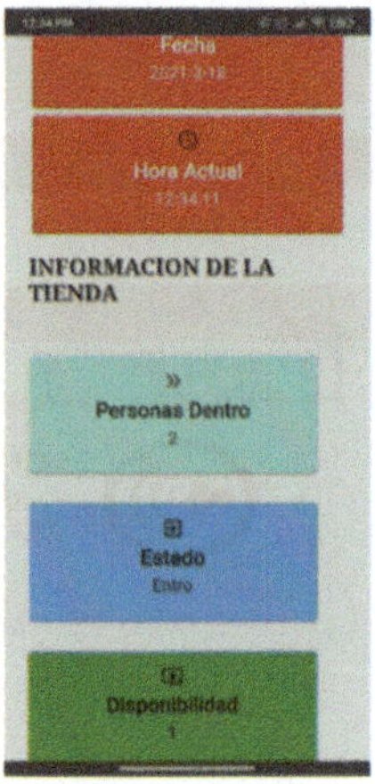

Fig. 8. Web page on an Android device.

With the implementation of this system, better control of the capacity of the stores are maintained by the users and the administrators of the stores.

In this way, the stores can provide greater confidence to their users, improving the reputation of the premises and avoiding the risk of infection.

6 Discussion

This prototype was developed to control the capacity of people in shopping centers, and in this way reduce the infection due to COVID 19. Low-cost infrared sensors were used for the implementation of the prototype, which is why it has a limited range, to test the system, an adequate distance must be had between the emitter and the receiver, which implies that people must be close enough to the input or output so that it comes into operation and can send the signal to the NodeMCU module, being that this, in turn, connects to an MQTT server to send and receive messages or alerts that will be reflected in the web page system, where all the information about the people who are in the place and how many more are displayed can enter.

There is a lot of reason by the proposal could fail, one of this is indiscipline of the people, because, they do not respect the entrance and exit respectively established by the stores, causing an alteration in the sensor at the time of accounting for people, and consequently display the wrong data on the screen. We also find that our project does not requires the user to have a mobile device when entering the store because the capacity will be displayed on a screen, and it will be visible to all users. On the other hand, if the user is somewhere outside the shopping center, wanting information in real-time of the capacity of the store, he can also find it through a web page, using a computer or mobile phone with internet access, and verify the information provided therein. Structurally referring to the project, we find that it requires WIFI to communicate with the server because it is designed with NodeMCU, an open-source development kit based on the ESP8266 chip (ESP-12E).

For the design and testing of the prototype, an infrared sensor HW-201 was used, this being a basic or simple sensor has certain limitations, and the main limitation is the short range of object detection which is around 3 to 4 cm.

Therefore, there could be several problems when it comes to detecting the entry and exit of people when they do not pass close enough to the sensors. To implement this system in a production environment, it is necessary to change the HW-201 sensor for a more powerful and better-quality sensor such as the EE-SPWD311 which has a detection range of up to one meter, taking into account where. The sensors are placed and the local conditions can also be used photoelectric sensors which have a greater detection range.

Depending on the needs of the stores, it is possible to increase input and output sensors, although this may require more NodeMCU modules and certain modifications to the code for optimal operation.

Regarding the web page, it is possible to improve its design and its alerts, it is possible to integrate it with other web pages in case the store already has one, you can also add more information such as temperature, number of ATMs, promotions, statistics, etc. This, in addition to providing security and confidence to customers, can be used as a marketing tool while people wait to enter the premises.

7 Conclusions

The objective of the project was to create a prototype, which fulfills the function of controlling the capacity of people in the premises of shopping centers, to contribute to the reduction of COVID-19 infections. Throughout the development of our proposal, we have managed to make a prototype, which works optimally in places with reduced space. To evaluate the functionality of the prototype, a simulation was carried out in a 9×9 room, which consists of an entrance door and an exit, in this way the project gave a solution to what was established, which was to provide information on a capacity to the users through a screen which is also available on a web page.

To achieve this objective, it was necessary to configure sensors and modules which communicate with an MQTT server, this server connects at the same time with a web page that is programmed under the REACT Framework which is in charge of displaying the information to the users.

Being that through infrared sensors the people are counted, establishing the number of people who are both inside and outside, checking its optimization in a reduced space of 9×9 square meters, giving a solution to what is established in the problem, which was to control the capacity of people and thus reduce the crowds, preventing the number of infections from rising, generating more control for entering these types of places.

8 Future Works

As future work, the development of an app for mobile devices that emits warning signals is proposed, this being very noticeable as a sound or in turn, an automatic voice. Additionally, it is proposed as future work the implementation of control utilizing automatic door locking when completing the maximum capacity established for each store in the shopping center, restricting access to people from outside who forcibly want to enter it.

References

1. Pedersen, S.F., Ho, Y.C.: SARS-CoV-2: hay una tormenta. Revista de investigación clínica **130**(5) (2020)
2. Orozco Polo, L.Á., Suárez Medina, J.I., Villaseca López, B.J.S.: Implementación de un plan de vigilancia, prevención (2020)
3. Berná Galiano, J.Á.: Sistema de detección de aglomeraciones de personas basado en una App para smartphones Android (2020)
4. Lyseen, A.K., et al.: A review and framework for categorizing current research and development in health-related geographical information systems (GIS) studies. Yearbook Med. Inform. **23**(01), 110–124 (2014). https://doi.org/10.15265/IY-2014-0008
5. Barrientos-Gutiérrez, T., Alpuche-Aranda, C., Lazcano-Ponce, E., Pérez-Ferrer, C., Rivera-Dommarco, J.: La salud pública en la primera ola: una agenda para la cooperación ante Covid-19. salud pública de México **62**(5), 598–606 (2020)
6. Boulos, M.N.K.: Descriptive review of geographic mapping of severe acute respiratory syndrome (SARS) on the Internet. Int. J. Health Geogr. **3**(1), 2 (2004). https://doi.org/10.1186/1476-072X-3-2
7. Dong, E., Du, H., Gardner, L.: Tablero interactivo basado en la web para rastrear COVID-10 en tiempo real. The Lancet. Recuperado de (2020). https://www.thelancet.com/journals/laninf/article/PIIS1473-3099(20)30120-1/fulltext
8. Ministerio de Salud Pública: Actualización de casos de coronavirus en Ecuador. Recuperado de (2021). https://www.salud.gob.ec/actualizacion-de-casos-de-coronavirus-en-ecuador/
9. Servicio Nacional de Gestión de Riesgos y Emergencias: Protocolo-centros-comerciales. Recuperado de (2021). https://www.gestionderiesgos.gob.ec/wp-content/uploads/2020/05/protocolo-centros-comerciales.pdf
10. Ruiz, G.P.I., León, A.C.I.: Evolución de la enfermedad por coronavirus (COVID-19) en Ecuador. La Ciencia al Servicio de la Salud **11**(1), 5–15 (2020)
11. Castrillón Santana, M., Lorenzo Navarro, J., Hernández Sosa, D.: Conteo de personas con un sensor RGBD comercial. Revista Iberoamericana de Automática e Informática industrial **11**(3), 348–357 (2014)
12. Guaycha Arreaga, B.R., Zambrano Sandoval, J.A.: Diseño de un prototipo electrónico escalable de crowd density y plataforma digital con notificaciones, para evitar multitudes en locales comerciales en tiempos de pandemias (Doctoral dissertation. Universidad de Guayaquil. Facultad de Ciencias Matemáticas y Físicas, Carrera de Ingeniería en Networking y Telecomunicaciones) (2020)
13. Ponce, P., Loaiza, V., del Río, M.D.L.C., Parra, L.B.: Efecto de la desigualdad y la actividad económica en el COVID-19 en Ecuador: un bosquejo de sus posibles determinantes económicos, sociales y demográficos. Contaduría y Administración **65**(4), 211 (2020)
14. Guerrero, S.: Coronavirus en ecuador: una opinión desde la Academia. LA GRANJA. Revista de Ciencias de la Vida **32**(2), 127–133 (2020)
15. Prado-Ortega, M.X., Grunauer-Robalino, G.R.: Salud Pública: Detección De Concentración De Personas Aplicando Big Data Para Evitar Brotes Epidemiológicos COVID-19. Identidad Bolivariana **4**(2), 5–19 (2020)
16. Maida, E.G., Pacienzia, J.: Metodologías de desarrollo de software [en línea]. Tesis de Licenciatura en Sistemas y Computación. Facultad de Química e Ingeniería "Fray Rogelio Bacon". Universidad Católica Argentina Disponible en (2015). https://repositorio.uca.edu.ar/handle/123456789/522
17. Ghandi, L., Silva, C., Martínez, D., Gualotuña, T.: Proceso de desarrollo de aplicaciones móviles: una experiencia práctica. En: 2017 XII Congreso Ibérico de Sistemas y Tecnologías de la Información (CISTI), pp. 1–6. IEEE, junio 2017

18. Lethbridge, T.C., Sim, S.E., Singer, J.: Studying Software engineers: data colletion techniques for Software field studies. Empir. Softw. Eng. **10**(3), 311–341 (2005)
19. Servicio Nacional de Gestión de Riesgos y Emergencias: Protocolo-centros-comerciales. Recuperate de (2021). https://www.gestionderiesgos.gob.ec/wp-content/uploads/2020/05/protocolo-centros-comerciales.pdf
20. Jimenez, J.: Descubre React. Recuperado en Octubre **2**, 2018 (2015)
21. Ubach Granados, C.: Programando directamente un ESP8266 (Bachelor's thesis, Universitat Politècnica de Catalunya) (2017)
22. Hunkeler, U., Truong, H.L., Stanford-Clark, A.: MQTT-S: un protocolo de publicación / suscripción para redes de sensores inalámbricos. En 2008 3ª Conferencia Internacional sobre Software y Middleware de Sistemas de Comunicación y Talleres (COMSWARE 2008), pp. 791–798. IEEE, enero 2008
23. Ríos, L.H., Bueno, M., Valencia, J.A.: Implementación de un sistema de medición de distancia con sensores infrarrojos utilizando dispositivos de lógica programable. Scientia et technica **2**(39) (2008)

Yaatree: A Mobile Application for Transportation Request Focused on People with Disabilities

Miquely Calvopiña, José Zamora, Sebastián Zúñiga, and Graciela Guerrero(✉)

Departamento de Ciencias de la Computación, Universidad de las Fuerzas Armadas ESPE, Sangolqui, Ecuador
{mecalvopina1,jazamora4,sazuniga,rgguerrero}@espe.edu.ec

Abstract. New technologies have the duty to help in the inclusion of people with disabilities in society. This article shows the design, development and evaluation of an application oriented to the transport of people with a motor disability. The application is developed for Android system, uses Google's geolocation services, maps and routes generation, in addition, it stores information in Firebase, also has a web interface that allows the validation of new users in mobile application. The results show that the application has a proper construction and a high level of usability. Finally, the long-term objective of the application is to be put into production to effectively help in the transportation of people with motor disabilities in Ecuador.

Keywords: Disability · Mobile application · Transportation · Real-Time Systems · Mobility

1 Introduction

In 2011, the World Health Organization [1] mentioned in its World Report on Disability that more than one billion people, i.e., about 15% of the world's population at that time, suffered or suffer from disability, of which about 20% experience integration problems, given the lack of regulations or actions regarding their integration into society. In Ecuador, by 2020, 485,325 people with some types of disability were registered, in addition, only 15% of them are active in the labor market, 10% are students from elementary school to high school and only 5% of the age group between 19 and 35 years old are in higher education [2]. At this point, according to Sánchez [3], Information and Communication Technologies (ICT) acquire a high social relevance, even though the digital gap in Latin America is much higher than the international average, therefore, their duty is to solve the best way possible problems of any kind; this includes disabilities. Da Cruz et al. [4] explained how people with disabilities feel the need to be independent and to be able to carry out their daily activities normally, without asking for constant assistance from their relatives or close friends. Despite this imperative need for independence, certain activities cannot be carried out alone, in most cases those that involve moving from one

M. Botto-Tobar et al. (Eds.): CIT 2021, LNNS 405, pp. 110–122, 2022.
https://doi.org/10.1007/978-3-030-96043-8_9

place to another, whether over short or long distances. This, this, in addition to the lack of inclusive elements such as ramps, braille displays or text-to-speech systems, prevents people with disabilities from moving around normally. According to Wasfi et al. [5], access to free mobility is fundamental, however, for those with a disability to find an accessible way to get to a specific place has become one of the basic daily challenges for people with disabilities.

At the moment, there are only applications that share Uber main objective, which consists in taking people from its origin point to a destination one, nevertheless, these applications do not allow to select an impairment filter, therefore, if a user with this characteristic uses this application, the transport will not have the necessary implements to perform the service. As a response to these facts, the purpose of the present project is focused on the design of a mobile application that helps people with disabilities to transport themselves with greater independence, using various technologies such as database, Global Positioning System (GPS) and Application Programing Interface (API) from various providers, focused on people with disabilities, in Ecuador.

The present work is structured as follows: (i) the second section describes the theoretical basis and characteristics of the related work that provided the necessary information for the research; (ii) the third section discusses the methodology, the architecture of the developed application, the mobile interface and the navigation method used by the application; (iii) the fourth section describes the results following the implementation of the application and its acceptance; (iv) Finally, the fifth section presents the conclusions of the research and future work.

2 Related Works

The inclusion of people with disabilities is one of the biggest challenges facing ICTs today, not only in the field of accessibility to digital tools, but also in all social aspects, including mobility. In view of this challenge, the related works to the present project are divided into two categories: i) Applications focused on providing the best route based on the user's needs and mode of transportation; ii) Applications focused on the link between public transportation services and users with disabilities.

2.1 Applications Focused on Providing the Best Route Based on the User's Needs and Mode of Transport

The transportation of people today relies mainly on GPS and the algorithms of different applications, such as Google Maps, to obtain the best routes. This idea can be extrapolated to people with disabilities, however, we are focusing on people with different transport needs, so the calculation of the best routes must be adapted to different requirements from the original ones.

Information integrating mobile services is being developed with the aim of increasing the accessibility of facilities and can be provided by authorities, volunteers and users [6]. According to Muñoz et al. [7], an application model is proposed in which the information providers are controlling or government entities, with the aim that they maintain a database with suitable spaces for disabled people. Alepsis and Nita [8], on the other hand, aim for the best route for the user based on their rating in the application, in addition to device-independent path storage. Finally, Vlahovic and Glažar [9] propose a cloud-based application model that allows the best routes to be established using the concept of a social network, allowing the users of the network identify and establish the best route, this information must be corroborated by volunteers to keep the application up to date.

Following this development line, Arellano et al. [10], propose an application for the mobility of people who use wheelchairs, whose distinction is presented in the generation of the best path, since it uses Dijkstra's algorithm and weights in different nodes of the trajectory to determine the one that presents less effort.

2.2 Applications Focused on Linking Public Transport Services to Users with Disabilities

Applications such as Google Maps, Waze, GPS navigation, among others, focus on providing information regarding the trip requested by the user to the different public transport stations, the problem lies in the lack of specialized guidance for people with disabilities. Mass public transportation is one of the preferred and reliable means of transportation for people with different needs, especially those who cannot drive [11].

Mobile applications have been built to solve the mobility of people with disabilities in public transport, these can be focused on people with any type of disability, Kishore et al. [11] explains how an application provides an interface suitable for any user with these difficulties, on the other hand, in 2017, Rodriguez et al. [12] propose a mobile navigation application aimed at people exclusively with visual disabilities, accessing it through voice commands or through a synthesizer activating GPS, consequently the application will indicate the nearest bus stop, bus routes and if necessary, send monitoring information in case of an emergency to a family member or friend.

Independence to perform daily activities is the biggest problem for a person with a visual impairment, so possible applications have been proposed to solve this problem, such as the one proposed by Landazabal et al. [13], which is based on providing support focused on the mobilization through public transport, using technologies that determine the geolocation and allow indicating the nearest optimal stops, using as a tool a cell phone that will receive an action either by touch or by voice commands, once the support is requested, the corresponding guide will begin.

In 2020, Wu et al. [14] propose a social business model based on a mobile platform to achieve barrier-free transportation, the model aims to indicate how ICTs can be integrated with providers that provide a transportation service and government entities, using their resources to meet the needs of people with disabilities when transporting by public transport services.

The different proposed solutions base their operation mainly on mobile devices, both to obtain information from users and generate the best routes [6], and to integrate them with public transport and deliver information about their characteristics [12]. In addition, not only these applications or proposals base their development on mobile devices, monitoring systems such as the one developed in [15] that uses IoT sensors linked to a central processing node to send the information collected by these to a mobile interface.

With the characteristics of the related works exposed, the present project will use the most relevant tools of these, such as mobile development, with the difference that the focus of the developed application seeks to provide a means of transportation suited to the needs of people with disabilities, rather than that these are adapted to a pre-existing environment.

3 Methodology and Architecture

This section details the architecture used, the development process and the evaluation of the system called Yaatree, whose objective is to provide a transport service for people with reduced mobility, with vehicles that have the necessary implements to fulfil the desired purpose.

Yaatree system has been developed for mobile devices that work under the Android operating system (OS), as this mobile OS is the most widely used worldwide [16] and allows a high degree of freedom when it comes to development [17], in addition, the system has a web environment whose access is exclusive to the application administrator.

3.1 Architecture Design

The proposed architecture of Yaatree system, as shown in Fig. 1, consists of the consumption of several APIs, a global positioning system (GPS) and a real-time storage system located in the cloud. The GPS will be used to obtain the current location of the users; therefore, drivers will be able to determine the trips that have been requested within a close range by taking their position as a reference.

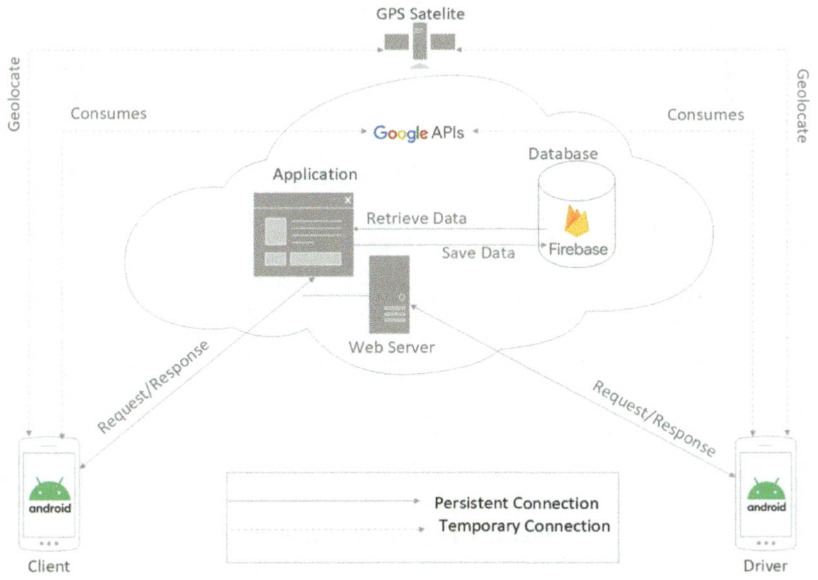

Fig. 1. Application architecture

The APIs required for the deployment of maps and routes will be provided by Google [18] and the storage system will be in a non-relational database, called Firebase [19].

The architecture has been designed to establish high availability independent of the type of network used by the user with a temporary connection to the GPS satellite to obtain its coordinates.

3.2 Development

The Yaatree system is designed for two types of users as shown in Fig. 2, the first will be the person with a disability, whose actions consist of: i) Register, ii) Login, iii) Request Trip & iv) Rate Trip. The second user is the driver whose actions are: i) Register, ii) Register vehicle, iii) Login, iv) Accept trip and v) End trip.

The above-mentioned users perform common activities such as login and registration, however, the information required for the registration of both is different. For the driver, in addition to his or her personal information, it has been validated the registration of the plate number of the vehicle with which he or she will provide the service and a section with checkboxes have been entered, with the objective of allowing the driver to check all the necessary characteristics for him or her to provide the service.

The disabled user will enter only the relevant data to identify him/her, in addition to verifying if he/she has certain characteristics within his/her own section with check boxes that will help the application to determine the appropriate vehicle and driver to perform the service. The proposed check boxes in the Yaatree user registration system are intended to be the guidelines for the documents that the person must upload for further validation.

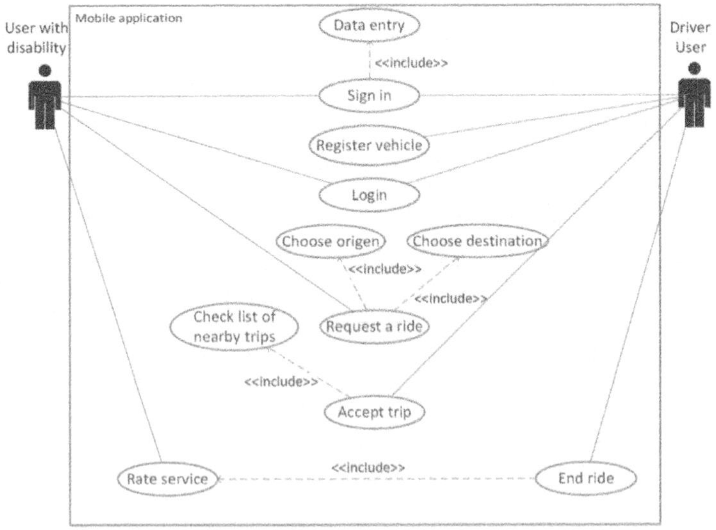

Fig. 2. Diagram of use cases

As for the driver user, the following parameters have been established: i) My car has the necessary implements and ii) I have followed a paramedic course. The driver user must upload images of his documents (ID card and driving licence) so that the administrator can verify the veracity of these and approve the user. The scenario is repeated for the disabled user with the following options: i) I have a CONADIS card, ii) I have a medical condition that could result in the need for first aid and iii) I need a wheelchair to get around.

The Yaatree system is divided into two applications: (i) mobile for users and (ii) web for the administrator, the latter being able to perform functions such as photo verification, user validation, login and enable user.

The Yaatree system administrator is responsible for verifying the veracity of the documents uploaded by users and, depending on this, must decide whether the user is suitable for the corresponding role or not, displaying this data in a table with the relevant fields as shown in Fig. 3.

State	Has CONADIS id	Medical condition that require assistance	Has wheelchair	Review
Rejected	No	No	No	Review
Rejected	Yes	No	No	Review
For review	No	No	No	Review

Fig. 3. Image control table in the web interface.

Firebase is used as an authentication system to perform different types of verifications and user registrations with email accounts. In addition, Firebase's own storage system is used to store user profile images, drivers and those requested at the time of registration. The administrator interface obtains these images directly from Firebase's storage system via asynchronous calls, to the functions provided by its Software Development Kit (SDK).

To generate a route, the user with a disability must choose the destination of his trip, using the Google Places API. The driver is assigned to the trip and gets the route, from his position to that of the user who requested it, with the help of the Google Directions API in conjunction with Android's own GeoFire API, the latitude and longitude of the location of both users is obtained. When picking up the disabled user, a new route to the selected destination is created, the longitude and latitude of the destination point are obtained from the location selected by the disabled user when requesting the service, taking the current location of the driver as the point of origin.

The driver assignment, at the moment a disabled user requests a trip depends on two factors, the distance and the matching of the requirements of the disabled user with the characteristics of each driver user. Regarding the distance, the functions provided by the Geofire API are used to perform geographical queries, using latitude, longitude and a search radius that increases up to a maximum of 30 km, to avoid memory overflow in the device. The second determining factor is the verification boxes, both driver and of the disabled user, by means of an asynchronous query to the database, the verification boxes of the disabled user and those of the potential driver are selected, in case the user's requirements match the characteristics of the potential driver, the latter is assigned the trip, otherwise the search will continue until finding the right one or until reaching the maximum radius previously defined.

4 Evaluation

The Yaatree system has been evaluated using videoconference and remote desktop control tools. These tools have been selected due to the current situation in the world caused by the SARS-CoV-2 pandemic and the restrictions imposed by the Ecuadorian state.

This section describes the evaluation of the system, which was carried out through surveys and observations of common users, who evaluated the interfaces of the different modules of the system and from whom the necessary information on usability was obtained.

4.1 Purpose of the Evaluation

To evaluate the usability of the Yaatree system, a simulated environment was used to verify, by obtaining results, that there is no problem with the interfaces corresponding to the mobile application. On the other hand, for the web interface, the expert users verified that there is no inconsistency in terms of usability, for which they were given the domain corresponding to the administrator's page, among with the necessary credentials to access it.

4.2 Scenarios

In order to evaluate the system, the division by scenarios was taken into account and these in turn are divided into sessions, the first corresponds to the process of induction to the use of the application and a survey to determine the demographic distribution of the participants, the second session consists of the observation process by the research team when using the application, while the third session was a usability survey based on the questions of the SUS questionnaire.

The three scenarios considered were evaluated by five participants each, this number was determined under the recommendations of Nielsen and Landauer [20], where it is explained that the minimum number of respondents should be 5 people and in [21], where Lee and Lin explain that the minimum number of participants in the observation process is 3.

In the first scenario, participants evaluated the mobile interface corresponding to a user with a disability in an average time of 12 min 14 s, and the second scenario, the mobile interface corresponding to a user with a disability in an average time of 12 min 14 s, while in the second scenario, participants evaluated the mobile interface of a driving user within the application for which, as in the first scenario, they made activities supervised by the research team, the observations were carried out in an average time of 12 min 24 s. The participants in the two scenarios described above were students from Ecuadorian universities who identify themselves as digital natives. Finally, in the third scenario, ESPE university teachers with a high level of experience in website management and construction were asked to evaluate Yaatree's administrative page, which took an average time of 20 min 48 s to complete.

4.3 Form of Evaluation

In order to evaluate the system, the following variables were taken in consideration: the attention given at the time of receiving the induction and when carrying out the requested actions, the usability of the application and the time taken to carry out these activities, the data was obtained by applying a survey delivered through the Microsoft Forms platform [22]. On the other hand, the observation technique evaluated the level of expertise to perform activities within the application, this was done with the use of Google Meet [23] and AnyDesk [24] for letting users take control of the computer of a member of the research group and perform the tasks mentioned in an emulator while they are observed, obtaining the necessary information for the evaluation.

4.4 Results

The results were obtained from the process mentioned in the previous section, so the information obtained from the scenarios in each session is detailed below. In the first session they were asked to carry out a survey, obtaining the results described in Table 1.

Demographic Survey and Induction. The first session was conducted by each of the participants in the different scenarios that make up the evaluation, this session was composed of an induction to the use of the application, in the first two scenarios, followed by a survey, focused on obtaining demographic information from the participants.

Table 1. Results of the demographic survey

Question	Mobile application users	Web application users
What is your age range?	18 to 24 years old (100%)	45 to 54 years old 60%) 35 to 44 years old (40%)
What is your academic level?	Third Level (60%) High School (40%)	PhD degree (100%)
What is your current employment situation?	Student (100%)	Full employment (100%)

Observation. The observation process was carried out in the second session of each of the scenarios, with the objective of collecting quantitative and qualitative data. During the observation, the research team proceeded to indicate to the participants the tasks they must perform and the time they took to perform them was recorded, in addition, elements such as the fluency of the participants when executing the tasks, the number of questions asked to the observers or if any problems were reported within the application were also collected.

Figure 4 shows the total time it took each participant to perform the tasks corresponding to the user and driver scenarios. In the first scenario, the time taken in the observation process resulted in an average of 2 min 50 s, within these data, an atypical case was found corresponding to observed 1, this participant had problems when executing the tasks "Register" and "Edit personal data". In the second scenario, corresponding to driving users, an average of 3 min 24 s was obtained, within these data, we found two outliers in relation to the mean, the first piece of information was from one of the participants that at the moment of starting the observation, mentioned that he has little technological knowledge and the second, contrary to what happened with the first one, he performed the tasks fluently, However, he mentioned aspects of the application that could be improved while he was working on them, which resulted in an unexpected extension of the time it took him to perform the activities.

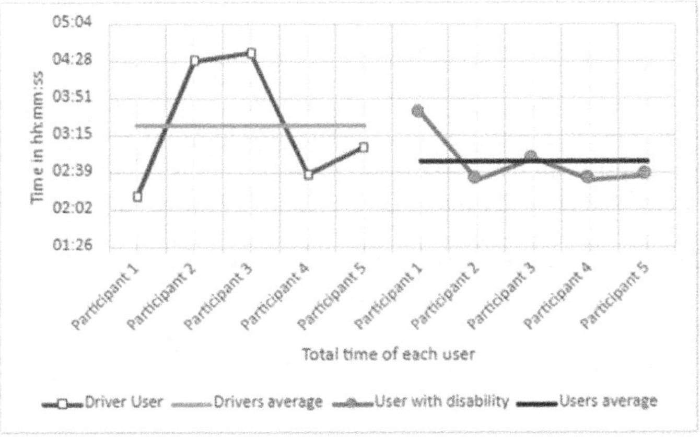

Fig. 4. Observation of users and drivers.

In Fig. 5, corresponding to the expert's scenario, an average of 1 min 55 s was obtained, within this we found only one outlier in relation to the mean, the cause of which is unknown to us, given that there was no anomaly in the observation, in addition to the fact that their level of expertise is the same as that of the other experts evaluated.

Based on the results obtained, the observation group was able to observe that users of both the mobile application and the website found it easy to perform different tasks due to the simplicity of both interfaces and familiarity with known applications and sites.

Usability Survey. In this last session for the three scenarios, a survey was applied to measure the usability perceived by the participants. Table 2 shows the distribution of the responses obtained, where the percentage represents the number of people who opted for a specific score for each question, out of five options.

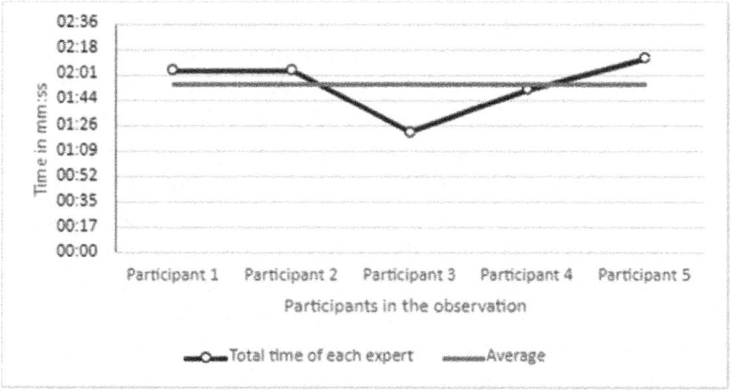

Fig. 5. Expert observation

Table 2. Usability survey results

Question	Mobile application users		Web application users
	User	Driver	
I think I would like to use this system frequently	4(80%); 5(20%)	5(80%); 4(20%)	3(40%); 4(40%); 5(20%)
I found the system unnecessarily complex	1(60%); 2(20%); 4(20%)	1(80%); 3(20%)	2(60%); 3(20%); 1(20%)
I think the system was easy to use	5(100%)	5(100%)	3(40%); 5(40%); 4(20%)
I think I need expert support to go through the system	2(60%); 1(40%)	2(80%); 1(20%)	2(80%); 1(20%)
I found that several of the system's functions were well integrated	5(60%); 4(40%)	4(60%); 5(40%)	3(40%); 4(40%); 5(20%)
I think there is too much inconsistency in the system	1(60%); 2(40%)	1(60%); 2(40%)	2(60%); 1(40%)
I imagine that most people will learn to use the system very quickly	5(100%)	5(100%)	4(60%); 5(40%)
I found the system very large as I went through it	1(60%); 2(40%)	1(100%)	1(80%); 2(20%)
I felt confident using the system	5(60%); 4(40%)	5(100%)	4(80%); 5(20%)
I needed to learn a lot of things before I could use the system properly	2(60%); 1(40%)	1(100%)	1(60%); 2(40%)

When evaluating the results detailed in Table 2, it can be seen that there are relationships between questions, the first focused on ease of learning, between the questions "I think the system was easy to use" and "I imagine that most people will learn to use the system very quickly" indicating that, in terms of the mobile application, the participants stated that it was easy to use and therefore, they think that someone else will also find it easy, in terms of the expert users, there were split results due to the perspective from which they analysed the website. The second relationship, regarding the consistency of the system, the questions "I found that several of the functions of the system were well integrated" and "I think there is too much inconsistency in the system" presented a reverse implication, as the majority of participants expressed that there are no inconsistencies in the system, which should be reflected in the results obtained regarding the correct integration of the functions in the system, which was satisfactory.

Subsequently, to verify the degree of usability of each of the Yaatree interfaces, the calculation indicated by the SUS questionnaire itself [25] was carried out to evaluate the questions, obtaining scores of 89, 95.5 and 80.5 in the scenarios of users with disabilities, drivers and website users, respectively, which indicated that the mobile application was highly accepted by users, while the website, because it was evaluated by experts in the area of usability, considered that there are elements that can and should be improved.

In addition to the questions taken from the SUS questionnaire, a question was added to the survey for all scenarios, as can be seen in Table 3, with the aim of finding out the probability of recommending the system to another person, in which the participant had to define their position on a scale with a minimum of one and a maximum of ten.

Table 3. Probability of recommendation

Question	Mobile application users		Web application users
	User	Driver	
How likely are you to recommend us to a friend or colleague?	10 (40%) 9 (40%) 8 (20%)	9 (80%) 10 (20%)	7 (40%) 9 (40%) 10 (20%)

Yaatree was positively accepted by the users that evaluated the system, who said they found the idea encouraging for the inclusion of people with disabilities in society and that the interfaces had an eye-catching but understandable design, although expert users told us that we need to make small changes.

5 Conclusions and Future Works

In this research work, a system with a mobile application was designed, developed and evaluated, the administrator is responsible for providing a transportation service to people with disabilities taking into account their needs and a web application for the administrator, in which he/she can accept or reject users by verifying the documents uploaded by them.

In the related works, it is observed that most of the applications oriented to the transportation of persons with motor disabilities, seek to match them to pre-established means of transport, although the idea is accurate, it does not take into account those people who for some reason are unable to do so. The present work was focused on providing a service, taking into consideration all persons affected by the mentioned problems, creating a suitable environment for their needs.

With regard to system development, using a non-relational database in real time provides many advantages over relational databases, For example, Firebase, as a database engine, is one of the most complete and provides a series of extra services that are very useful and represent a time saving in the development process. However, despite the advantages of real-time databases, a specific problem that arose in the development process is related to the use of "listenners", these need to be closed once they have fulfilled their function, because if they have not been closed the right way, they may cause a lack of performance and correct functioning of the system.

At the end of the evaluation of the system, scores were obtained for the SUS questionnaire, which once were processed showed positive results, indicating that users feel comfortable and find it easy to use the mobile application. On the other hand, the results of the web interface, evaluated by experts in the field, indicate that the functionality of the site is correct, however, there are elements at the interface level that require improvement, such as the display of data, with respect to its organization and quantity, which were verified by following the recommendations of the experts during the observation process carried out.

Yaatree is initially developed to be made up of volunteers, however, there are plans to add a module to allow payment for the service provided, as well as to improve the generation of routes in the mobile application. On the other hand, if the user has been rejected by the system administrator, there is no possibility to rectify the invalidated evidence, so it is planned to add this functionality, allowing the user to re-enter this information until they are accepted, which would imply more tasks for the administrator, who should have the possibility to edit the users' data in the system.

References

1. Organización Mundial de la Salud: Informe Mundial sobre la Discapacidad (2011)
2. Consejo Nacional para la Igualdad de Discapacidades: Consejo Nacional para la Igualdad de Discapacidades, June 2020. https://www.consejodiscapacidades.gob.ec/estadisticas-de-discapacidad/. Accessed 12 July 2020
3. Sánchez Montoya, R.: TIC y discapacidad en América Latina y el Caribe. Comunicación y pedagogía: Nuevas tecnologías y recursos didácticos, pp. 34–40 (2006)
4. Cesar da Cruz, D.M., Figueiredo e Silva, V., Ferigato, S., Agostini, R.: Personas con discapacidad y sus roles ocupacionales: trabajo, familia, independencia y participación social. Revista Chilena de Terapia Ocupacional **16**, 107 (2016)
5. Wasfi, R., Steinmetz-Wood, M., Levinson, D.: Measuring the transportation needs of people with developmental disabilities: a means to social inclusion. Disabil. Health J. **10**, 356–360 (2017)
6. Kulakov, K.A., Shabaev, A.I., Shabalina, I.M.: The route planning services approach for people with disability. In: 2015 17th Conference of Open Innovations Association (FRUCT), Yaroslavl (2015)

7. Muñoz, E., Serrano, M., Vivó, M., Marqués, A., Ferreras, A., Solaz, J.: SIMON: assisted mobility for older and impaired users. Transp. Res. Procedia **14**, 4420–4429 (2016)
8. Alepis, E., Nita, S.: Mobile application providing accessible routes for people with mobility impairments. In: 2017 8th International Conference on Information, Intelligence, Systems and Applications (IISA), Larnaca (2017)
9. Vlahovic, N., Glazar, D.: Urban accessibility application for users with mobility impairments: conceptual model based on social networking and volunteered geographic information. In: 2018 41st International Convention on Information and Communication Technology, Electronics and Microelectronics (MIPRO), Opatija (2018)
10. Arellano, L., Del Castillo, D., Guerrero, G., Tapia, F.: Mobile application based on Dijkstra's algorithm, to improve the inclusion of people with motor disabilities within urban areas. In: Rocha, Á., Adeli, H., Reis, L.P., Costanzo, S. (eds.) WorldCIST 2019. AISC, vol. 931, pp. 219–229. Springer, Cham (2019). https://doi.org/10.1007/978-3-030-16184-2_22
11. Kishore, A., et al.: CENSE: a cognitive navigation system for people with special needs. In: 2017 IEEE Third International Conference on Big Data Computing Service and Applications (BigDataService), Redwood City (2017)
12. Rodriguez, A., De la Cruz, D., Tobar, J., Mejia, P., Paredes, N., Olmedo, G.: Voice—TOUCH GPS: navigation and mobility assistant for people with visual disability in Ecuador. In: 2017 12th Iberian Conference on Information Systems and Technologies (CISTI), Lisbon (2017)
13. Landazabal, N.P., Rivera, M.A., Martinez, M. H., Nates, C.R., Uchida, B.T.: Design and implementation of a mobile app for public transportation services of persons with visual impairment (TransmiGuia). In: 2019 XXII Symposium on Image, Signal Processing and Artificial Vision (STSIVA), Bucaramanga (2019)
14. Wu, Y.J., Liu, W.-J., Yuan, C.-H.: A mobile-based barrier-free service transportation platform for people with disabilities. Comput. Hum. Behav. **107**, 105776 (2020)
15. Abeledo, M.C., Bruschetti, F., Aguilera, G., Iriso, P., Marsicano M., Lacapmesure, A.: Remote monitoring of elderly or partially disabled people living in their homes through the measurement of environmental variables. In: 2016 IEEE Congreso Argentino de Ciencias de la Informática y Desarrollos de Investigación (CACIDI), Buenos (2016)
16. Statcounter: Mobile and Tablet Operating System Market Share Worldwide, June 2020. https://gs.statcounter.com/os-market-share/mobile-tablet/worldwide#monthly-201906-202006-bar. Accessed 12 July 2020
17. Goadrich, M.H., Rogers, M.P.: Smart smartphone development: iOS versus android. In: Proceedings of the 42nd ACM Technical Symposium on Computer Science Education - SIGCSE 2011, Dallas (2011)
18. Google: APIs de Cloud, July 2020. https://cloud.google.com/apis/. Accessed 12 July 2020
19. Google: Firebase, July 2020. https://firebase.google.com/. Accessed 12 July 2020
20. Nielsen, J., Landauer, T.K.: A mathematical model of the finding of usability problems. In: Proceedings of the SIGCHI Conference on Human Factors in Computing Systems - CHI 1993, Amsterdam (1993)
21. Lee, Y.-J., Lin, C.-J.: Evaluation and satisfaction survey on the interface usability of online publishing software. Math. Probl. Eng. **2014**, 1–10 (2014)
22. Microsoft: Microsoft Forms, July 2020. https://support.microsoft.com/es-es/forms. Accessed 12 July 2020
23. Google: G Suite Essentials: la manera más simple de trabajar en equipo, July 2020. https://gsuite.google.com/intl/es-419/essentials/. Accessed 12 July 2020
24. AnyDesk (2020). https://anydesk.com/es. Accessed 12 July 2020
25. Martins, A.I., Rosa, A.F., Queirós, A., Silva, A., Rocha, N.P.: European Portuguese validation of the system usability scale (SUS). Procedia Comput. Sci. **67**, 293–300 (2015)

Lifelong Learning as a Strategy for Digital Literacy in University Teachers

Margarita Alejandra Aucancela[1(✉)], Ana Cecilia Andrade[1], David Antonio Ureña[1], and Jéssica Pamela Torres[2]

[1] Escuela Superior Politécnica de Chimborazo, Riobamba EC060155, Ecuador
maucancela@espoch.edu.ec
[2] Facultad Latinoamericana de Ciencias Sociales, Quito EC170201, Ecuador

Abstract. The outbreak of the COVID-19 pandemic generated the transition from a face-to-face teaching modality to a virtual teaching modality. For continuity deliverance to the university educational system's activities, it was crucial to generate permanent training courses to manage virtual classrooms, virtual educational resources, and video conferencing platforms to establish contact with students. This research aims to corroborate whether permanent training constitutes a meaningful and valuable digital literacy strategy for university teachers. Following the paradigm of a qualitative research design, a survey application before and after starting the permanent training courses explicitly selected for the teachers from Escuela Superior Politécnica de Chimborazo of the Public Health faculty. As a result, there is a significant difference between the digital skills that teachers initially possess and those developed after participating in permanent training courses. The data was corroboree by obtaining a p-value of 0.007. Lifelong learning acquires a leading role when digital literacy and digital skills development are necessary for an emerging situation.

Keywords: Digital literacy · Permanent training · Digital educational resources · University

1 Introduction

According to primary journalistic sources [1], in Ecuador, by June 2020, due to the COVID 19 pandemic, around 50,000 students from the 91 public technical and technological institutes are participating in classes virtually. In another article of the same newspaper publications, around 40,000 students from the Central University of Ecuador began their academic activities by opening 7,000 virtual classrooms by using the Microsoft Teams platform for delivery class, valuation of tasks, as well as the development of virtual works evaluation with other teachers [2].

The transition from a face-to-face modality to a virtual modality, due to the disruptive event caused by the COVID 19 pandemic, has mustered a series of digital, pedagogical, and organizational implications to give continuity to the university educational activities system [3]. This upheaval process required endeavors to produce permanent training

M. Botto-Tobar et al. (Eds.): CIT 2021, LNNS 405, pp. 123–134, 2022.
https://doi.org/10.1007/978-3-030-96043-8_10

courses that contribute to digital skills development, such as the mandatory management of virtual classrooms, digital educational resources, apps, and videoconferencing platforms. Concerning the management of virtual classrooms, universities in Ecuador use the b-learning modality. They combine virtual environments such as learning management systems (LMS) and physical environments such as classrooms and laboratories for learning. Development of the teaching-learning process both in the face-to-face and blended modalities [4]. In the virtual modality, the use of virtual environments (LMS) is mandatory, in compliance with regulations such as the Regulation for Academic Programs and Careers in Online, Distance and Blended Modalities or Media Convergence approved by the Higher Education Council [5], which in a certain way, strengthened the development of digital skills in Ecuadorian teachers.

The present research work aims to corroborate whether permanent training constitutes a digital literacy strategy in university teachers in the health area. The research analyzes the digital skills that teachers initially possess to plan continuous training activities that contribute to reducing the negative effect caused by the change from face-to-face mode to virtual mode due to the COVID 19 pandemic after digital competencies were evaluated after continuous training activities to determine if they achieved the expected effect. The survey evaluated the management of digital resources and their use in the development of the teaching-learning process.

1.1 Importance of Digital Skills Development

The volume of digital educational resources and the existence of models for the provision of services for online such as distance, open education, mixed education, short courses based on the acquisition of skills such as massive online courses, open access (MOOC), and Open Educational Resources (OER), leads to the need of improving and guarantee the quality and recognition of programs. Requiring higher education systems and institutions to take innovative measures related to the development of educational programs, curricula, and learning processes, to guarantee equality and accessibility [6].

On the other hand, the European Union Commission raised eight key competencies that all people need to fulfill and personal development, employability, social integration, sustainable lifestyle, success in life in peaceful societies, and way of life. Healthy life and active citizenship, these competencies respond to literacy skills, the use of different languages adequately and effectively for communication (multilingual), mathematical competence and competence in science, technology and engineering, digital competence, personal, social, and learning to learn competence; citizen, entrepreneurial competence and competence in cultural awareness and expression [7].

Digital disruption in the education sector is inevitable [8]. Thus, high education institutions guarantee the quality, equality, and accessibility for ad-dressing critical competencies within their educational programs, even more so when they try to maintain and expand the provision of learning services mediated by technology. In this way, digital competence gains interest in the teaching staff since it allows them to use a large volume of digital educational resources when programming the learning processes of their students [6].

Digital competence implies the safe, critical, and responsible use of digital technologies for learning, work, and participation in society and interaction with them. Including information and data literacy, communication and collaboration, media literacy, digital content creation (including programming), security (including digital well-being and cybersecurity-related competencies), property-related issues, intellectual, problem solving, and critical thinking [6].

People must use digital technologies and access, filter, evaluate, create, schedule, and share digital content. Besides, they must manage and protect the information, content, data, and digital identities and recognize and interact effectively with software, devices, artificial intelligence, or robots [7]. Competencies must allow students' effective participation in the knowledge society through creativity, problem-solving, lifelong learning, collaboration, and critical thinking [9]. Skills related to technology-mediated learning should allow:

- Read, write, speak, and listen in real time.
- Identify, locate, evaluate, and use information in real time.
- Understand and produce visual messages from teachers to students and vice versa.
- To create question that challenge and evaluate the meanings and purposes of texts.
- To create question, to analyze, interpret, evaluate, and create media messages.
- Use tools to manage, consume and create information.
- Use digital technology, communication tools, or networks to locate, evaluate, use, and create information.

Digital competencies developed with digital literacy at the university level seek to encourage academic staff to use ICT Information and Communication Technologies for online learning.

Digital literacy is an essential process through which we acquire the necessary skills to use new technologies. A person can perform different tasks in a digital environment [10]. Digital literacy is considered an emerging priority [11]. It increases the efficiency of human capital for sustainable development [8].

Lifelong learning is a continuous process that makes it possible to expand the knowledge, interpretive frameworks, teaching methods, and values of teachers in a constant, meaningful, appropriate, and appropriate way to the contexts, needs, changes, or advances of society [12, 13]. Ongoing training follows up from training actions that imply a close link between theory and practice [13]. Lifelong learning in universities is a continuous educational process regards any learning activity throughout life to improve and update a person's knowledge, skills, and abilities [14].

1.2 Superior Education Primary Educational Digital Resources

ICTs have a series of digital resources which, used in an educational environment, facilitate the development of learning activities [15].

Digital educational resources (RED) are multimedia content based on technology used in learning and training. The author designed, developed, and administered digital educational resources with a specific purpose. Thus, using them in the learning activities planned by the teacher to provide experience and create opportunities for students to (a) build and (b) use knowledge and, at the same time, (c) develop new skills in the use of technology [9].

Microsoft 365

Microsoft 365, also known as Office 365, generated a market share of 30.37% for 2019, the second software with the highest penetration at the business level after Google [15]. With the COVID 19 pandemic, the use of Teams increased such that Microsoft added 95 million users in 2020, and more than 183,000 educational institutions use Microsoft Teams [16].

Microsoft 365 is a platform that everyone needs to know, not only because of its market share but also because its applications are familiar to all users. Conceptually, Microsoft 365 (also known as Office 365) is a series of tools, which collaboratively allow to store resources in the cloud, edit text files, spreadsheets, and presentations, send emails, improve communication be-tween the employees of an organization through chats and videoconferencing and help groups or work teams in their daily tasks. The tool created by the giant Microsoft allows to create, access, and share documents online between different users in Word, Excel, PowerPoint, and OneNote, among others. The only need is access to the Internet and has the OneDrive program [17]. Microsoft versions for the home, business, large companies, and educational institutions [18]. The products that Microsoft 365 owns are visible in Graphic 1.

Learning Management Systems (LMS)

A learning management system (LMS) is software that allows you to cre-ate, offer, reuse and manage digital educational content [19]. One of the most used LMS is the Moodle platform since it has advantages such as operating without modifications in any operating system compatible with PHP. Be-sides, it works as a set of modules and allows adding or removing elements flexibly in different stages.

An LMS creates virtual classrooms, a digital environment through which academic staff and students have various telematic tools that facilitate teaching and learning processes [21].

Teachers use it to upload the course syllabus, post materials, deliver notes, request and collect assignments [22] (Fig. 1).

Videoconferencing Platforms

Videoconferencing platforms are systems that establish multidirectional, face-to-face communications between attendees. Video conferencing is an interactive system that allows several users to have a virtual conversation through the real-time transmission of video, sound, and text over the Internet [21]. Among the most used videoconferencing platforms are Microsoft Teams, Zoom, Skype, and Google Meet.

The main functionalities that a videoconferencing platform has are [22]:

- Friendly management for the teacher-student user.
- Autonomy to schedule, modify or cancel classes.
- Ability to share PowerPoint presentations in presenter mode.
- Autonomy to decide what to record and how to do it.
- Possibility of working with groups of students and simulating movements.

Fig. 1. Microsoft 365 applications **Elaborated by:** Margarita Aucancela

- Ability to share control of the session by a co-host.
- Space to manage questions/answers (Q & A).
- Easy integration with other resources.
- Facilitates synchronous e-learning [23].

For instance, Microsoft Teams integrates to its videoconferencing system functions to create Class-type teams, which allow publishing information and documents, uploading class materials in files, requesting, collecting, and most importantly, for a teacher to review. Tasks, which facilitate the delivery of notes displayed in the graphic (Fig 2).

Other Digital Educational Resources

Other digital educational resources that from experience as a university teacher is to use scientific databases, which are compilations of publications with scientific-technical content, such as magazine articles, books, theses, conferences, and many others, with thematic content, whose objective is to gather all the possible bibliographic production on an area of knowledge [26], the leading scientific databases in the health area are: Cochrane, Science Direct, Pub-med, among others.

Using Educational Blogs and Search Engines such as Google, Bing, Altavista allows us to overview the information, deepened when using scientific databases.

The use of videos such as Youtube, Vimeo provides visual information for the development of procedures. On the other hand, specialized software such as clinical and surgery simulators contribute to procedures. There are websites for the creation of visual, media, and information messages. Some of these sites allow the creation of diagrams, other presentations or concept maps, other timelines, and other flow diagrams, some of these sites are Lu-acid chart, Genially, Canva, Draw.IO, and Timeline.

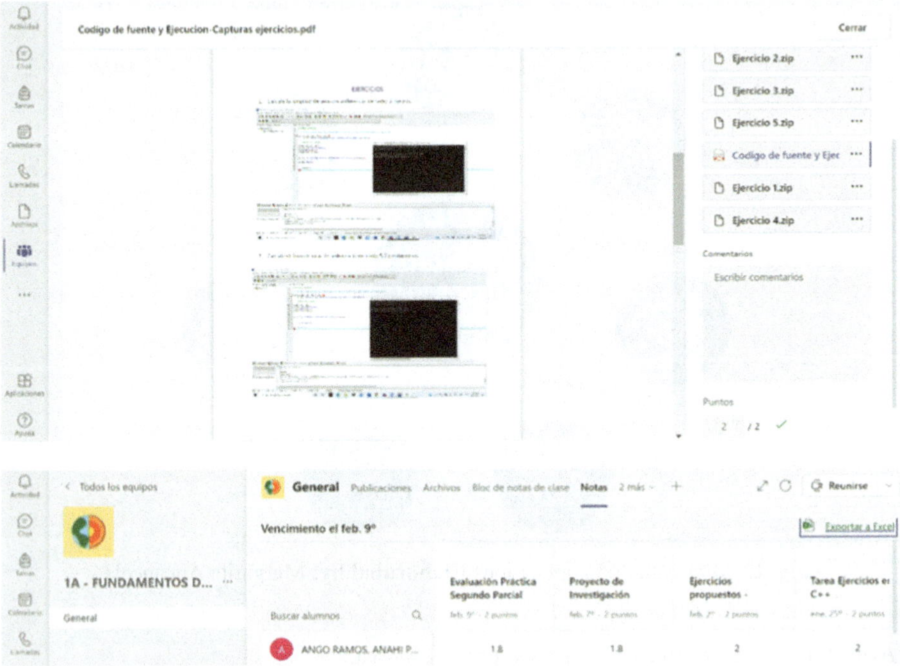

Fig. 2. View assignments and grades using Microsoft Teams **Source (s):** Microsoft Teams

The increase in mobile technology has motivated smartphone applications, which have enormous potential in all areas of medical practice, offering promising opportunities to improve the scope and quality of health services. Some applications are iResus, iDoctus, Omnio, Medscape, Medicalc, Pubmed Mobile, Epocrates [27].

The use of the different virtual resources will depend on the approaches and methodologies used by the teacher, the use of the REDs may depend on the activities planned by the teacher [9], such as the learning result that the teacher sees necessary to reach [24].

2 Methods and Materials

The research corresponds to a quantitative research design. The design used search strings for the bibliographic review, survey outline, digital competence and education, digital resources, and education. Google Scholar platform and Springer's Scientific allow exploring databases, newspapers, and websites of organizations such as the European Union and UNESCO.

For evaluating the effect of continuous training, the development of several activities was necessary to cope with data; application of a survey before starting the training processes and another at the end of the training processes was a must. Thirty-eight teachers from Escuela Superior Politécnica de Chimborazo, Public Health Faculty, participated. The survey has eight questions used to know the gender and age of the teacher, as well as the domain of knowledge in the management of digital educational resources such as Email (Outlook, Gmail), Database scientific data (Cochrane, Science Direct), Cloud storage (Onedrive, Google Drive, Dropbox), Educational Blogs, Search engines (Google, Bing, Altavista), Videos (Youtube, Vimeo), Specialized soft-ware (Simulators), Office automation (Word, Excel, Powerpoint), Websites (Healthline, Medscape) using a Likert scale. Besides, digital educational re-sources develop their academic activities evaluation, using virtual classrooms and videoconferencing platforms for virtual learning.

For data collection, it was essential to use the Microsoft Forms platform. SPSS program carried the treatment of the collected data.

3 Results and Discussions

In general, the teachers who participated in the development of this study have that 52.63% are male and 47.37% are female, the age of the teachers is between 26–35 years old, 36–45 years old, 46–55 years old, and more than 56 years old (Table 1).

Table 1. Teachers' gender and age data. **Source**: Survey **Elaborated by**: Margarita Aucancela

Gender	Age				Total
	26–35 years	36–45 years	46–55 years	More than 56 years	
Female	1	9	5	3	18
Male	2	6	4	8	20
Total	3	15	9	11	38

3.1 Teachers' Digital Skills

To evaluate whether the effect of continuous training helped the development of digital competencies in health teachers, that is, the domain of knowledge in the management of digital resources, the following hypothesis was raised:

H0: The average knowledge about the management of digital resources be-fore continuing education is equal to the average knowledge managing digital resources after continuing education.

H1: The average knowledge about digital resources management before continuous training is not equal to the average knowledge about digital resources management.

Once established, the hypothesis and significance level of 5%. The normality test determined the type of statistic to be used, with the result that the data have a normal distribution because the p-value for the data collected before the training activities is equal to 0.130. At the same time, the p-value for the data collected after the training activities was 0.408. The following table displays the data (Table 2).

Table 2. Normality test - Shapiro-Wilk **Source**: Analysis SPSS **Elaborated by**: Margarita Aucancela

	Statistics	gl	P
After	0,95502013	38	0,13036739
Before	0,97061742	38	0,40852715

Once applied the normality tests, the implementation of the test statistic by applying the student's t-test for related samples was the next step. As a result, the p-value is 0.007, which means that it is less than 0.05. Therefore, the alternative hypothesis is accepted: H1, in this way, shows a significant difference between the competencies that initially the teacher dominated and the teacher's competencies when receiving the permanent training courses. The following table exhibits the results (Table 3).

Table 3. T student for paired samples **Source:** Survey **Elaborated by:** Margarita Aucancela

	Statistics	gl	P
After-Before	−2,867	37	0.07

3.2 Use of Digital Resources in the Development of Academic Activities

It is one thing to handle digital resources and use these resources to develop teachers' academic activities. To evaluate the use of digital resources such as Email (Outlook, Gmail), Scientific databases (Cochrane, Science Direct), Cloud storage (Onedrive, Google Drive, Dropbox), Blogs Educational, Search Engines (Google, Bing, Altavista), Videos (Youtube, Vimeo), Specialized Software (Simulators), Office Automation (Word, Excel, Power-point), Websites (Healthline, Medscape), a Likert scale was indispensable.

For knowing the use of digital resources in the development of teachers' academic activities, the hypothesis was:

H0: The average use of digital resources before continuing training is equal to the average use of digital resources after training.

H1: The average use of digital resources before continuing training is not the same as the average use of digital resources after training.

A significance level of 5% was applied. Then the normality test was carried out, which gave a p-value of 0.505 for the data collected before the training activities and a p-value of 0.448 after the training activities. The following table demonstrates the statistical data (Table 4).

Table 4. Normality test-Shapiro-Wilk **Source:** Survey **Elaborated by:** Margarita Aucancela

	Statistics	gl	P
Before	0,974	38	0,505
After	0,972	38	0,448

Once applied the normality tests, fulfillment of the statistic test by applying the student's t-test for related samples was a must. As a result, the p-value is 0.000, which means that it is less than 0.05, rejecting the null hypothesis. The following table displays the analysis results (Table 5).

Table 5. T-student for paired samples **Source:** Survey **Elaborated by:** Margarita Aucancela

	Statistics	gl	p
Before-After	−4,588	37	0

For evaluating the use of virtual classrooms, the authors directly asked whether the participants used them or not. Continuous training contributed to their use, increasing by 39.5%, representing the data in the following graphic (Fig. 3).

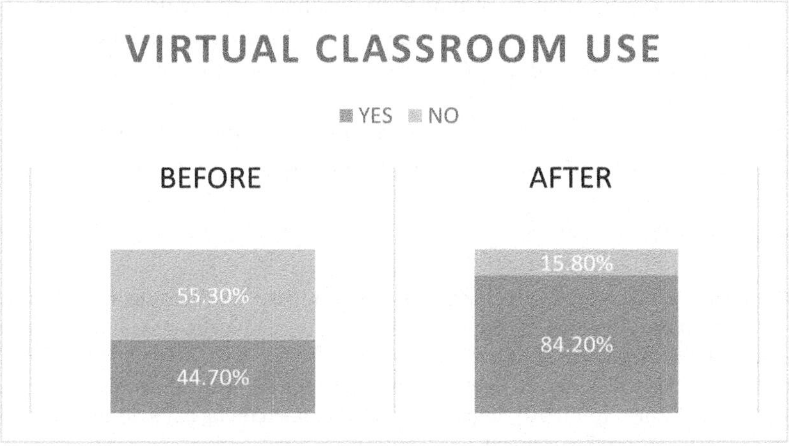

Fig. 3. Virtual classroom analysis use **Source**: Survey **Elaborated by:** Margarita Aucancela

The use of videoconferencing platforms evaluated by requesting one indicated it used to perform most academic activities (Fig. 4).

As shown in graphic 4, the use of video conferencing platforms such as Microsoft Teams and Zoom increased considerably, but not via Skype or others.

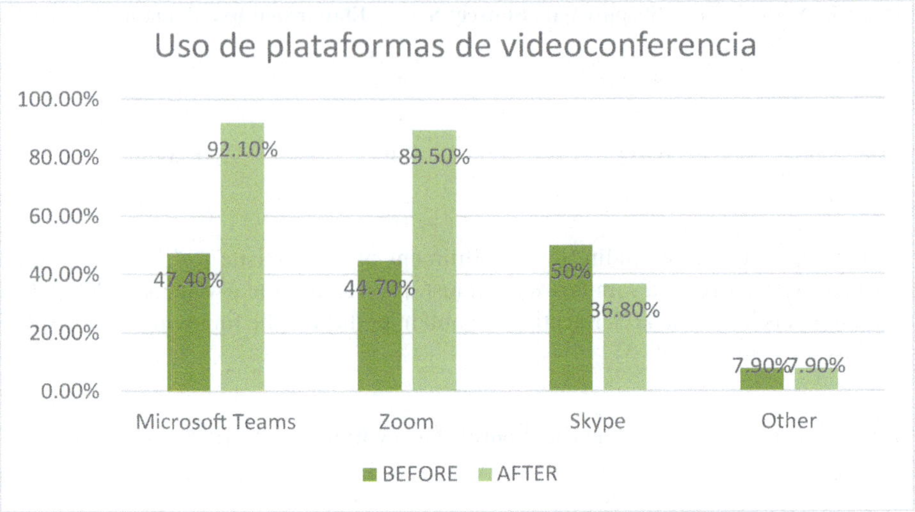

Fig. 4. Use of videoconferencing platforms **Source**: Survey **Elaborated by:** Margarita Aucancela

4 Conclusions

Permanent training has taken on a leading role when changing from face-to-face to virtual mode, not only when it is necessary to overcome an emerging situation such as the disruptive event caused by the COVID 19 pandemic, but also because through it has been achieved digital literacy in university teachers and the development of digital skills. The use of digital educational resources allows directing digital literacy to students and promoting in them the development of learning activities that contribute to their professional training.

Future research should focus on deepening synchronous e-learning, the activity-based learning approach, and the effect of video conferencing platforms.

References

1. El Comercio: 50 000 estudiantes empezaron clases online en institutos tecnológicos | El Comercio (2020). https://www.elcomercio.com/actualidad/estudiantes-clases-institutos-tecnicos-educacion.html. Accessed 22 Mar 2021
2. El Comercio: 'MI UCE en casa' es el curso para iniciar a los estudiantes en el manejo de aulas virtuales | El Comercio (2020). https://www.elcomercio.com/actualidad/plataforma-uce-clases-aulas-virtuales.html. Accessed 22 Mar 2021
3. García-de-Paz, S., Santana Bonilla, P.J.: Transición a entornos de educación virtual en un contexto de emergencia sanitaria. Rev. Educ. Distancia **21**(65) (2021). https://doi.org/10.6018/red.450791
4. Reinoso, G.L., Castro, A.C., Izquierdo, J.E., Cornejo, A.N.: El B-learning y su aplicación en la enseñanza universitaria del Ecuador (2020). https://doi.org/10.37954/se.v5i2.146
5. Correa Peralta, M., Vinueza Martinez, J., Lazo Sulca, R.: Fundamentos de aplicación de blenden-learning para las Universidades del Ecuador. Dominio las Ciencias **3**(2), 401–423 (2017)

6. UNESCO: Educación superior digital (2020). https://es.unesco.org/themes/educacion-superior/digital. Accessed 31 Mar 2021
7. Consejo de la Unión Europea: Recomendación del Consejo, de 22 de mayo de 2018, relativa a las competencias clave para el aprendizaje permanente. D. Of. la Unión Eur. **C189/1**(2), 1–13 (2018)
8. Anthonysamy, L., Koo, A.C., Hew, S.H.: Self-regulated learning strategies in higher education: fostering digital literacy for sustainable lifelong learning. Educ. Inf. Technol. **25**(4), 2393–2414 (2020). https://doi.org/10.1007/s10639-020-10201-8
9. Churchill, D.: Educational reforms, learning-centred education and digital resources for learning. In: Digital Resources for Learning. STE, pp. 1–17. Springer, Singapore (2017). https://doi.org/10.1007/978-981-10-3776-4_1
10. UNIR: Alfabetización digital: ¿Qué es?, ¿Cuál es su importancia? (2019). https://www.unir.net/educacion/revista/alfabetizacion-digital/. Accessed 13 Apr 2021
11. Pérez-Escoda, A., García-Ruiz, R., Aguaded, I.: Dimensions of digital literacy based on five models of development/Dimensiones de la alfabetización digital a partir de cinco modelos de desarrollo. Cult. Educ. **31**(2), 232–266 (2019). https://doi.org/10.1080/11356405.2019.1603274
12. Barba, R.: Tesis Doctoral : La Investigación-Acción Participativa Desde La Mirada De Las Maestras Participantes En Un Proceso De Formación Permanente Del Profesorado Sobre Educación Inclusiva, pp. 83–95 (2019). http://uvadoc.uva.es/handle/10324/37920
13. Fonseca, S., Hernández, V., Forgas, J.: Formación Continua Y Formación Permanente Desde El Desarrollo De Competencias Docentes En Las Instituciones De Educación Superior. Mikarimin. Rev. Científica 55–62. ISSN 2528–7842 (2017)
14. UNED: ¿Qué es la Formación Permanente? (2021). https://portal.uned.es/portal/page?_pageid=93,37646506,93_37646507&_dad=portal&_schema=PORTAL. Accessed 13 Apr 2021
15. Zapata, M.: Recursos educativos digitales: Conceptos básicos (2012). https://aprendeenlinea.udea.edu.co/boa/contenidos.php/d211b52ee1441a30b59ae008e2d31386/845/estilo/aHR0cDovL2FwcmVuZGVlbmxpbmVhLnVkZWEuZWR1LmNvL2VzdGlsb3MvYXp1bF9jb3Jwb3JhdGl2by5jc3M=/1/contenido/. Accessed 17 Apr 2021
16. Statista: Software de productividad para oficinas: cuota de mercado a nivel mundial en 2019 (2019). https://es.statista.com/estadisticas/1060714/cuota-de-mercado-de-software-de-productividad-de-oficina-a-nivel-mundial/. Accessed 16 Mar 2021
17. Curry, D.: Estadísticas de uso e ingresos de Microsoft Teams 2021-Negocio de aplicaciones (2021). https://www.businessofapps.com/data/microsoft-teams-statistics/. Accessed 19 Mar 2021
18. BeServices: ¿Qué es y para qué sirve Microsoft Office 365? (2020). https://www.beservices.es/microsoft-office-365-que-es-n-5441-es. Accessed 01 Apr 2021
19. Microsoft 365: Microsoft 365 con las aplicaciones de Office | Microsoft 365 (2021). https://www.microsoft.com/es-ww/microsoft-365?ocid=oo_support_mix_marvel_ups_support_smcuhfm365&rtc=1. Accessed 01 Apr 2021
20. Manrique-Losada, B., Zapata Cárdenas, M.I., Arango Vásquez, S.I.: Virtual environment to co-create digital educational resources in higher education. Campus Virtuales **9**(1), 101–112 (2020)
21. Universidad de Murcia: ¿Qué es el aula virtual? - Ayuda Aula Virtual (2021). https://www.um.es/aulavirtual/primeros-pasos/que-es-el-aula-virtual/. Accessed 17 Apr 2021
22. Cabero-Almenara, J., Arancibia, M.L., Del Prete, A.: Technical and didactic knowledge of the moodle LMS in higher education. beyond functional use. J. New Approaches Educ. Res. **8**(1), 25–33 (2019). https://doi.org/10.7821/naer.2019.1.327
23. Learning, E.-A.: ¿Qué es la Video Conferencia? (2021). https://www.e-abclearning.com/queesvideoconferencia/?lang=en. Accessed 13 Apr 2021

24. Vera, F.: Impacto de las plataformas de videoconferencia en tiempos de COVID-19, pp. 23–24 (2020). https://fernandovera.cl/wp-content/uploads/2020/10/Impacto-de-las-plataformas-de-video-conferencia-en-tiempos-de-COVID-19.pdf
25. Reinoso-González, E.: La videoconferencia como herramienta de educación: ¿qué debemos considerar? Rev. Española Educ. Médica **1**(1), 60–65 (2020). https://doi.org/10.6018/edumed.426421
26. Universidad de Extremadura: Bibliogufas: Cómo buscar en las bases de datos de forma eficaz: QUÉ SON LAS BASES DE DATOS (2021). https://biblioguias.unex.es/c.php?g=572073&p=3944411. Accessed 17 Apr 2021
27. Iglesias-Posadilla, D., Gómez-Marcos, V., Hernández-Tejedor, A.: Apps y Medicina Intensiva. Med. Intensiva **41**(4), 227–236 (2017). https://doi.org/10.1016/j.medin.2017.01.003
28. Bolaño, M.: Herramientas multimedia interactivas Aplicadas en la educación en tiempos de pandemia por Covid-19 y sus desafíos. I Simp. Int. Implementación las herramientas tecnológicas para la Educ. durante el tiempo pandemia por COVID-19 (2020)

Simulation System of a Tomato Sorting Process Using Artificial Vision

Sandro Balarezo[1(✉)], Xavier Arias[2], Kevin Espín[2], Miguel Aquino[3], and Guillermo Novillo[3]

[1] Universidad Nacional de Colombia, Medellín, Colombia
sfbalarezog@unal.edu.co
[2] Universidad de Las Fuerzas Armadas ESPE, Sangolquí, Ecuador
{axarias,kaespin2}@espe.edu.ec
[3] Escuela Superior Politécnica de Chimborazo, Riobamba, Ecuador
{socrates.aquino,gnovillo}@espoch.edu.ec

Abstract. This present research consists of a design of a virtual plant to simulate the automatization of a tomato sorting process by means of artificial vision. This sorting process starts with the entry of tomatoes on a conveyor belt, the fruits move at a constant speed until they are under the scope of the camera. An image processing system developed in LabVIEW software is in charge of detecting three main characteristics: green color, red color and defects on the surface of the fruit. The simulation was performed in Factory I/O software in conjunction with TIA Portal to obtain a real simulation environment.

Keywords: Machine vision · Tomatoes · Processing · Imaging

1 Introduction

The tomato commercialization process depends directly on the size and color of the fruits, these factors intervene in the quality of its derivative products and subsequently in the percentage of consumer acceptance; for example, a series of sensory tests carried out on ketchup have revealed that brightness and hue angle are the elements responsible for consumer acceptance. The pigment taken up by tomato fruits during the ripening process allows differentiation of color changes in a progression from green to red tomato [1]. Farmers must make many decisions regarding the process of planting and harvesting tomatoes based on the empirical knowledge of the people who harvest the fruits, and this method is one of the most inefficient [2]. Currently, most growers sort tomatoes manually, meaning that there are people who, due to their experience, identify the color of each tomato and detect the presence of any defects on the surface of the harvested fruit. The problem with this method of sorting arises in large-scale production, because as the fruits increase, several experienced workers are needed to separate the good fruits from the bad ones, and this increases the cost of production. An adequate quality control system would allow the integration of programs or systems within the tomato production process

M. Botto-Tobar et al. (Eds.): CIT 2021, LNNS 405, pp. 135–146, 2022.
https://doi.org/10.1007/978-3-030-96043-8_11

to improve productivity and ensure that the company seeks continuous improvement in the quality of its products [3].

Nowadays technology offers new computer vision systems that allow quantitative characterization of various physical features such as size, morphology, color and textural properties using a digital image. In addition, these systems are often coupled with systems that can replace human manipulation in order to make the processes more efficient.

1.1 Contaminants in Tomatoes

In Ecuador tomato is the fourth most important fruit in production, there are 3333 hectares of tomato plantation in the country and the annual production is around 61426 tons per year. The tomato fruit has a large amount of vitamins and mineral salts and is considered an excellent element to add to the diet [4].

According to agronomists specializing in tomato planting and cultivation, the causes of the presence of contaminants are due to pests or diseases that affect these fruits, the most relevant in the area being anthracnose or commonly known as chicken eye or tomato moth and calcium deficiency. Some of the contaminants and pests in tomatoes are:

Gray Rot. This disease shows its first traces on the leaves and flowers of the tree, while in the fruit there is a rotting in which hairs of the fungus are observed, as can be seen in Fig. 1.

Fig. 1. Tomato with gray rot [10].

Tomato Moth. Figure 2 shows the attack of the pest that affects any stage of development of tomato, that is, from its formation the fruits can already be attacked. The larvae enter the fruits to feed on them, creating holes and mines.

Calcium Deficiency. This disease begins with the presence of brown areas on the skin of the fruit, which with the passage of time dries out as the cells in the affected area die. The effect of this disease can be seen in Fig. 3 [5].

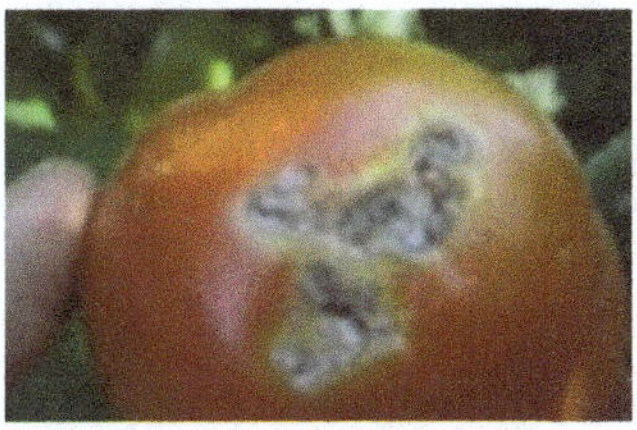

Fig. 2. Tomato with moth [10].

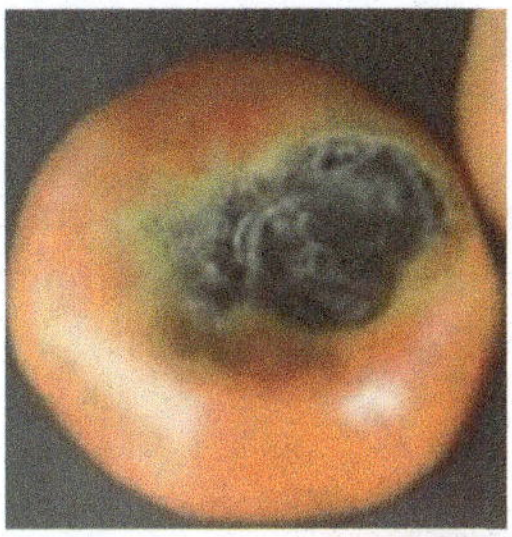

Fig. 3. Tomato with calcium deficiency [10].

1.2 Artificial Vision

The use of artificial vision is aimed at extracting information from the physical world from digital images. This technique appeared with the evolution of cameras, which are currently used not only to capture souvenir photographs, but also for quality control in industry, which is one of its most important applications [6].

After capturing an image or video, which is basically a sequence of images, processing must be applied to extract the desired information by filtering colors, shapes, textures, etc. A camera with excellent pixel quality can obtain very specific data regarding failures on mechanical surfaces exposed to the effect of hostile environments and analyze these data in an in- report that details the results obtained for the personnel in charge to perform preventive or corrective maintenance as appropriate [7].

Stages of an Artificial Vision System. Artificial vision attempts to reproduce the behavior of the human eye, which captures light through the eyes, and this information passes through the optic nerve to the brain where it is processed.

- Acquisition of digital images using some type of sensor, typically a camera.
- Image processing, removing non-useful parts and/or enhancing areas of interest.
- Isolation of the elements of interest for later interpretation.
- Recognition, or distinguishing the isolated elements based on their characteristics [8].

1.3 Image Processing

This process consists of the modification of a digital image captured from the real world to obtain information about some specific aspect of an object of study, such as colors, shapes, textures, etc. [9].

Image. According to Vélez, Moreno & Sánchez (2016) a two-dimensional image is defined as: "A function that to each pair of coordinates (x, y) associates a value relative to some property of the point it represents (its brightness or hue)". It also defines the digital image as: "A two-dimensional matrix, each element of the matrix is called a pixel". The pixel stores the brightness with 256 levels of both grayscale and color images, with 256 levels in each layer (red, green and blue).

RBG Image. It is the one that associates to each point of the image a triad of intensity values (red, green and blue). Figure 4 shows the image of a tomato in good condition with all its colors [8].

Fig. 4. RGB image of a tomato [10].

Grayscale Image. It is the representation of an image in which each pixel is drawn using an individual numerical value, on a scale that can be between black and white. In Fig. 5 shows the same image of Fig. 5 but transformed to grayscale [10].

Fig. 5. Image of tomatoes in grayscale [10].

2 Materials and Methods

2.1 Materials

Industrial process simulation software was used to develop a tomato sorting machine, which consists of four important stages and is shown in Fig. 6.

Dosing. A spiral was designed as a sorting machine feeder, in which all the tomato production will be placed to maintain the same output direction and positioned on a conveyor belt.

Transport. The conveyor belt is in charge of transporting the tomato from the starting spiral to the sorting stage, passing through the sensors.

Sensors. The camera located on the conveyor belt detects three characteristics in the tomatoes for their respective classification: green color, red color and tomatoes that have some type of surface damage.

Sorting. Once the image processing stage is complete, three rotating actuators direct each of the tomatoes onto different ramps. The ramps are used to separate the red ripe tomatoes from the green tomatoes and the tomatoes that are damaged.

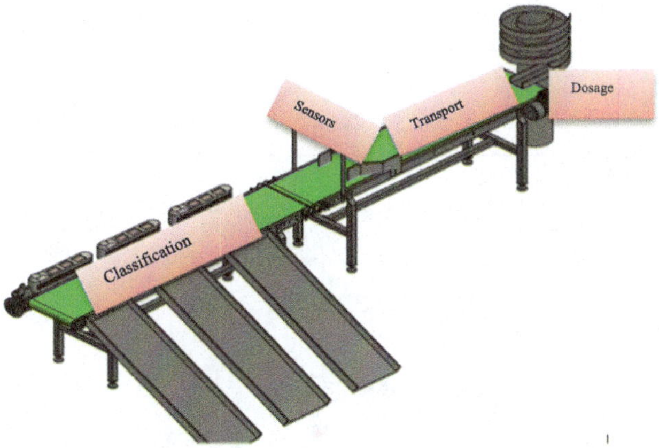

Fig. 6. Design of the tomato sorting process with artificial vision.

2.2 Methods

Color Detection. For the development of the control system and image processing, LabVIEW software was used for its ease of programming and the Vision Assistant tool. The procedure for detecting red tomatoes, green tomatoes and damaged tomatoes is described below.

1. The initial process with obtaining an image of the tomato by accessing a camera connected to the computer. The blocks used for this step are found in NI-Imaq-dx and must be connected in the following order: Open Camera - Configure Grab - Grab and finally Close Camera as shown in Fig. 7.

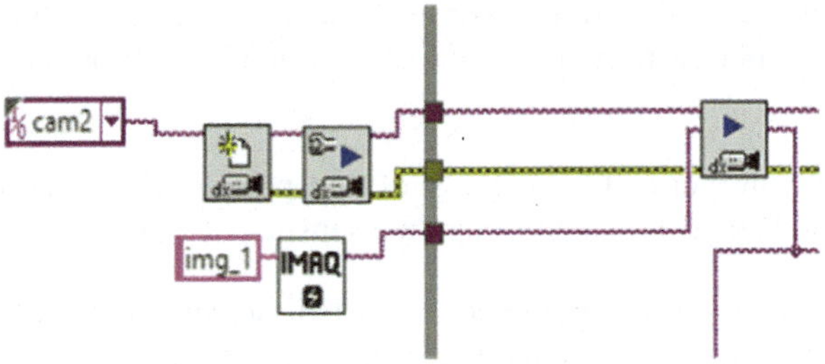

Fig. 7. Image acquisition with NI-IMAQ-dx

2. Once the desired image is obtained, image processing is applied to detect all tomatoes that are red and in good condition. This requires taking a template of a tomato that represents the red color desired by the user and the computer assigns it a label called red. Figure 8 shows the commands used, which are applied to all the tomatoes circulating on the conveyor belt.

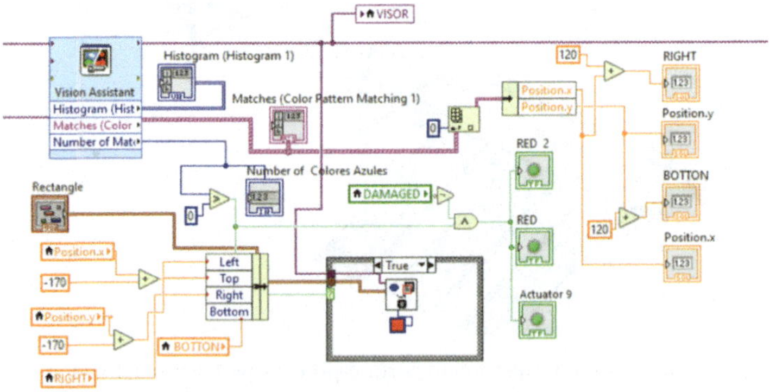

Fig. 8. Image processing for red tomatoes.

3. The Vision Assistant tool is configured to obtain as much information as possible from each image and to draw a circle each time the desired tomato color appears. To verify the correct operation of this tool, a Case instruction was applied so that a box is drawn on the image only when the color is identified as the desired color as shown in Fig. 9.

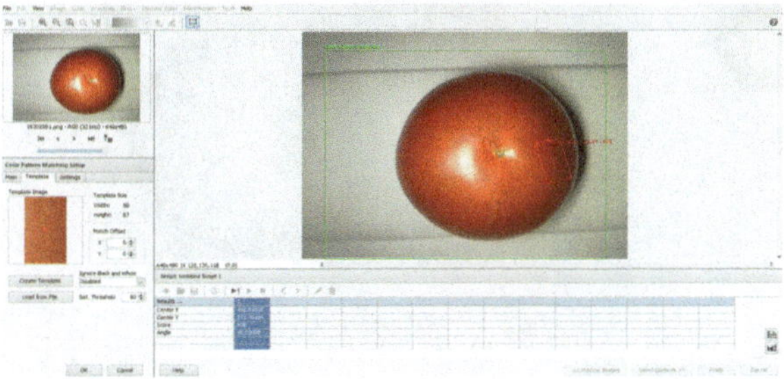

Fig. 9. Vision assistant configuration to detect the red color.

4. Repeat the same process for green tomatoes, taking into account that the Vision Assistant must be configured for green tomatoes.
5. Figure 10 shows a set of three LEDs that will light up as a warning when the artificial vision system detects the color that has been previously configured by the user.

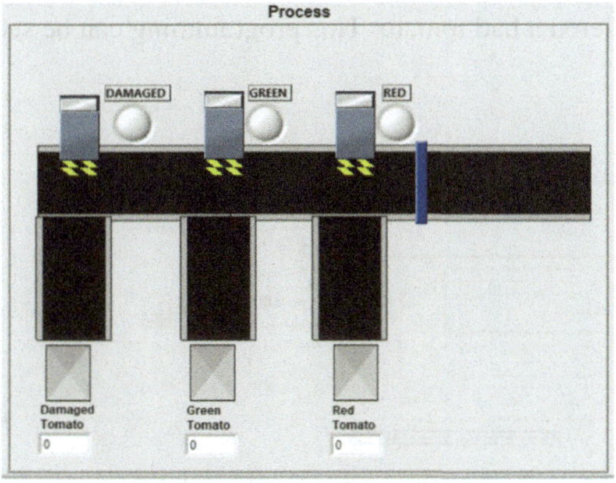

Fig. 10. Color identification LEDs.

Detection of Imperfections. Quality control corresponds to the analysis of each of the tomatoes to detect the presence of pits or deformations on the surface of the tomato. To achieve this, the following steps were followed:

1. The main function of the Features Detection tool is to detect points of interest in an image. The points of interest in an image are those that have a well-defined and robust position on the surface of the tomato. Figure 11 shows the configuration of the Vision Assistant to detect defects in tomatoes.

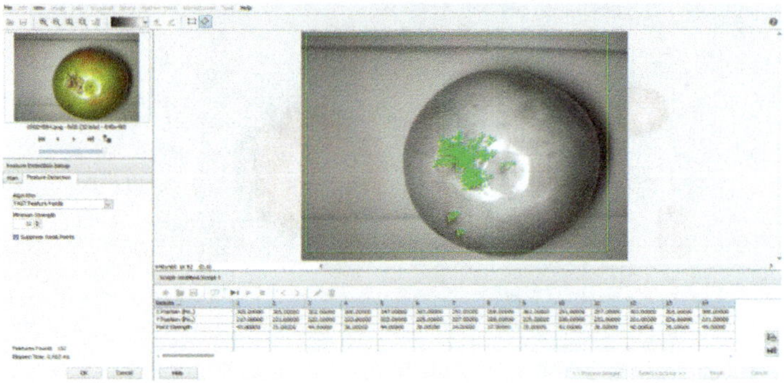

Fig. 11. Detection of tomato surface damage.

2. The Vision Assistant gives the number of coincidences that in turn represent the number of defects found in a tomato, these defects are more common when the tomato is red in color and the higher the number of points detected, the greater the defect in the fruit, which means that the tomato is damaged.
3. To determine whether a tomato is defective or not, an acceptance range is used, i.e., when the number of defects is in a range that exceeds the limit entered by the user, it will be considered a bad tomato. This programming can be seen in Fig. 12.

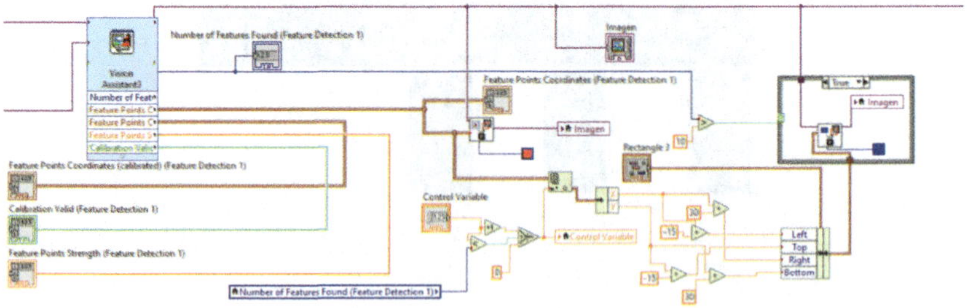

Fig. 12. Image processing for detection of damaged tomatoes.

4. Figure 13 shows the tomato quality indicators as follows: Poor, Fair and Good quality to be distributed in the different ramps.

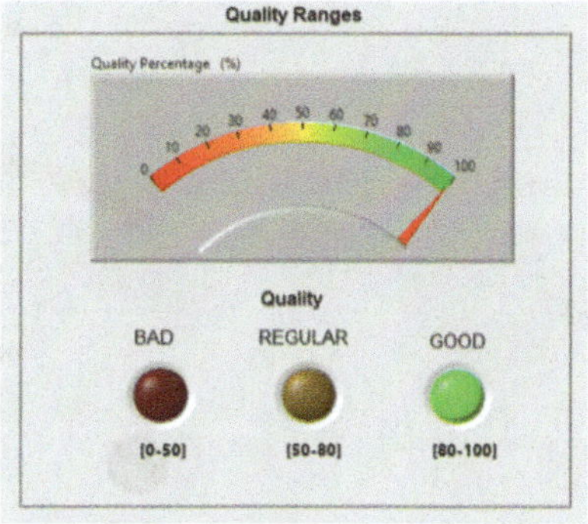

Fig. 13. Tomato quality indicators.

3 Results and Discussion

The testing protocol consisted of placing different tomatoes under the camera and first identifying the color of each one of them, then the quality control was applied to determine if this tomato is good, fair or bad. Three case studies are described below:

3.1 Case 1

This case corresponds to the analysis of a red ripe tomato with no damage, considered as an ideal case of tomatoes. Once the system has detected the color of the tomato, the red LED is activated, which in turn activates the corresponding actuator to separate the tomato on the ramp, as shown in Fig. 14.

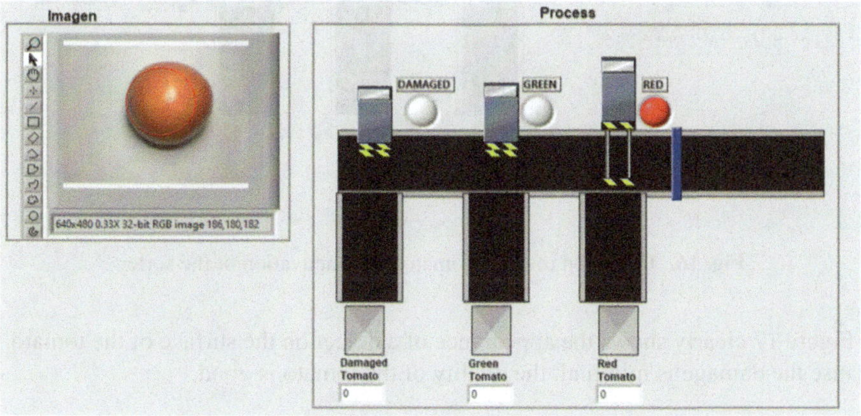

Fig. 14. Red tomato detection and activation of the sorter

The quality control of the tomato has determined that it is of good quality because no defects are found on the surface of the tomato. Figure 15 shows the results of this process.

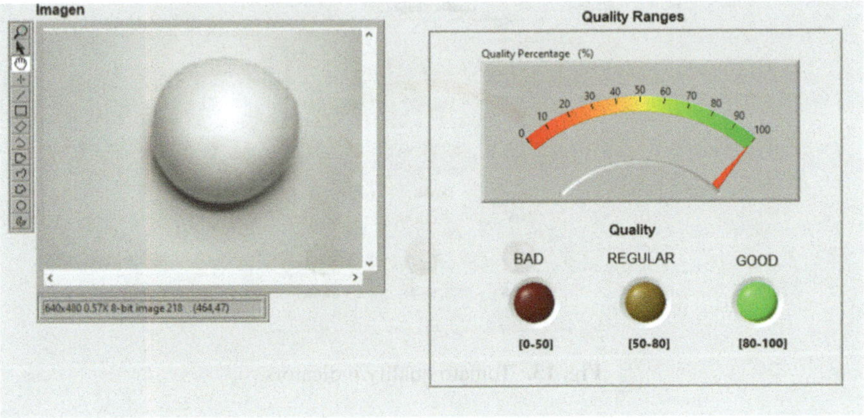

Fig. 15. Percentage of detected red tomato quality

3.2 Case 2

In this case, the results of a green tomato were analyzed. Figure 16 shows the test tomato in the top left part and its recognition by means of the LED located in the process stage.

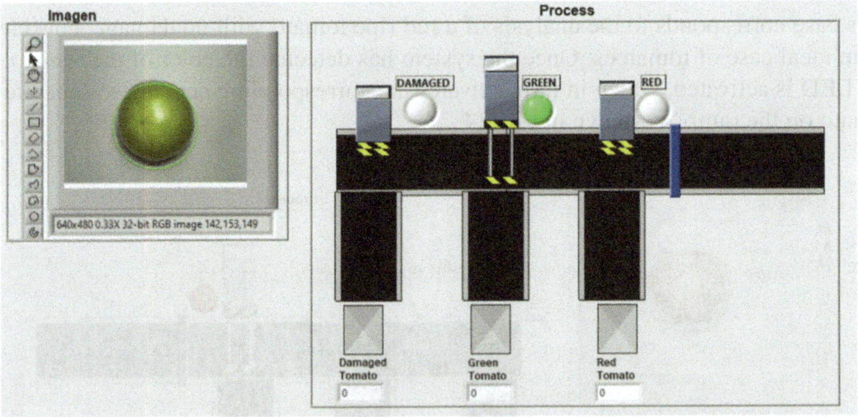

Fig. 16. Detection of green tomatoes and activation of the sorter.

Figure 17 clearly shows the appearance of a defect on the surface of the tomato, but because the damage is minimal, the quality of the tomato is good.

3.3 Case 3

The last test case consists of the analysis of a green tomato with several damage. Figure 18 shows the activation of the sorting LED corresponding to the ramp of damaged tomatoes.

Figure 19 shows the quality analysis of a tomato with poor quality due to the number of defects found.

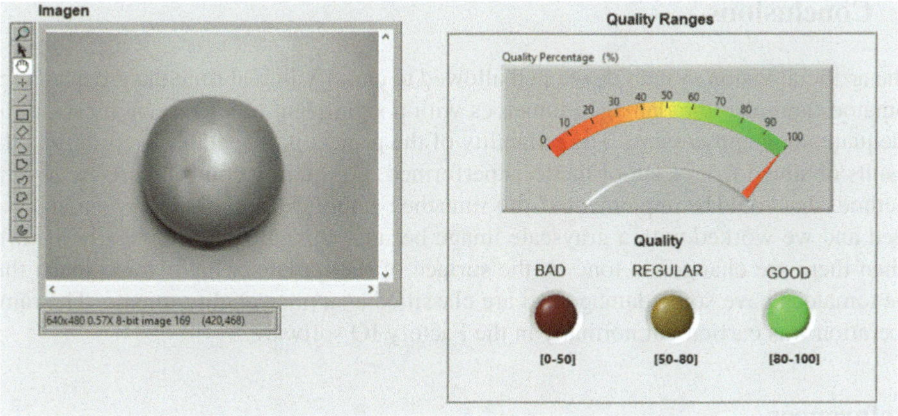

Fig. 17. Percentage of green tomato quality detected.

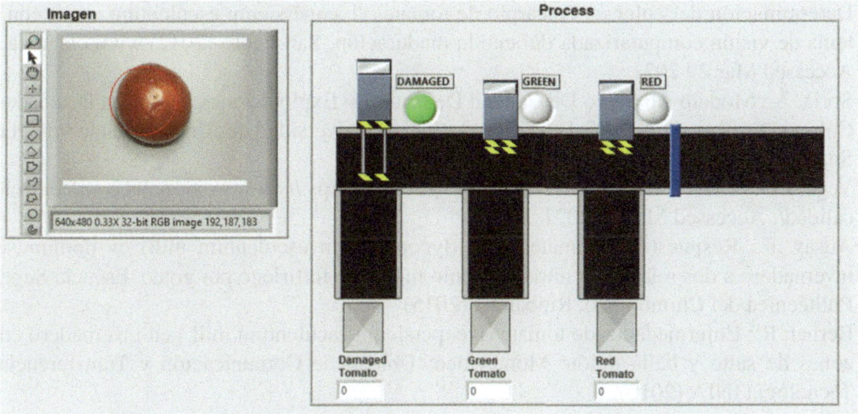

Fig. 18. Detection of damaged tomatoes and activation of the sorter.

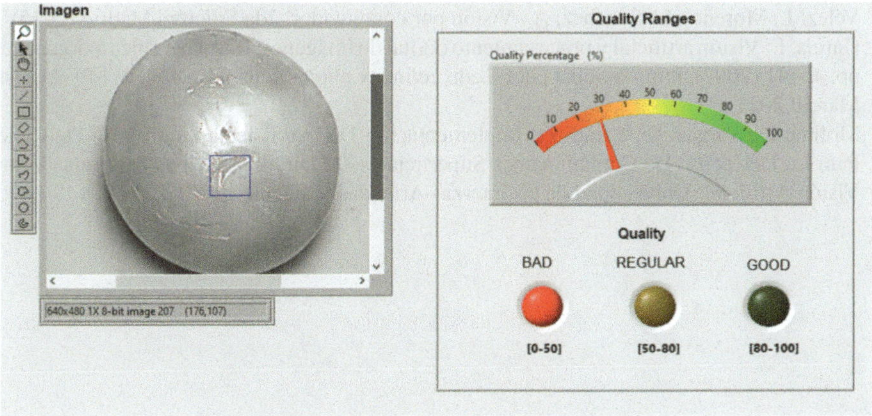

Fig. 19. Percentage of tomato quality detected.

4 Conclusions

The artificial vision system developed allowed to classify in real time the green and red tomatoes and also the damaged tomatoes with a significant efficiency by means of an adequate image processing. The reliability of the project is acceptable according to the results obtained from each of the tests performed. For quality control there are several methods that could be implemented, this time the Feature Detection Setup command was used and we worked with a grayscale image because this allows us to clearly identify when there are changes in tone on the surface of the tomato, which would mean that the tomatoes have some damage and are classified as a poor-quality tomato. The ramp operation was carried out normally in the Factory IO software.

References

1. Padrón Pereira, C.A., Padrón León, G.M., Montes Hernández, A.I., Oropeza González, R.A.: Determinación del color en epicarpio de tomates (lycopersicum esculentum mill.) con sistema de visión computarizada durante la maduración, San Pedro (2012).www.cia.ucr.ac.cr. Accessed Mar 29 2021
2. Soria, Á.: Modelo Analítico De Calidad De Fruto En Explotaciones Agrícolas Dedicadas Al Cultivo Del Tomate A Partir De Visión Artificial. Universidad de Alcalá Escuela Politécnica Superior, Alcalá (2020)
3. Vega, J.O.: Control de calidad *Gestiopolis* (2020). https://www.gestiopolis.com/control-de-calidad/. Accessed Mar 30 2021
4. Ausay, E.: Respuesta de tomate riñón (lycopersicum esculentum mill) cv dominic bajo invernadero a dos relaciones nitrato/amonio mediante fertiriego por goteo. Escuela Superior Politécnica del Chimborazo, Riobamba (2015)
5. Bernal, R.: Enfermedades de tomate (lycopersicum esculentum mill.) en invernadero en las zonas de salto y bella unión. Montevideo: Unidad de Comunicación y Transferencia de Tecnología INIA (2010)
6. Bilbao, J.: Fotografía y emigración. 1era edici, Salamanca (2008)
7. Viera-Maza, G.: Procesamiento De Imágenes Usando Opencv Aplicado En Raspberry Pi Para La Clasificación Del Cacao. Universidad de Piura, Piura (2017)
8. Vélez, J., Moreno, A., Sánchez, Á.: Visión por computador, 2da Edición. Madrid (2015)
9. García, I.: Visión artificial y procesamiento digital de imágenes, Rev. Científica axioma, no. 3, pp. 40–41 (2007). http://axioma.pucesi.edu.ec/index.php/axioma/article/view/149. Accessed Mar 30 2021
10. Molina, L., Vargas, C.: Estudio E Implementación De Un Sistema De Control De Calidad Para La Detección De Contaminantes Superficiales De Diferentes Tipos De Frutas Usando Visión Artificial. Universidad de las Fuerzas Armadas, Latacunga (2019)

A Prediction Model to Prevent Phishing Attacks on E-Mails Using Data Mining Techniques

Johanna Mishell Rosero, Walter Fuertes(✉), Theofilos Toulkeridis, and Henry Cruz

Department of Computer Sciences, Universidad de las Fuerzas Armadas – ESPE, P.O. Box 15-17-231-B, Sangolquí, Ecuador
{jmrosero1,wmfuertes,ttoulkeridis,hocruz}@espe.edu.ec

Abstract. The current study presents a design and implementation of a prediction model that prevents phishing attacks on emails using data mining techniques. Hereby, we collected appropriate data sets in order to identify the characteristics of the infected emails, which allow a phishing attack to be successful from three significant sources, being Monkey.org, Enron.org, and PhishTank.com. For the design and implementation of the model, we used the CRISP-DM methodology in order to generate the detection model based on the characteristics that recognize an email as phishing. Likewise, we developed a descriptive analysis using Decision Trees, Bayesian Networks, and Random Forest as data mining algorithms, whose most precise has been Random Forest. Once the email was detected as infected, the algorithm quarantines the email. Subsequently, it sends a notification message through WhatsApp. Finally, we evaluated the model in a controlled environment. The results demonstrate the functionality of the model since it achieved a precision score greater than 97.28% in the detection of phishing emails.

Keywords: Data mining · Phishing · Prediction model · Random Forest

1 Introduction

Phishing is one of the most common and severe threats on the Internet for the last 20 years. This activity is performed by using instant messaging or e-mail services [17] when taking advantage of companies and users' vulnerabilities in cyberspace. In such cyberattacks, the phishers attempt to steal confidential user information by posing as a trusted entity [6] and carrying out fraud, generally for economic purposes.

According to the Verizon Data Breach Investigations Report (DBIR) [14], phishing is the main threat action variety in security breaches. This report also indicates that users are more susceptible to become phishing victims if they

M. Botto-Tobar et al. (Eds.): CIT 2021, LNNS 405, pp. 147–162, 2022.
https://doi.org/10.1007/978-3-030-96043-8_12

receive the attack through mobile devices. Furthermore, as claimed by the Anti-Phishing Working Group (APWG) on their 4th Quarter 2020 Report, the financial institution, webmail, and SaaS site category was the one most frequently victimized by phishing in that quarter. Business e-mail compromise scams are becoming more costly for victims, which requires the implementation of protection tools for cybersecurity based on data science to detect them and subsequently establish countermeasures. Unfortunately, the current phishing detection methods may be inaccurate, as phishers use innovative methods, taking advantage of users' ingenuity while using e-mails.

Within this context, the industry tries to decipher such issues constantly. Baykara et al., in [1] presents a complete analysis of the meaning of phishing and its characteristics which allow an attack through e-mails. Moradpoor et al., in [10] proposed new methods to counteract phishing, and Bikov et al., in bikov2019phishing focuses on the losses and damages caused by such attacks. Furthermore, Chatterjee and Namin [4], centers on phishing websites. Backes et al. [16] consider the time, costs, and profitability in mitigation strategies. Chin [5] focuses on phishing signature classification. Lastly, Oña et al. [11], and Espinosa et al. [7] developed anti-phishing prototypes using classification algorithms and machine learning methods.

The present study aims to design and implement a prediction model in order to detect social engineering attacks such as phishing in infected e-mails using data mining techniques. To achieve such scope, we first identify the characteristics of infected e-mails which admit the phishing attack to be successful. Therefore, we performed analyses of different phishing e-mails from three essential sources as Monkey.org, Enron.org, and PhishTank.com, which are identical ones, allowing the generation of a dataset. Then, we used the CRISP-DM methodology for the design and implementation of the model [9]. The detection model was generated based on the characteristics that recognize an e-mail as phishing. Lastly, we evaluate the model through proofs of concept in a controlled environment. The results indicate the functionality of the model since it reached a degree of precision of being more significant than 97.28% in detecting e-mails infected with phishing.

The main contribution of this study is to implement a phishing-type attack prediction model using data mining. This contribution achieves to detect such attacks in e-mails without requiring additional investments, thus detecting them and alert users via WhatsApp.

The remainder of this article is organized as follows: Sect. 2 describes the research design and focuses on constructing and evaluating the prediction model using data mining. Section 3 enlightens the evaluation of the results while it also contains the discussion of its central findings. Section 4 finalizes the conclusions and the potential lines of future development.

2 Materials and Methods

The present chapter reports the application of data mining using several levels of abstraction organized so that the phases, general tasks, specialized tasks, and process instances of the methodological guide allow the construction of the phishing attack prediction model. For this purpose, we use the Cross-Industry Standard Process for Data Mining (CRISP-DM) methodology, which is a practical guide to these ends. CRISP-DM has six phases, whose sequence is not rigid, as it allows bidirectional movement between each phase, as deemed necessary [13]. A description is given in detail.

2.1 Understanding the Data

In order to characterize the emails infected with phishing, we collected the URLs and email addresses from the PhishTank website in a .CSV format. Furthermore, the API that PhishTank provides to verify if a link is in its database and if it is phishing or not [2]. Likewise, we used Monkey.org and Enron.mbox to develop tests, the same ones, which provide emails with phishing and non-phishing features. The community, dedicated to studying such attacks, ensures new data and characteristics for improved results while continuously updates such data. This information allows us to identify a set of previously defined characteristics stored in a dataset in the emails. Through data mining processes, they would sustain the identification of infected emails.

Furthermore, it was possible to group and synthesized effectively all those main characteristics which define that an email is phishing, although some previous studies enabled analyzing and determining them [8,15]. Additionally, groups of words that appear regularly in phishing-infected emails were used as characteristics. Hereby, the entire mining process was facilitated, as described in the following section.

2.2 Predictive Phishing Detection Model

Predictive analysis is considered an area in data mining that focuses on extracting adequate data and predicting behavioral trends. These can be applied to any unknown event, whether in the past, present, or future. In order to conduct predictive analysis, it is essential to count with a considerable amount of data, where specific characteristics generate the problem by treating and extracting it. We performed the predictive analysis with the datasets obtained with the data mining technique.

Regarding the development of the predictive model, we considered the following techniques which allowed us to fulfill the stated objective, such as:

- **Bayesian techniques**: this technique uses Bayes' theorem, which can be determined as parameter-based learning, that given a data structure, the probability can be determined before the event occurs and under required conditions;
- **Techniques based on decision trees and rules learning system**: this technique classifies an object based on a series of rules, which belong to the "divide and conquer" type of algorithms;

- **Techniques based on artificial neural networks**: this technique requires training based on the weight of each neuron that interconnects with each other, based on the topology of the network and its weights in the different connections, determines a pattern classification.

Data Collection Process: The data mining process aims to extract information from a data set and transform it into an understandable structure for later use. Given this principle, we listed the sources initially collected for the development of this project (see Table 1).

- **PhishTank database**: We have extensive records updated every hour, which allowed technicians to fasten and up-to-date the phishing detection, being ideal for integrating into the model to be designed. This results in a page for collaborative clearing of data and information on phishing on the Internet [2]. The developers of this page have an open API so that developers and researchers can integrate anti-Phishing codes. For this project's data collection, we downloaded the records in two formats: CSV and JSON.
- **Monkey.org**: is a community page where a group of people is dedicated to the study of phishing attacks on emails uploads infected emails daily. Additionally, it is a complete database being fully accessible to anyone who needs to perform tests or studies based on the emails stored in its database;
- **ENRON email database**: There are more than 5000 regular or non-phishing emails, which will allow a better classification and learning of the given model.

Table 1. Amount of data collected in each source.

Data sources	Before ETL	After ETL
PhishTank	14857	13368
Monkey.org	3900	2000
Enron	5000	2000
TOTAL	23757	17368

Characterization of Phishing in Emails: We compared legitimately versus other infected emails with phishing to define the characteristics that distinguish an email infected with phishing. This process allowed us to create the dataset of characteristics, the same that would be the input for the learning algorithms of the model to be developed.

Data Exploration: We determined if the collected data is complete, has consistency, and will meet the objective. The exploration was performed through searches and basic queries of the .txt and .CSV documents. The exploration also allowed to identify and confirm the most frequent words in emails with phishing, so they are considered characteristic of being used in the code for the detection model. After determining the number of times each word is repeated, the relative frequency was obtained in order to identify the difference and influence of its use in emails with and without phishing.

Data Preparation and Cleaning: We proceed to prepare the data to adapt it correctly to the selected data mining techniques. The two obtained databases, being private-Phishing4.mbox and emails-Enron.mbox were cleaned, as they contained infected and uninfected phishing emails, respectively. From these databases, a subset of data was selected to clean them and improve their quality, allowing the extraction of characteristics in the required format so that the modeling tools can process them. Table 2 lists the selected characteristics.

Table 2. Extraction of representative characteristics for modeling

Item	Selected characteristics
1	Empty subject
2	Detection of @ character in email URLs and subject
3	Presence of URL in the body of the email
4	Detection of HTML content and forms in email
5	Detection of IP addresses in URL, instead of domain name to mask the name of the site to which the link is redirected
6	Presence of JavaScript code that can be embedded in the body of the email or in a link
7	Number of points in the domain must not exceed 3
8	Http presence in URLs
9	Https presence in URLs
10	Sender's mail format
11	URL size >74
12	Using CSS in the body of the email
13	Presence of the OnClick event

Data Integration: The characteristics extracted from each email are consolidated into a matrix in which its row represents the record, and each column is a characteristic field extracted. At the end of the analysis and determination process, this matrix is saved within a CSV file for later analysis in the prediction algorithms used in Data Mining.

Formatting the Data: Once we have a consolidated matrix with the characteristics which determine or not phishing in an email, it is mandatory to identify them somehow. The column where the source file's name is found is used, and this information is replaced. This occurs with the records, which contain the name of the private-Phishing4.mbox file whose content refers to infected emails. Then, they will be replaced with the number "1", and the records containing the name of the emails-Enron.mbox file, which refers to non-presence emails phishing, will also be replaced with the number "0". In this way, we have the information required for the prediction models.

2.3 Modeling Phase

In this phase, we have chosen which of the data mining models is the most appropriate. Therefore, a test plan was conducted based on metrics that will define each model's percentage of precision. Finally, it will be evaluated how this model meets the established success criteria.

Choosing the Modeling Technique: The appropriate technique selection was based on those whose function contributes to meet the data mining objectives. This occurred besides the fact that the "sklearn" library is available for the specific Python language since it provides functions related to data mining [3]. Furthermore, as the data's cleaning and construction were performed in this language, the model's construction is facilitated. The techniques defined and available in the library are Naive Bayes, Decision Trees, and Random Forests, which are adjusted to predict whether or not an email contains phishing based on its multiple characteristics.

Generate the Test Plan: The test plan begins with the data matrix containing 4789 emails containing or not phishing features. They were separated under Pareto's theory in a proportion of 80% of the training data and the residual 20% used for testing. Hereby, the "train-test-split" function was used, which allowed us to separate the percentage mentioned above and in a random way if the data content or not any phishing. It is worth noting that the information needed to be more considerable for the training since it covers the most significant number of characteristics related to what the objective requires to achieve. Its efficiency level will be tested for each technique based on the following four metrics, being Accuracy score, Precision score, Recall score, and F1, as defined on [12].

The **Accuracy Score** is a metric that directly represents the accuracy of the model of a ratio of correctly predicted observations to the total observations, which is calculated as follows (see Eq. 1).

$$AccuracyScore = \frac{VP + VN}{VP + VN + FP + FN} \tag{1}$$

The **Precision score** is a metric representing the relationship between the cases calculated as true positives and the total positive cases encountered in the test (see Eq. 2).

$$PrecisionScore = \frac{VP}{VP + VN} \tag{2}$$

The **Recall score** is a metric, which represents if relationships exist between the observed cases as truly positive with the sum of these and those that were positive but have been observed as negative (see Eq. 3).

$$RecallScore = \frac{VP}{VP + FN} \tag{3}$$

The **F1 score** is the weighting and relationship between Precision Score and Recall Score, resulting in both false positive and false negative values obtained within the observation. This value is used in cases where the case distribution differs. It is helpful in the current project since the data division is random (see Eq. 3).

$$F1Score = \frac{PRECISIONSCORE * RECALLSCORE}{PRECISIONSCORE + RECALLSCORE} \tag{4}$$

Based on these measures, the success obtained in predicting whether or not an email contains phishing can be specified. For the calculation of these mentioned metrics, we need the following values [18]:

- **True positive (TP)**, which is the number of cases which the test declared as positive and truly positive;
- **False positives (FP)**, which is the number of cases which the test declares as positive and is true negative;
- **True negative (TN)**, which is the number of cases which the test declares as negative and a true negative;
- **False Negative (FN)**, which is the number of cases which the test declares as negative and which are true positive.

2.4 Construction and Employment of Model

Since the three techniques' inputs are similar, only one parameter set must be ready and processed. Adjusting and separating the data is subsequently performed. Hereby, the data is separated into characteristics in a single matrix called "X" or an independent variable. The column, which identifies whether or not the email has phishing, is placed in a vector called "Y" or the dependent variable. Once the variables are separated, the following parameters applied a compute confusion matrix in order to evaluate the accuracy of classification based on [12]:

- **X_train**: Variable that contains 80% of the data which characterizes an email containing or not phishing;
- **Y_train**: Variable that contains 80% of the data which indicates whether or not an email is a carrier of phishing;

- **X_test**: Variable that contains 20% of the data which characterizes an email containing or not phishing;
- **Y_test**: Variable that contains 20% of the data which indicates whether or not an email is a carrier of phishing.

Once the data is adjusted and ready to be processed by the different chosen Data Mining techniques, the analysis mentioned above is conducted for the three chosen data mining techniques. All three techniques use the variables X_train and Y_train as inputs for it to use for training. Once the network has been trained, the tests are used with the variable X_test. When the three techniques obtain the prediction data, it is stored in the variable y_pred as a vector. This contains the information of the prediction determined, which contrasted with the variable Y_test, and it can be evaluated if it belongs to the group of TP, TN, FP, or FN.

Model Evaluation: For the evaluation of the model, the metrics proposed in the present study's test plan were used (i.e., Accuracy score, Precision score, Recall score, and F1 score). These arguments are provided by the high-level Python language and specifically the scikit-learn 0.24.1 library. When comparing the real values and the predicted values, the model estimates the percentages based on the formulas proposed for each of the metrics used to evaluate the proposed model. Table 3 presents a summary of each of the metrics' values and their respective model.

Table 3. Values calculated by the metrics defined per each model

Model	Accuracy score	Precision score	Recall score	F1 score
Decision Tree	97,15%	96,57%	96,96%	96,77%
Naive Bayes	91,91%	90,40%	91,31%	90,85%
Random Forest	97,60%	96,70%	97,77%	97,28%

Model Implementation: In the first instance, there are three essential subprocesses: reading non-phishing emails, reading phishing emails, and reading emails; this achieves the generation of the dataset with the extraction of characteristics.

Figure 1a illustrates the flowchart for the reading sub-process. The algorithm reads the file private-phishing4. mbox where non-phishing emails are found. Afterward, a repetitive For cycle is entered, which ends when the emails contained in the file have been analyzed. Within this For cycle, several functions are conducted to treat each email and extract existing characteristics in the PhishTank repository database through a restful service API, where each of the URLs contained is sent with the mail. Then, an XML file returns to determine if the URL exists within the PhishTank database and if it is defined as phishing.

The algorithm then analyzes the data in each email to later proceed with the extraction of characteristics. It is worth noting that each character will be represented by a binary value, with "1" being phishing detection and "0" non-phishing. All this data is saved in the datasetTesisFinal.csv file for each scanned email.

Once this cycle is completed, the rest subprocesses are executed, similar to the one presented, although with slight differences (e.g., name of the database and the number of emails). It is essential to point out that the generated dataset creates 25 columns, of which the first 24 represent the defined characteristics. Column 25 identifies which emails are phishing and not, represented by a "1" or "0," respectively. After performing these processes, it generates the dataset.csv file, which allows the generation of the model and the tests in an effective way.

Employment of the Predictive Model: This stage of the project begins with reading the generated dataset, followed by the definition of the variables "x", "y" which allow the training and prediction of the model. The 24-phishing features are assigned to "x" and to "y" the target variable, which defines whether the characterization belongs to an email with phishing or not.

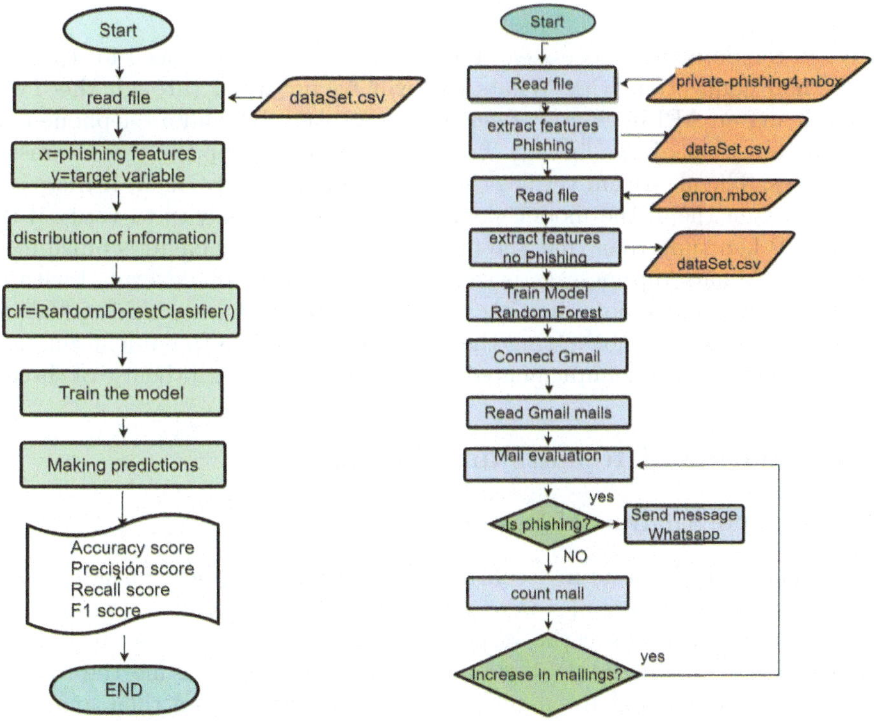

(a) Prediction model generation (b) Prediction model process

Fig. 1. Flowcharts of the generation and functioning of the prediction model.

Once these variables are declared, the percentage of data for training and testing is divided. The selected data were performed randomly. This division is generated based on the Pareto principle, considering 80% of data for training and 20% for testing. This is conducted by declaring the variable random_state = 1. Then, the model is trained with the RamdonForest algorithm, which is in charge of performing the respective process so that the model learns from the characteristics and objective variable previously declared. Furthermore, the corresponding predictions are created. This process allows us to obtain the metrics (Accuracy, Precision, Recall, and F1 scores), which are in charge of validating the precision and reliability of the model regarding the detection of phishing in analyzed emails. Once the model is generated, it is applied in the controlled environment of Gmail for the respective analysis.

After recognizing how each of the model's sub-processes works, we may obtain a better overview of the project's general process and its operation. In Fig. 1b, we are able to appreciate in a general way which is the flow of the generated algorithm fulfilling the defined objectives. Once each detailed sub-process has been performed, the model is put into operation under a controlled environment in order to test its efficiency. These tests are performed through the Gmail connection.

When the detection process can be visualized without the need for the code, the Python connection with WhatsApp will be produced through the Twilio library, being an API in charge of generating a token, ID, and a telephone number, which is what allows linking the code with WhatsApp, by the submission of messages. The algorithm verifies the number of emails in Gmail. If there is an increase within them, the model yields the respective analysis to verify if it is phishing. When supposing that the model detects that the email is phishing, in that case, a WhatsApp message is sent. This delivery will be executed with the subject, sender, and a note alerting that this email contains phishing features, allowing mitigation by not letting the user expose itself to fraudulent emails. It is worth noting that it is implemented every five minutes with the use of threads.

3 Evaluation of Results and Discussion

3.1 Test and Analysis of Results

We conducted all the tests in a controlling environment, where phishing emails were generated and where files were obtained from reliable sources, as described above. In order to perform the respective analysis, we divided such a process into three stages, being model training, predictions, and applying the model in a controlled environment in emails of Gmail. The results and their analysis of the model are detailed below:

First Stage - Model Training: In this phase, a total of 3,375 emails were used to train the model, of which 1505 were infected, and 1870 are emails that lacked contain phishing data. It is essential to point out that these data represent 80%

of the total data obtained in the dataset. As illustrated in Fig. 2a, 1125 data pieces represent the residual 20% of the data in order to allow the predictions to be distributed in 495 emails with and 630 emails without phishing. The 3375 emails were used to conduct the respective learning process.

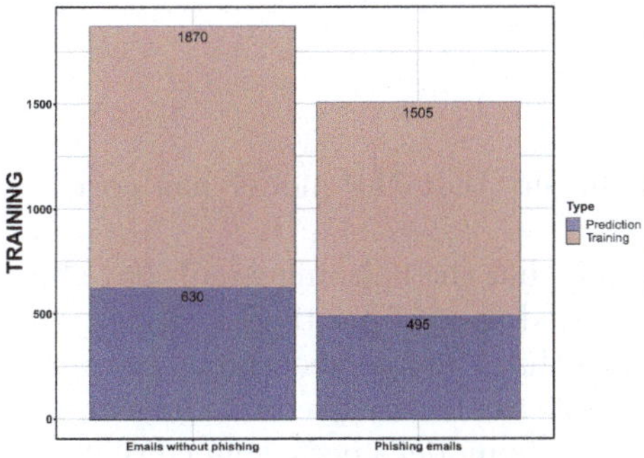

(a) Model training phase

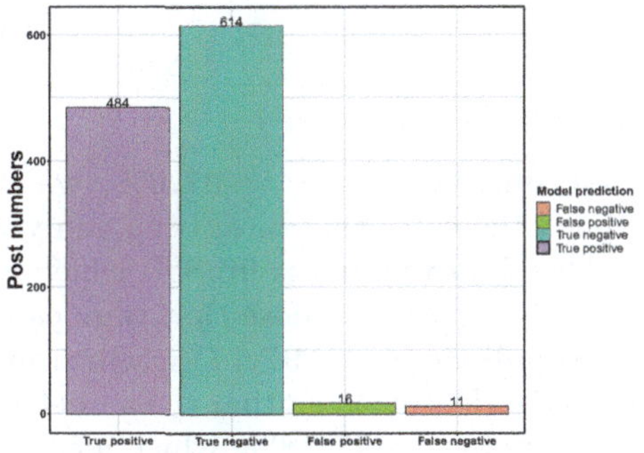

(b) Prediction phase of the model

Fig. 2. Bar charts of tests and analysis of model results.

Second Stage - Prediction of the Model: As described previously, 1125 emails were used for the prediction phase, using the RamdonForest algorithm. This process uses the scikit-learn 0.24.1 library, which provides the necessary metrics to evaluate the results of the model's predictions. Hereby, the variables True positive (VP), True negative (VN), False Positive (FP) were defined, as well as the False negatives (FN). All these variables are defined by the library, which works with it in its internal functions (scikit learn, 2020). The exact amount of emails in each variable may be visualized according to the model's predictions.

According to the number of emails in each variable, the model's prediction shows that 484 emails were correctly categorized as true positives (TP) because they were marked as phishing. This value differs from the true negatives (TN) since it exceeds all categories by reaching 614 emails, determining those marked as not phishing when they are not. Unlike the false positives (FP), which count 16 correct records and the false negatives (FN) of only 11 emails (see Fig. 2b).

To assess the model's performance in this phase, we used the data described in Table 4, which are necessary in order to determine the value of the established metrics. Such process allowed us to analyze the model statistically. Using these values, with the application of the Eqs. 1 to 4 of each metric were calculated, which allows to establish the level of precision when training and learning this model. The results obtained in each evaluation of each metric are displayed below, as stated numerically as well as graphically (see Fig. 3a):

Table 4. Number of emails in each variable

Variables	y_pred	y_test	Total emails
TP	1	1	484
TN	0	0	614
FP	1	0	16
FN	0	1	11

Accuracy Score = 97.6%, which means that the model has correctly defined the cases that have been analyzed.

Precision Score = 96.8%, which means that the phishing emails have been correctly identified: those that were categorized as true positives (VP).

Recall Score = 97.77%, which means that the infected emails have been correctly identified and that the model was able to identify, predict, and estimate.

F1 Score = 97.28%, which means that the model's performance that they have marked with phishing has really been.

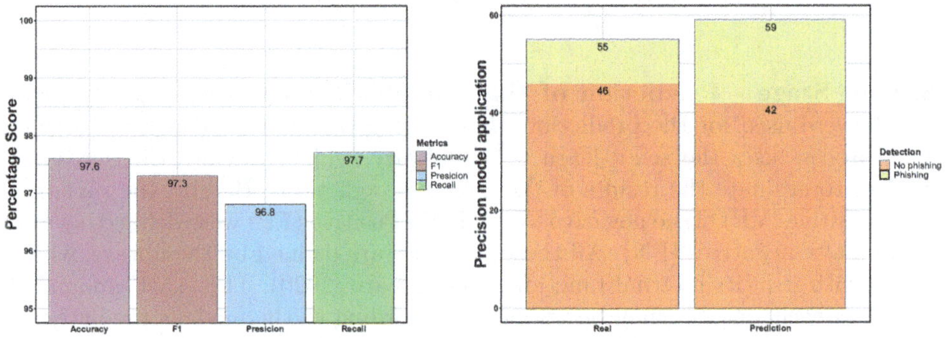

(a) Percentage score for each metric (b) Prediction score of the model

Fig. 3. Bar charts of percentage scores of each metric and the accuracy of the model prediction.

Finally, in Eq. 5, the error rate (ER) is determined. That is the percentage in which the model lacked to classify the emails correctly. The classification of incorrect emails, i.e., they have been categorized as phishing when they are not and vice versa, is 2.4%.

$$ErrorRate = \frac{FN + FP}{VN + FN + FP + VP} \quad (5)$$

Stage 3 - Application of the Model: This stage performs an analysis of the results, referring to the number of emails detected with phishing in a controlled (i.e., real) environment. Therefore, we used a Gmail account, where there are 101 emails, of which 55 are phishing, and the residual 46 are not.

When analyzing these emails, the prediction model categorizes 59 emails with phishing and 42 emails without phishing. These results in percentage terms give a precision of 96.03%. It should be noted that this analysis is performed not only with generated emails but also with real emails sent to the public and commercial entities (see Fig. 3b).

Finally, we compared the aforementioned with other analyzed data mining models in order to determine the selected model's efficiency. Therefore, through statistical values, we supported to select the Random Forest model, which was also correct. In Fig. 4, we observed the three models (i.e., the decision tree, Naive, Bayes, and Random Forest). The latter will be contrasted with the previous ones, thus determining their efficacy.

Concerning the results obtained from each of the metrics, we were able to determine that the Random Forest Model has a higher percentage of precision when it is estimated with accuracy (97.6%), F1 (97.2%), Precision (96,7%), and Recall Score (97.7%). Thus, this model has greater precision when categorizing and predicting an email and having a higher degree of complexity concerning learning. Also, the models' average is estimated. In that case, it is confirmed that the Random Forest occupies the first place with 97.3%, followed by the decision tree (96.7%), and finally, the Naive Bayes model with about 90.9%.

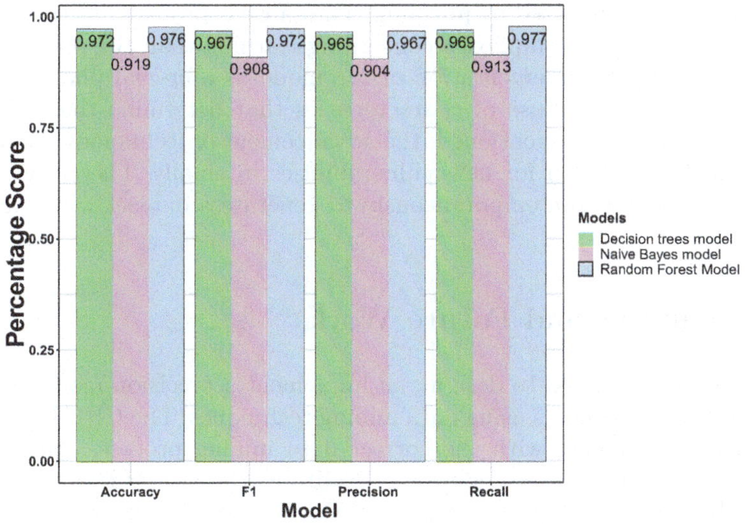

Fig. 4. Bar chart of the comparison with the rest data mining algorithms.

3.2 Discussion

After exploring the data, we were able to deduce that the data sources obtained are complete. They provide the information necessary to obtain specific characteristics, allowing us to generate the model with high precision. As for the emails with phishing, they originated from a reliable source where several studies used it. These data are up-to-date, ensuring that new results can be obtained in this model to be generated. Also, some emails lack complete information in these databases, whose solution is to eliminate through the data mining process since this data will neither contribute nor consider them.

The database with emails without phishing is of great support when learning the model. Hereby, identifying emails with phishing can be given greater validity as it provides greater characteristics to distinguish between infected emails.

The PhishTank database is a reliable source, which helped when analyzing URLs and domains in an email. All the data obtained provide essential and accurate information, ensuring no risk of noise in the process and will be positive and accurate.

Python is a high-level, open-source programming language with many libraries, tools, and functions that allow the development of data mining solutions to prevent phishing attacks.

Based on the analysis of the results in the tests of the three selected algorithms, i.e., Decision Tree, Naive Bayes, and Random Forest, we can deduce that Random Forest is the model with the highest precision when categorizing and predicting an email infected by phishing. The cause is probably as this algorithm has a higher degree of complexity concerning learning than the previous ones.

Maintaining the implementation of the model is a crucial phase due to the constant change or increase of characteristics that determine that an email is phishing. As previously mentioned, the advancement of technology allows phishers continually to search for new vulnerabilities in emails. Therefore, the code needs to be updated contemporaneously to generate a dataset with much more precision.

4 Conclusions and Future Work

The current study aimed to design and implement a precision model to prevent phishing attacks on emails using data mining techniques. The CRISP-DM phases were applied to comply with the correct data mining process, combined with Python. Based on the analysis of the results in the selected models' tests, it was possible to conclude that Random Forest resulted in being the model with the highest precision when categorizing and predicting an email infected with phishing. We also concluded that the model's percentage of precision depends on the correct selection of characteristics and the amount of data used during training and prediction.

As future work, we plan to analyze users' behavior when deciding to open an infected email, focusing on cognitive security.

Acknowledgment. The authors would like to thank the Universidad de las Fuerzas Armadas ESPE of Sangolquí, Ecuador, for the resources granted for the development of the research project entitled: "Detection and Mitigation of Social Engineering attacks applying Cognitive Security, Code: PIC-ESPE-2020-SE" and "Design and Implementation of the IT infrastructure and service management system for the ESPE Academic CERT", Code: PIC-ESPE-2020-ESPE-CERT.

References

1. Baykara, M., Gürel, Z.Z.: Detection of phishing attacks. In: 2018 6th International Symposium on Digital Forensic and Security (ISDFS), pp. 1–5. IEEE (2018)
2. Bell, S., Komisarczuk, P.: An analysis of phishing blacklists: Google safe browsing, OpenPhish, and PhishTank. In: Proceedings of the Australasian Computer Science Week Multiconference, pp. 1–11 (2020)
3. Buitinck, L., Louppe, G., Blondel, M., et al.: API design for machine learning software: experiences from the Scikit-learn project. arXiv:1309.0238 (2013)
4. Chatterjee, M., Namin, A.S.: Detecting phishing websites through deep reinforcement learning. In: 43rd Annual Computer Software and Applications Conference (COMPSAC), vol. 2, pp. 227–232. IEEE (2019)
5. Chin, T., Xiong, K., Hu, C.: Phishlimiter: a phishing detection and mitigation approach using software-defined networking. IEEE Access **6**, 42516–42531 (2018)
6. Chuenchujit, T.: A taxonomy of phishing research (2016)
7. Espinoza, B., Simba, J., Fuertes, W., et al.: Phishing attack detection: a solution based on the typical machine learning modeling cycle. In: International Conference on Computer Science and Computational Intelligence (CSCI), pp. 202–207. IEEE (2019)
8. Gangavarapu, T., Jaidhar, C.D., Chanduka, B.: Applicability of machine learning in spam and phishing email filtering: review and approaches. Artif. Intell. Rev. **53**(7), 5019–5081 (2020). https://doi.org/10.1007/s10462-020-09814-9
9. Macas, M., Lagla, L., Fuertes, W., Guerrero, G., Toulkeridis, T.: Data mining model in the discovery of trends and patterns of intruder attacks on the data network as a public-sector innovation. In: Fourth International Conference on eDemocracy & eGovernment (ICEDEG), pp. 55–62. IEEE (2017)
10. Moradpoor, N., Clavie, B., Buchanan, B.: Employing machine learning techniques for detection and classification of phishing emails. In: 2017 Computing Conference, pp. 149–156. IEEE (2017)
11. Oña, D., Zapata, L., Fuertes, W., Rodríguez, G., Benavides, E., Toulkeridis, T.: Phishing attacks: detecting and preventing infected e-mails using machine learning methods. In: 2019 3rd Cyber Security in Networking Conference (CSNet), pp. 161–163. IEEE (2019)
12. Pedregosa, F., et al.: Scikit-learn: machine learning in python. J. Mach. Learn. Res. **12**, 2825–2830 (2011)
13. Schröer, C., Kruse, F., Gómez, J.M.: A systematic literature review on applying crisp-DM process model. Procedia Comput. Sci. **181**, 526–534 (2021)
14. Sharma, T., Ferronato, P., Bashir, M.: Phishing email detection method: leveraging data across different organizations (2021)
15. Sonowal, G.: Phishing email detection based on binary search feature selection. SN Comput. Sci. **1**, 1–14 (2020)

16. Speicher, P., et al.: Formally reasoning about the cost and efficacy of securing the email infrastructure. In: 2018 IEEE European Symposium on Security and Privacy (EuroS&P), pp. 77–91. IEEE (2018)
17. Vijayalakshmi, M., Shalinie, S.M., Yang, M.H., et al.: Web phishing detection techniques: a survey on the state-of-the-art, taxonomy and future directions. IET Netw. **9**(5), 235–246 (2020)
18. Yacouby, R., Axman, D.: Probabilistic extension of precision, recall, and F1 score for more thorough evaluation of classification models. In: Proceedings of the First Workshop on Evaluation and Comparison of NLP Systems, pp. 79–91 (2020)

Anomaly Detection Method for a Local Area Network

Ángel Ramón Párraga-Palmar[1(✉)], Marely del Rosario Cruz-Felipe[2], and José Párraga-Valle[2]

[1] Instituto de Postgrado, Universidad Técnica de Manabí, Portoviejo, Ecuador
aparraga4693@utm.edu.ec

[2] Facultad de Ciencias Informáticas, Universidad Técnica de Manabí, Portoviejo, Ecuador
{marely.cruz,jose.parraga}@utm.edu.ec

Abstract. This paper shows the development of an anomaly detection method for a local area network. This research work aims to use an unsupervised machine learning algorithm that integrates different network management tools for anomaly detection. The experiment includes the description of three essential modules: variable definition and protocol configuration, network monitoring and data collection, and fault prediction. SNMP, ICMP, and WMI communication protocols were selected to collect information about the current state of the network, and variables such as latency, packet loss, availability, traffic volume, and speed were defined. The network was polled by temperature sensors, CPU consumption, memory consumption. Network probing was done through the PRTG network monitor and access to the collected data was done through APIs, which were formatted to a template with the structure corresponding to the defined variables. In addition, a method was developed using an unsupervised machine learning algorithm in Python. The tools used for the development of this research were PRTG, Python, Pycaret Library, Collaboratory, Power BI for monitoring, development, and simulation respectively. For the analysis of the model efficiency, the parameters of accuracy (Accuracy), area under the curve (AUC), completeness (Recall), precision (Prec), F-value (F1-score), Matthews correlation coefficient (MMC) were used. The model was implemented through the Power BI tool. It was obtained as a result of measurements and simulation in different scenarios that the implementation of an anomaly detection method reduces degradation, improving the quality and availability of services offered in a local area network.

Keywords: Detection method · Network monitoring · Machine learning

1 Introduction

The growth of local area networks leads to new challenges in ensuring performance. Anomaly detection is rapidly becoming a key element of communication

M. Botto-Tobar et al. (Eds.): CIT 2021, LNNS 405, pp. 163–177, 2022.
https://doi.org/10.1007/978-3-030-96043-8_13

systems. Accurate detection of such anomalies has become a major problem for the networking community. Needing to distinguish anomalies that change traffic abruptly or slowly [1].

As organizations automate their processes and become more dependent on technologies, the impact of degradation of telematic services and non-compliance with operating levels increases, caused by failures in the IT infrastructure that affect services and data integrity. For this reason, the management of anomalies must be carried out early, making corrective decisions based on the evaluation of the impact of damage to the organizations' equipment. Likewise, various efforts are made around fault management, which includes detection, isolation, and control actions, to correct problems and effectively reduce the degradation of services in the active equipment of a local area network.

One of the techniques used today to provide solutions to anomaly detection is through the use of machine learning, which has been evolving in recent years in the branch we know as artificial intelligence so that we can have machines capable of predicting a large amount of information based on what has already been stored and processed previously [2].

As networks become larger and more complex, the need for advanced fault management capabilities becomes critical. Faults are inevitable in large and complex communication networks, but rapid detection and identification can significantly improve network reliability. Packet monitoring has become a standard technique in network fault detection, but when applied to a large-scale network, it produces a large volume of traffic. To overcome this problem, some techniques are proposed. However, the proposed techniques are based on the popular SNMP agent, RMON technology, which is characterized by centralization and does not provide high scalability and local processing capability [3].

There are several studies on the detection of faults and anomalies in local area networks, ranging from methods, policies, algorithms for data analysis, and analysis of fault detection procedures. Among which is a study [4] where the proposal of a system for early detection and isolation of faults in the active equipment of the network, minimizing the impact on users is evidenced. Through a system that incorporates a Policy-Based Management architecture that allowed the automated correction of failures, reducing the impact on services. In addition, it allowed speeding up the localization of failures through the application of a change detection algorithm to the analysis of parameters on the variables of the MIBs of the active network equipment.

Likewise, the objective of the precedent [5] is to design the procedures of a model for failure management of communications equipment (routers). For this purpose, it worked with the CACTI network monitoring tool from which the main failure indicators per equipment were obtained. The Fault Management model was elaborated according to the description of each management function, such as criticality levels, classification of criticality levels by indicator, values for reporting a fault indicator, and the processes defined for Fault Management were logically tested.

Some solutions to the problem of fault detection are realized by using neural networks as shown in [6], where deep neural networks were applied to the problem of fault detection and classification. In the case of fault detection, they investigated the effects of two hyperparameters (number of hidden layers, number of neurons in the last hidden layer) on the performance of the networks, and concluded that increasing the network size does not improve the fault detection accuracy. Then, they show that increasing data may be key to further increase the fault detection accuracy.

Another study of fault detection employing neural networks is proposed in [7], which appears promising. First, the features and faults need to be analyzed in detail to understand how the neural network classifier works and improve its performance. Second, the effects of data augmentation need to be further investigated. Finally, different types of neural networks should be tested to see if they are better suited to the problems of fault detection and classification. Also in [8], neural networks are used for pattern recognition resulting in fault classification.

Rui Zhang, Shaoyan Zhang, Yang Lan, Jianmin Jiang present a class support vector machine (OCSVM)-based method to detect network anomalies. For which they used the performance data of a telecommunication network. First, the raw data is preprocessed to produce the vector sets required by the OCSVM algorithm. After preprocessing, the vector set of the training data is used to train the OCSVM detector, which can learn the nominal behaviors of the data. Then, the trained detector is applied to the test data to detect the anomalies [9].

From the solutions analyzed, it is observed that most of them employ neural networks, and use data that provide network performance parameters, however, they do not contain log-hosted data, such as physical failures coupled with performance parameters.

This research presents a study of anomaly detection taking into account physical failures or network performance such as power failures, cabling, connection, interfaces, service, temperature, storage, availability, speed. To achieve the objective it was necessary to review the literature of the different methodologies used in different investigations, these have served as a guide to propose a method of anomaly detection that combines different network management tools through an unsupervised algorithm (IForest) of machine learning, using the pycaret library of Python, communication protocols, PRTG, Power BI to collect information and to detect faults on the services available on a local area network.

The research has been developed in three stages determined by the authors, which are: definition of variables and protocol configuration, network monitoring and data collection, and finally the prediction stage.

2 Materials and Methods

This research work has been developed in three stages determined by the authors, which are: the definition of variables and protocol configuration, network monitoring and data collection, and finally the prediction stage.

This methodological development has a great contribution to the elaboration of the fault detection method for a network, which includes the interaction between different tools and software dedicated to the administration or management of networks, which are detailed below:

- **Communication protocols. -** A communications protocol is a set of rules that all equipment or software involved in a data communication must comply with.
- **PRTG. -** This tool is a network monitor that facilitates the administration of infrastructure, it is easy to use in different web environments and free for up to 100 sensors, it works through SNMP, ICMP, WMI protocols, among others. This software offers the possibility of collecting information through the APIs included in the installation package.
- **Machine Learning. -** It is a branch within the field of Artificial Intelligence. Thanks to the enormous increase in computing power and the large volume of data that companies are beginning to handle. What Machine Learning provides us with is a set of algorithms cataloged as Machine Learning whose objective is to provide computers with the capacity to learn without the need to be explicitly programmed. Machine Learning algorithms are classified into three types of classes which are: Supervised, semi-supervised and unsupervised. In this research, an unsupervised algorithm is used, based on the selected data that does not have a label that defines whether there is any failure or not, which requires the use of an unsupervised algorithm.
- **Python. -** It is a multipurpose programming language that facilitates the development of machine learning models with artificial intelligence.
- **Colaboratory. -** It is a programming IDE that allows you to run and program Python functions, requires no additional configuration, provides free access to GPUs, and allows easy content sharing.
- **Pycaret. -** It is an open-source library that can be used to develop and implement a large-scale machine learning model.
- **IForest. -** It is an unsupervised machine learning algorithm for anomaly detection used exclusively on large amounts of data.
- **Power BI. -** It is a big data tool that allows you to analyze and share information through a variety of visualization and data management components under a simple and friendly appearance. This application is capable of facilitating the deployment of models developed in Python, thanks to the integration that the new versions of Power BI currently have with the Python programming language.

Having mentioned the materials involved in the research, it is still necessary to describe each of the phases of the proposed methodology as shown in the Fig. 1.

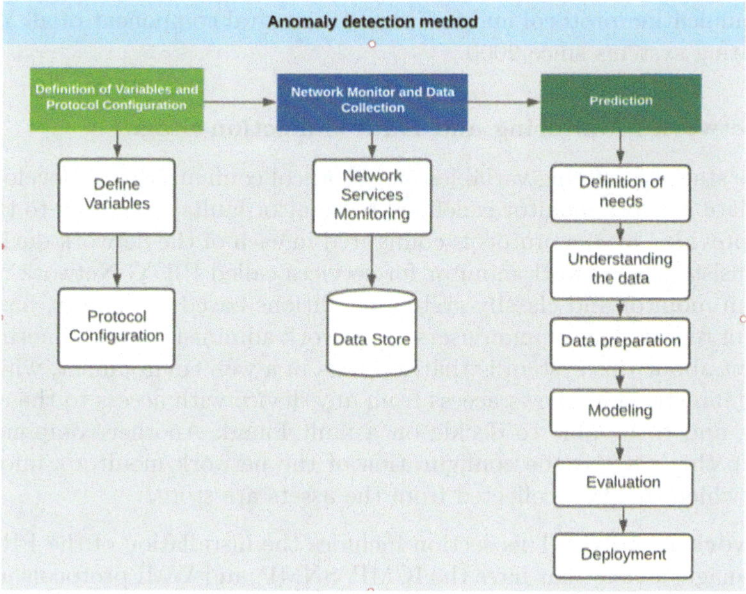

Fig. 1. General diagram of the anomaly detection method. **Source:** Own elaboration.

2.1 Variable Definition and Protocol Configuration Stage

The first part of this research is to define the input variables that can achieve the expected accuracy of the method, such as latency, packet loss, availability, traffic volume, traffic speed, temperature sensors, CPU consumption, memory consumption, among others. Once this process has been carried out, the SNMP (Simple Network Administration Protocol), ICMP (Internet Control Message Protocol), and WMI protocols are configured in the different network devices, to obtain real-time information on the current state of the medium, for which it is necessary to know the structure of the data returned by the protocols used, and thus be able to feed the corresponding higher stages.

- **Define variables:** this module has been developed to determine the largest number of variables to be measured in the higher modules.
- **Configure the protocols in the devices:** To ensure the operation of this module, three very important protocols have been enabled, such as ICMP, SNMP, and WMI in each of the elements to be monitored. The SNMP protocol is an agent that collects information about a device on the network through the MIB (Management Information Base), and this is processed and stored in the next stage with a coherent structure to be interpreted and presented to end-users. In addition, it manages the ICMP protocol that is used to send error notifications to the network layer about the delivery status of packets or datagrams and the Windows Management Instrumentation (WMI) protocol that is based on the DCOM (Distributed Component Object Model)

communication protocol and has been an integral component of all Windows operating systems since 2000.

2.2 Network Monitoring and Data Collection Stage

Once the stage of defining variables and protocol configuration is developed, the appropriate network monitor is defined to monitor faults, according to the information provided by the protocols configured in each of the network devices, this stage consists of a network monitor for services called PRTG Network Monitor, which can monitor and classify system conditions based on sensors, has a large number of reports for common users or network administrators. Something very important about this system is that it works in a web environment, which facilitates administration, allows access from any device with access to the analyzed network, and to be able to decide on a fault found. Another component that makes up this stage is the configuration of the network monitor's information base, in which the data collected from the assets are stored.

- **Network monitor:** This section includes the installation of the PRTG tool to manage devices that have the ICMP, SNMP, and WMI protocols activated in their local configuration through sensors, which are activated and allow storing data of the events occurring in the network, which must be configured in an information base.
- **Configuration of the information base:** Guarantees the operation of the entire system, from the sensors activated in the network monitor, we proceed to configure the information base, once the data is stored, LOG records were obtained through the PRTG APIs, which are used as input data for the abnormality detection methods developed in this research.

2.3 Prediction Stage

This process was developed after the collection of the information by the network monitoring, thanks to the definition of the variables it was possible to obtain the template of the historical data contained in reality, through the detection of anomalies and predictions we can anticipate any disaster that may intervene in the operation of the network, for this reason, this stage was based on the CRISP-DM methodology as shown in Fig. 2 in which the stages of successful anomaly detection are defined, first the needs are defined, an understanding of the data is performed, the data is prepared, the modeling of the algorithm is performed, then the algorithm is evaluated and finally the deployment of the algorithm is performed.

Prediction: this component comprises a small system that acts as an internal agent to detect and predict faults that may occur in the network, and is made up of six modules:

- **Definition of requirements:** this module includes the structuring of the historical data collected by the network monitor, which defines the classification of the input data.

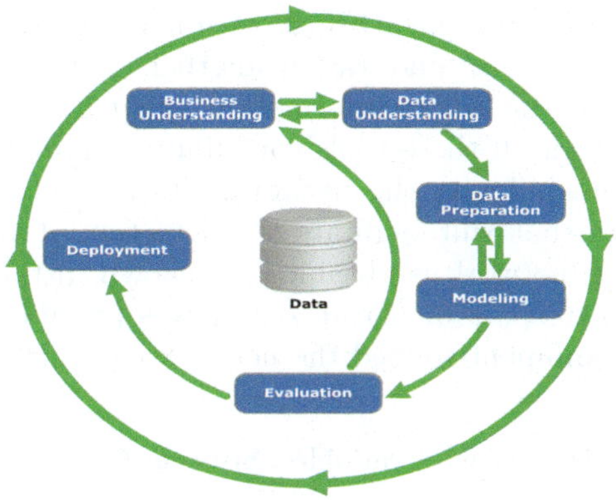

Fig. 2. CRISP-DM methodology. **Source:** Own elaboration.

- **Understanding the data:** This module includes familiarization with the data to identify quality issues to gain preliminary insights.
- **Data preparation:** This module covers all the activities necessary to build the final data set, as well as transformation and cleaning to avoid junk data that may interfere with the deviation of the final results.
- **Modeling:** In this phase, Pycaret's iForest modeling techniques are applied and their parameters are calibrated to optimal values. Therefore, it is almost always necessary to return to the data preparation and calibration phase.
- **Evaluation:** Once the iForest continuous learning model has been built, an in-depth understanding is made, the steps taken are reviewed and the results are compared with the objectives set.
- **Deployment:** For the deployment of the iForest model it is necessary to increase the knowledge about the data, where the knowledge obtained must be organized and presented through the Power BI tool so that the client can use it. Depending on the requirements, this phase can be as simple as generating a report or as complex as automating a data analysis process in the organization.

3 Results and Discussion

As a result of the monitoring of each of the stages, an anomaly detection method was obtained to maintain control of the assets of a LAN (Local Area Network) through the implementation of a continuous learning model that facilitates the detection of anomalies. For the implementation of the proposed model, the deployment was done through the Power BI tool, facilitating the work of network administrators. The results obtained from the developed experiment show the success of the model deployment, showing improvements in the performance and availability of the services offered to the end-user. As illustrated in Fig. 1, the architecture of this study is formed by three main components, organized in such a way that the communication between them effectively prevents the degradation of the performance of the managed elements.

The creation of this experiment brings with it the need to combine different tools originally designed to in one way or another facilitate the administration of a network and avoid the loss of information, in such a way that a model has been developed in which all these tools contribute in part to the success of the work of detecting anomalies in the network. At the end of the first phase, a list of input variables that will facilitate the detection of anomalies within the active equipment of a network is obtained, as shown in the Table 1. Likewise, by configuring the corresponding protocols, it was possible to access the basic information of the equipment to feed the network monitor.

Table 1. Definition of variables. **Source:** Own elaboration.

ID	**Name**
1	Latency
2	$\text{Lost}_{package}$
3	Availability
4	Traffic_{volume}
5	Traffic_{speed}
6	$\text{Temperature}_{1006}$
7	$\text{Temperature}_{1007}$
8	$\text{Temperature}_{1008}$
9	CPU
10	Memory

Once the appropriate tool to monitor and collect the information was defined, it was decided to install and configure a monitor called PRTG to obtain baseline information from the devices and services, as shown in the Fig. 3.

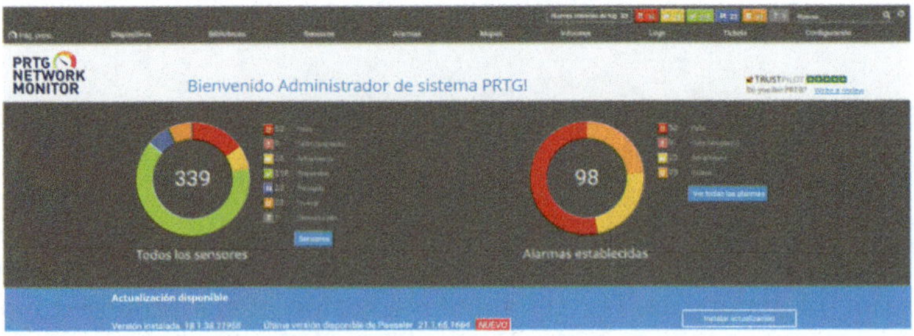

Fig. 3. PRTG network monitor home page. **Source:** Own elaboration.

Similarly, Fig. 4 shows the list of configured devices with different types of sensors added to each of the assets found within the network infrastructure.

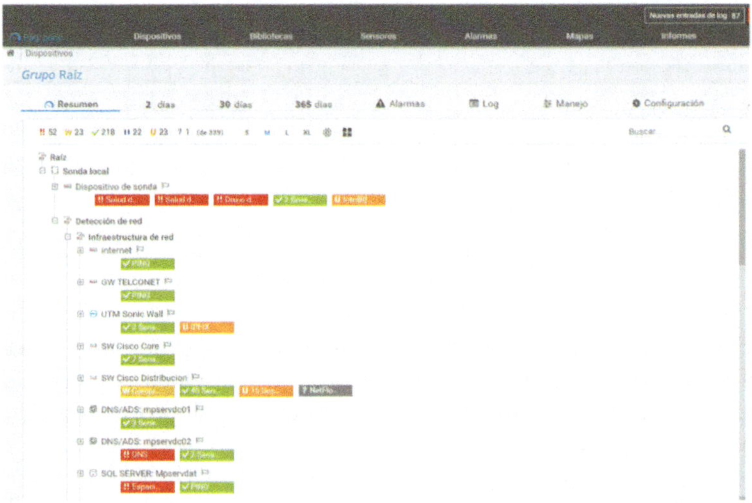

Fig. 4. Device groups are configured in the PRTG network monitor. **Source:** Own elaboration.

Another of the facilities that PRTG provides for administration is a network mapping, where it can graphically report any type of fault that is occurring as shown in the Fig. 5.

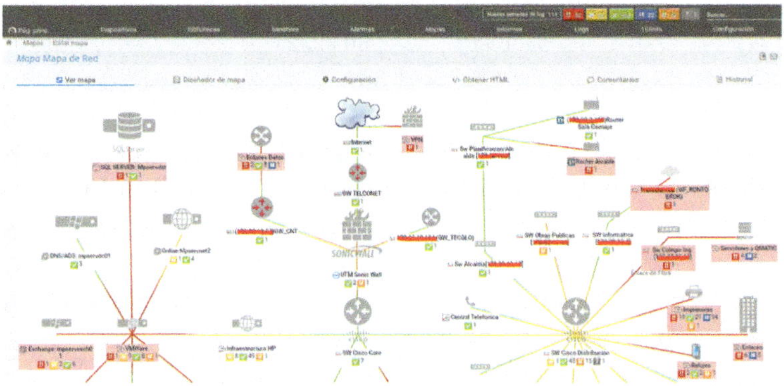

Fig. 5. Network map. **Source:** Own elaboration.

Within this phase, it is necessary to give importance to the collection of the information that was performed in CSV format in the local area network, through the consumption of the APIs that the PRTG network monitor has, in addition to the need to analyze each of the columns of the data returned by the API, a template was also defined with the variables established in the first phase as shown in the Fig. 6.

	A	B	C	D	E	F	G	H	I	J	K
1	Fecha Hora	Tiempo de Pi	Perdida de paque	Disponibilidad de siste	Trafico suma (volum	Trafico sur	Temperature 100	Temperature 10	Temperature 10	CPU 10	Available Memo
664	30/10/2020 10:00:00 - 11:00:00	2.4194	0.8333	0	7724.0147	2.1456	4294968332	4294968332	4294968332	3.1667	356599887.6
665	28/10/2020 14:00:00 - 15:00:00	2.4565	0	0	1782.9872	3.3707	4294968360	4294968360	4294968360	4.6406	355899705.1
666	28/10/2020 15:00:00 - 16:00:00	2.3117	0	0	12644.0128	3.5124	4294968356	4294968356	4294968356	4.15	355938629.1
667	28/10/2020 16:00:00 - 17:00:00	2.215	0	0	13008.96	3.6138	4294968356	4294968356	4294968356	3.1667	355997059.2
668	28/10/2020 17:00:00 - 18:00:00	2.1933	0	0	8422.02	2.3397	4294968356	4294968356	4294968356	3.35	356001120.5
669	28/10/2020 18:00:00 - 19:00:00	2.275	0	0	6360.0163	1.7668	4294968356	4294968356	4294968356	2.8333	356010526.2
670	28/10/2020 19:00:00 - 20:00:00	2.2967	0	0	6671	1.8532	4294968356	4294968356	4294968356	2.4167	356010424
671	28/10/2020 20:00:00 - 21:00:00	2.3017	0	0	6468.98	1.7971	4294968356	4294968356	4294968356	2.7167	356010803.5
672	28/10/2020 21:00:00 - 22:00:00	2.145	0	0	7571.0137	2.1032	4294968356	4294968356	4294968356	2.7	356008363.6
673	28/10/2020 22:00:00 - 23:00:00	2.1633	0	0	6517.9745	1.8107	4294968356	4294968356	4294968356	2.6667	355999542.3
674	28/10/2020 23:00:00 - 0:00:00	2.2517	0	0	6622.0237	1.8396	4294968356	4294968356	4294968356	2.7333	355996335.9
675	29/10/2020 0:00:00 - 1:00:00	2.2267	0	0	7620.0018	2.1168	4294968356	4294968356	4294968356	2.8	355997152.7
676	29/10/2020 1:00:00 - 2:00:00	2.1183	0	0	7669.0027	2.1304	4294968356	4294968356	4294968356	2.6667	355996308.5

Fig. 6. Template resulting from PRTG API consumption. **Source:** Own elaboration.

After defining the variables, installing the monitoring tool, and collecting the data, we proceeded to develop the prediction phase, which is composed of the CRISP-DM methodology comprising a 6-step flow as shown in the Fig. 2. Once known, understood the needs, and performed the preparation of the collected data, the result obtained is a division of the data with 95% destined to the modeling of the algorithm and 5% as validation data as shown in the Table 2.

Table 2. Data division. **Source:** Own elaboration.

Destination	%	Valor
Modeling	95	7432
Validation	5	392
Modeling		
Train	70	5108
Test	30	2324

Next, a matrix was obtained containing the set of input variables to the model with their respective statistical analysis as described in Fig. 7.

	Latencia	Perdida_paquetes	Disponibilidad	Trafico_volumen	Trafico_velocidad	Temperature_1006	Temperature_1007	Temperature_1008	CPU	Memory
count	7.823000e+03	7823.000000	7.823000e+03	7.823000e+03	7.823000e+03	7823.000000	7823.000000	7823.000000	7.823000e+03	7.823000e+03
mean	3.235855e+05	17.899271	1.052198e+08	6.457048e+06	2.103380e+07	122772.298613	89023.722332	177768.173156	4.372953e+05	3.565131e+08
std	5.831605e+05	191.371250	7.616127e+08	2.217720e+09	1.491791e+08	151294.060221	164440.768655	211230.809452	7.588965e+05	6.405974e+05
min	2.016700e+00	0.000000	3.603000e+03	0.000000e+00	0.000000e+00	26.000000	33.000000	35.050000	1.016700e+00	3.539263e+08
25%	2.240000e+00	0.000000	3.137044e+06	0.000000e+00	0.000000e+00	30.450000	38.000000	42.025000	2.066700e+00	3.561119e+08
50%	3.735000e+03	0.000000	5.579297e+06	0.000000e+00	0.000000e+00	34.000000	39.200000	44.000000	4.166700e+00	3.563253e+08
75%	4.056345e+05	0.000000	9.329383e+06	0.000000e+00	0.000000e+00	300167.000000	44.000000	421333.000000	5.234540e+05	3.565745e+08
max	2.952288e+06	8333.000000	9.817183e+09	9.990511e+09	6.692510e+09	409833.000000	469667.000000	490333.000000	2.897530e+06	3.593339e+08

Fig. 7. Data analysis. **Source:** Own elaboration.

Once the result of the processed data was obtained, a comparison of models was made to choose the best algorithm according to the working environment as shown in Fig. 8, where the accuracy value (Accuracy), area under the curve (AUC), Exhaustiveness (Recall), precision (Prec), F-value (F1-score), Matthews correlation coefficient (MMC) are evidenced.

	Model	Accuracy	AUC	Recall	Prec.	F1	Kappa	MCC	TT (Sec)
ada	Ada Boost Classifier	0.9851	0.9967	0.7882	0.9105	0.8374	0.8297	0.8361	0.2755
lightgbm	Light Gradient Boosting Machine	0.9825	0.9947	0.7514	0.8859	0.8075	0.7985	0.8044	0.1665
rf	Random Forest Classifier	0.9821	0.9934	0.7596	0.8742	0.8068	0.7976	0.8030	0.8770
gbc	Gradient Boosting Classifier	0.9812	0.9934	0.7209	0.8875	0.7884	0.7788	0.7872	0.7885
et	Extra Trees Classifier	0.9810	0.9953	0.7074	0.9018	0.7831	0.7735	0.7848	0.5765
lr	Logistic Regression	0.9749	0.9581	0.6027	0.8540	0.6976	0.6852	0.7011	0.1220
dt	Decision Tree Classifier	0.9723	0.8336	0.6797	0.7554	0.7057	0.6914	0.6976	0.0395
lda	Linear Discriminant Analysis	0.9696	0.9857	0.5442	0.7806	0.6272	0.6125	0.6300	0.0240
qda	Quadratic Discriminant Analysis	0.9683	0.9862	0.7173	0.6849	0.6916	0.6751	0.6802	0.0190
nb	Naive Bayes	0.9644	0.9550	0.7030	0.6368	0.6619	0.6434	0.6477	0.0185
ridge	Ridge Classifier	0.9570	0.0000	0.1703	0.8312	0.2707	0.2593	0.3507	0.0175
knn	K Neighbors Classifier	0.9528	0.8534	0.3255	0.5591	0.3936	0.3721	0.3947	0.1365
svm	SVM - Linear Kernel	0.9443	0.0000	0.3593	0.5285	0.3760	0.3523	0.3833	0.0325

Fig. 8. Comparison of models. **Source:** Own elaboration.

After comparing the models, it was decided to work with the unsupervised IForest machine learning model, which allows isolating the observations selected as anomalous according to the percentage of contamination corresponding to 5% as shown in the Fig. 9.

```
iforest = create_model('iforest')
iforest

IForest(behaviour='new', bootstrap=False, contamination=0.05,
    max_features=1.0, max_samples='auto', n_estimators=100, n_jobs=-1,
    random_state=123, verbose=0)
```

Fig. 9. Creation of the model. **Source:** Own elaboration.

Next, the evaluation of the model was performed through a statistical plot indicating the data considered anomalous that were found by the IForest model as shown in Fig. 10.

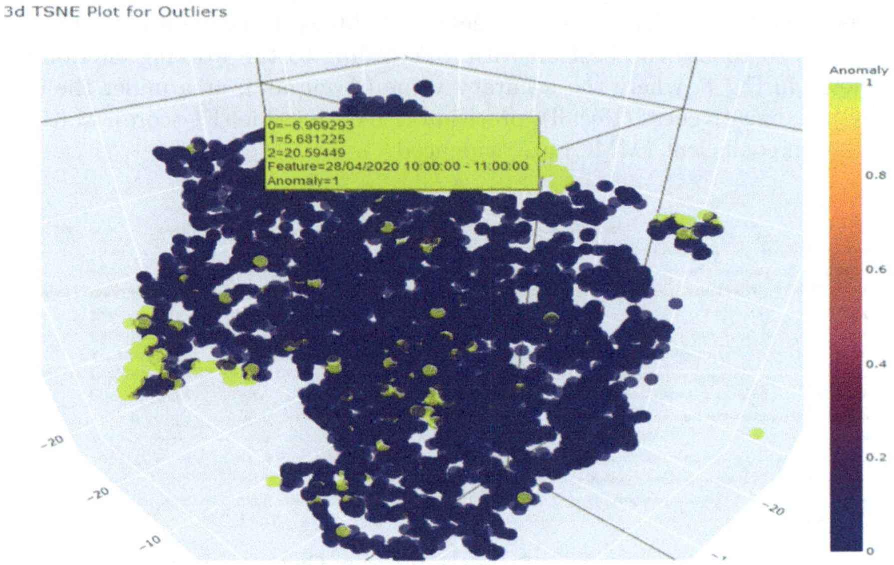

Fig. 10. Anomaly detection. **Source:** Own elaboration.

Before finalizing the model, a prediction was made with the 30% data set corresponding to 2324 records with 13 columns as shown in Fig. 11.

	Model
0	Isolation Forest

	Latencia	Perdida_paquetes	Disponibilidad	Trafico_volumen	Trafico_velecidad	Temperature_1006	Temperature_1007	Temperature_1008	CPU	Memory	Anomaly	Label	Score
0	2.176300e+00	0.0	6104897.0	0.000000e+00	0.0	31.000000	39.000000	42.099998	44231.00	356226128.0	0	1	0.9579
1	2.751405e+06	0.0	3934097.0	0.000000e+00	0.0	292833.000000	38.000000	42.000000	2.00	356476992.0	0	1	0.9711
2	6.051640e+05	0.0	3646796.0	0.000000e+00	0.0	35.000000	42.000000	45.000000	2.95	356591648.0	0	1	0.9895
3	5.051640e+05	0.0	3495696.0	0.000000e+00	0.0	32.000000	40.000000	437667.000000	706047.00	356720384.0	0	1	0.9895
4	2.138300e+00	0.0	2130321.0	9.211048e+09	25587592.0	51.000000	59.000000	43.000000	2.95	357132544.0	0	1	1.0000
...	...	...	...	...	...	...	...	...	...	...	...	...	...
2319	9.755300e+04	0.0	5518097.0	0.000000e+00	0.0	305667.000000	50.299999	42.099998	97561.00	356242648.0	0	1	0.9895
2320	6.819410e+05	0.0	5435996.0	0.000000e+00	0.0	36.900001	44.000000	462667.000000	2.15	356042400.0	0	1	0.9711
2321	2.305000e+03	0.0	6785298.0	0.000000e+00	0.0	31.000000	39.000000	43.000000	2.45	356179648.0	0	1	0.9797
2322	3.034800e+04	0.0	2407696.0	0.000000e+00	0.0	29.000000	37.000000	413333.000000	858668.00	356480064.0	0	1	0.9895
2323	1.208600e+04	0.0	6289183.0	0.000000e+00	0.0	32.000000	40.000000	43.000000	2.00	356120352.0	0	1	0.9711

2324 rows × 13 columns

Fig. 11. Prediction with test data. **Source:** Own elaboration.

Furthermore, the prediction was done on the validation dataset and was able to simulate a real environment with 391 rows of 13 columns equivalent to 5% of the original data as shown in Fig. 12, thus was validated the model before taking it to production.

	Latencia	Perdida_paquetes	Disponibilidad	Trafico_volumen	Trafico_velocidad	Temperature_1006	Temperature_1007	Temperature_1008	CPU	Memory	Label	Score
0	1.523390e+05	0.0	12912307.0	0.00	0.0	33.05	40.45	438333.00	1.071259e+06	353952124.9	0	0.6915
1	4.810580e+05	0.0	12933907.0	0.00	0.0	32.00	40.00	431833.00	1.071259e+06	353939452.8	0	0.6801
2	6.512610e+05	0.0	12941107.0	0.00	0.0	31.95	39.00	43.00	4.431700e+04	353937923.3	0	0.7663
3	2.165000e+03	0.0	12959107.0	0.00	0.0	30.00	38.00	42.05	1.801744e+06	353942713.0	0	0.6718
4	2.158300e+00	0.0	13099507.0	0.00	0.0	32.00	39.00	43.00	1.250000e+00	353958657.1	0	0.8074
...	...	...	...	...	...	...	...	...	...	...	...	...
386	2.345000e+03	0.0	6327890.0	23373144.24	6493027.0	33.00	41.00	44.00	1.984424e+06	356406097.0	0	0.8300
387	2.180000e+00	0.0	6396290.0	26443582.24	73460313.0	32.00	400833.00	44.00	3.183300e+00	356401309.5	0	0.7244
388	2.163300e+00	0.0	6468290.0	9905182.62	27516689.0	31.00	40.00	43.75	2.532260e+06	356402685.1	0	0.7547
389	6.464170e+05	0.0	6497090.0	22531690.90	5269342.0	33.10	41.00	44.05	2.471657e+06	356401654.7	0	0.7373
390	2.179435e+06	0.0	6507890.0	11032842.71	30649254.0	33.00	41.00	442833.00	2.714881e+06	356415278.5	0	0.5445

391 rows × 12 columns

Fig. 12. Prediction with validation data. **Source:** Own elaboration.

Finally, the deployment of the model could be carried out in conjunction with the Power BI tool by executing a python script, as shown in Fig. 13 where the instructions used that allowed the deployment of the model.

Ejecutar script de Python

Especifique los scripts de Python en el editor para transformar y modelar los datos.

Script

```
from pycaret.anomaly import *

load_iforest = load_model('C:/Users/I7/Downloads/iforest_20210329')

dataset = predict_model(load_iforest, data= dataset)
```

Fig. 13. Read the saved model. **Source:** Own elaboration.

A report was also produced as an example that demonstrates the effectiveness of the model implementation as shown in Fig. 14.

After analyzing the results of the research, it is necessary to mention that the work done by [4] that proposes a policy-based system that allows the automated correction of faults to reduce the impact on the services of active network equipment, where the scope of this work is very similar to the development of the anomaly detection method that also aims to reduce the degradation of services and improve network performance, it should be noted that the development process of these investigations is immersed in different environments, but even so, they can share their scope. One of the stages with the highest degree of difficulty in the development of the research was the collection of data in the correct format; necessary so that they can serve as input parameters to Pycaret's IForest model and thus ensure the success of the prediction of the method as indicated in Fig. 10. It should be noted that the method developed has been an unsupervised

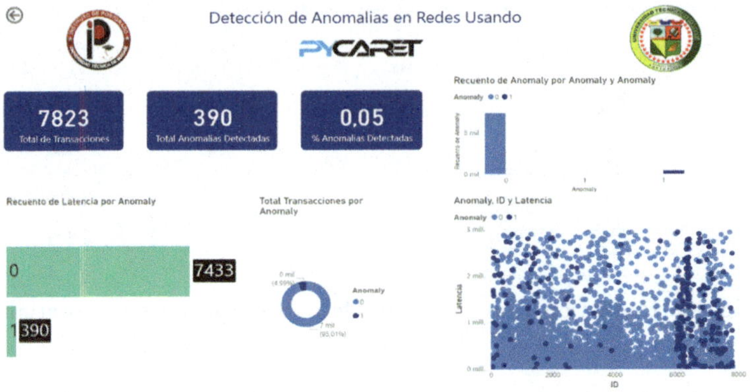

Fig. 14. Unattended IForest machine learning model deployment within Power BI. **Source:** Own elaboration.

model where it is not necessary to have a characterization or tag on the input data, this being a strength of the method, that means that new parameters can be easily incorporated to help detect anomalies within any LAN.

4 Conclusions

After carrying out the study and its respective analysis, the current problems in the study network were identified. The data obtained from the data collection instrument allowed us to reach the following conclusions:

- From the analysis of the different methods associated with the subject under study that is used to maintain control of the assets of a network, systems based on policies for the detection and isolation of faults are appreciated, as well as procedures for a fault management model, traffic predictions through neural networks that point to the particularity of decision making, thanks to the feedback acquired it can be determined that no method allows the combination of the different tools used in the research.
- The researcher was able to corroborate that the existing methods for detecting failures in the literature do not fit the general objective of the research, presenting the need for a new method.
- The development of the fault detection method is based on three levels of the process, which have a horizontal relationship, where everything starts with the configuration of the protocols in the network assets, followed by the installation of the monitoring and data collection tools, reaching the prediction stage in which a model was developed for the detection and prediction of anomalies, obtaining the new anomaly detection method.
- Within the method obtained, one of the phases with the highest degree of complexity to achieve the expected result was the stage of installation of the monitoring and data collection tools to feed the prediction stage.

- The results obtained show the effectiveness of the implementation of the method within a LAN, contributing to decision making, reliability, availability of services offered to end-users within the network.

5 Recommendations

To give continuity to this research, it is recommended:

- Analyze new monitoring tools and adjust them in terms of centralized information gathering to feed the anomaly detection model.
- Implement additional safety or service variables to the predictive model data to increase the level of asset control.
- Evaluate and adjust different deployment environments as needed.
- Focus on predictive techniques through agents installed directly on network equipment and raise the level of supervision.

References

1. Kavitha, P., Usha, M.: Detecting anomalies in WLAN using discrimination algorithm, pp. 1–6 (2013). https://doi.org/10.1109/ICCCNT.2013.6726547
2. Zamorano Ruiz, J.: Comparativa y análisis de algoritmos de aprendizaje automático para la predicción del tipo predominante de cubierta arbórea, p. 134 (2018)
3. Voronin, V.V., Davydov, O.A.: Intermittent connection fault diagnosis for CAN using data link layer information. IEEE Trans. Ind. Electron. **64**, 2286–2295 (2017). https://doi.org/10.1109/TIE.2016.2624261
4. Peña Casanova, M., Lauriano da Silva, J., Febles Díaz, O., Anías Calderón, C.: Sistema para detección y aislamiento de fallas. Rev. Cubana de Ciencias Informáticas. **12**, 58–73 (2018). http://scielo.sld.cu/scielo.php?script=sci_abstract&pid=S2227-18992018000200005&lng=es&nrm=iso&tlng=es
5. Cruz Villa, J.A.: Desarrollo de procedimientos para un modelo de gestión de fallas de la red para la plataforma ISP de la CNT EP. Pontificia Universidad Católica del Ecuador (2015). http://repositorio.puce.edu.ec:80/xmlui/handle/22000/10008
6. Heo, S., Lee, J.H.: Fault detection and classification using artificial neural networks. IFAC-PapersOnLine. **18**, 470–475 (2018). https://doi.org/10.1016/j.ifacol.2018.09.380
7. Voronin, V.V., Davydov, O.A.: Local area network failures types, consequences and criticality analysis, pp. 184–187 (2017). https://doi.org/10.1109/RPC.2017.8168095
8. Usman, M.U., Ospina, J., Faruque, M.O.: Fault classification and location identification in a smart distribution network using ANN, pp. 1–6 (2018). https://doi.org/10.1109/PESGM.2018.8586471
9. Zhang, R., Zhang, S., Lan, Y., Jiang, J.: Network anomaly detection using one class support vector machine, vol. 1, pp. 19–21 (2008)

Analysis of a Network Implementing ROIP Technology

Alex Gilces-Zambrano[1(✉)], Dannyll Michellc Zambrano[2], and Lorena Bowen-Mendoza[2]

[1] Instituto de Postgrado, Universidad Técnica de Manabí, Portoviejo, Ecuador
agilces9551@utm.edu.ec
[2] Facultad de Ciencias Informáticas, Universidad Técnica de Manabí, Portoviejo, Ecuador
{mdzambrano,lbowen}@utm.edu.ec

Abstract. Communications are a relevant aspect in emergency response institutions when coordinating activities before, during, and after an incident. Radio over IP (ROIP) technology allows to increase the coverage of different radio communications systems, regardless of their technology, taking advantage of the benefits provided by voice over IP systems, but with the use of two-way radios, with Push To Talk (PTT) transmissions, which is an attractive option for communication solutions with greater interoperability, flexibility, and rapid deployment, using an existing IP infrastructure, as well as the Internet, to allow a more scalable communication and reducing implementation costs. This research work is oriented to the development of a network design implementing ROIP technology, as well as the evaluation of the most important parameters to determine the quality of data transmission, allowing to propose a communication alternative between different geographical areas, to ensure coordination, effective reporting, and communication of events in the field.

Keywords: Radio over IP · SIP · Push to talk · VOX

1 Introduction

Currently, there are several radio frequency technologies such as high frequency radio systems (HF), very high frequency (VHF), ultra high frequency (UHF), analog or digital technologies, which makes it difficult to have an integrated system that allows the convergence of all these types of equipment [7]. In Ecuador, in recent years a national trunked network has been implemented, allowing the communication of groups of users, regardless of the channel or frequencies available, which is used by institutions such as National Police, Armed Forces, Ministry of Health, National Service of Risk Management and Emergencies, among other security institutions, with a coverage of 95% nationwide, but with high implementation and maintenance costs, making it difficult to use in all response institutions in the country.

M. Botto-Tobar et al. (Eds.): CIT 2021, LNNS 405, pp. 178–191, 2022.
https://doi.org/10.1007/978-3-030-96043-8_14

RoIP technology was developed in 2002 by David Cameron in British Columbia, Canada, and is based on voice over IP, which generally consists of capturing voice sound waves with a microphone, digitizing and encoding the signal to convert it into bits, transmitting it in real-time, decoding and converting it from digital to analog, and reproducing it through a peripheral; but unlike traditional VoIP systems, it includes two-way radio systems instead of telephone calls. RoIP also transmits certain signals that are specific to Land Mobile Radio systems, such as control signals for PTT (Push To Talk) to enable transmission and COR (Carrier Operated Relay) to indicate the receive mode. In a RoIP system, the PTT and COR signals are sent over the LAN/WAN network via the RoIP gateway in IP format [6].

Tabsombat [10] tells us that the SIP-based ROIP system is an IP-based communication that uses SIP, an application layer protocol based on TCP/IP, with a syntax similar to HTTP. SIP generally uses port number 5060 or 5061 for both UDP and TCP and can implement other transport protocols. The Radio over IP (ROIP) system offers good voice quality both in transmission and reception between radio-radio or radio-telephone over limited bandwidth. The system can allow interconnection between different types of equipment, such as cellular phones, two-way radios, PSTN, Internet phones, VoIP phones, etc.

According to Riofrio [6], the use of Radio Frequency over IP (RoIP) technologies allows to increase the coverage of different radio communication systems, regardless of their technology, taking advantage of the benefits offered by Voice over IP systems, but with the use of two-way radios, with Push To Talk (PTT) transmissions, which are converted into IP packets and transmitted over the network, offering advantages such as greater scalability, reduced implementation costs, greater interoperability between equipment.

Communications are a relevant aspect in emergency response institutions when coordinating activities before, during, and after an emergency or incident. In this work, a network proposal was made to integrate the communications of the radio frequency equipment of the different fire stations of the Portoviejo city with the existing telephony over IP system, allowing interoperability between different technologies, generating high scalability and rapid deployment.

In 2015, research was conducted in Ecuador for the "Solution based on ROIP (Radio over IP) as an alternative or complementary system for mission-critical trunked services", in which it is concluded that it is possible to communicate between radio frequency systems and systems that support IP networks at low cost, in this case, the study was established communication between a private LAN network without internet access with a trunked system [6].

In 2016 an investigation and design of an electronic communications solution for places without radio coverage of the Fire Department in the province of Imbabura was carried out, where the electronic communications solution was oriented to meet three essential requirements, easy installation/operation, coverage, and low cost of implementation, Thus, an analysis of the most appropriate technologies was carried out, without forgetting the concept of coverage in the place of care, which refers to providing a connection with all the actors involved in emergency care, mobile and portable radios, always complying with the concept of PTT, because it is the most appropriate for emergency care [11].

In 2018, an Analysis of radio over IP system to strengthen communication between the Fire Brigades of the province of Manabí was carried out, which concludes that to solve basic problems such as communication in a society currently involves the use of technology, integrating radio and voice over internet technologies to agree on an autonomous and efficient system for timely communication between the different points where it can be implemented [15].

In 2018, Kester [4] proposed an integrated RoIP communication network for effective collaboration during emergency and disaster management, which allows integration with other networks for effective communications, deployment, and information management, obtaining as results that the ease of use was effective and is suitable for easy implementation and management of disasters and emergencies.

Likewise, in 2019, research was developed on a design for the evolution of the Air Force air traffic control voice network using Radio over IP technology, where it was concluded that the use of RoIP and IP has allowed achieving convergence and interoperability through better management of communications networks [5].

With the exposed background, it was determined that it is possible to implement a communication system that integrates the technologies of Radio Frequency over Internet Protocol, as a complement to the VHF system, to extend the reach to the rural parishes of the Portoviejo city. However, it was necessary to design and implement a VPN (Virtual Private Network) connection to ensure the transmission of communications over the Internet and to evaluate the performance of the transmission of ROIP packets to determine the efficiency of the network.

2 Materials and Methods

For the realization of the project, it was decided to implement a methodology proposed by the author, which consisted of four phases: Analysis of the current situation, design of a communications network implementing ROIP technology, implementation of the test environment, and analysis of results. Each of the phases is detailed below:

2.1 Phase 1: Analysis of the Current Situation

Initially, the information was collected. Figure 1 shows the topology diagram of the current network, which due to a considerable geographical separation between the fire stations and the administrative office, there is no radio frequency communication between each of the facilities, which makes it difficult to perform the activities of operational personnel optimally. In addition, the Table 1 details the information obtained from the equipment that makes up the radio frequency communications system currently used in the various fire stations in Portoviejo city, as well as the voice over IP equipment used in the administrative office.

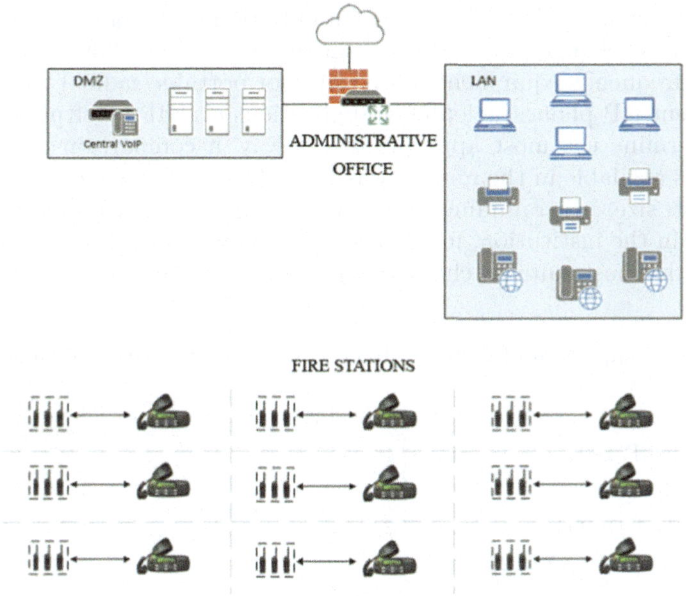

Fig. 1. Current CBP network topology. **Source:** Own elaboration

2.2 Phase 2: Design of a Communications Network Implementing ROIP Technology

Once the current state of the entity's communication system was known and the analysis of other works related to the research was done, we proceeded to redesign the communications network for the Portoviejo Fire Department, which allows the integration of radio frequency equipment.

Table 1. CBP radiofrequency and voice over IP equipment. **Source:** Own elaboration

Type	Manufacturer	Model
Repeater	Motorola	MOTOTRBO SLR 5100 (VHF)
Base radio	Motorola	PRO 5100 (VHF)
Vehicle base radio	Motorola	EM200 (VHF)
Portable radio	Motorola	DEP550E (VHF)
Portable radio	Motorola	EP450S (VHF)
Telephone exchange	Grandstream	Serie UCM6200
IP phone	Grandstream	GXP2140
IP phone	Grandstream	GXP1615

In this design, it was determined the need to implement a ROIP gateway and a Gain Balancer for the conversion of push to talk (PTT) audio to IP data packets, compatible with the session initiation protocol (SIP) supported by the telephone exchange available in the institution, which allows communication with radio frequency equipment (base radios or portable radios) with voice over IP equipment (IP phones, telephony applications over IP - Softphones).

To determine the most appropriate gateway, a comparison of some of the alternatives available in the market was made (Table 2), considering as parameters its cost, size, programming interface, as well as compatibility with existing equipment in the institution, for this comparison we worked with a scale of 1 to 5, being 5 the most suitable characteristics for the choice of equipment.

Table 2. Comparison of Gateway RoIP equipment. **Source:** Own elaboration

Device	Cost	Size	Interface	Compatib.	Average
RoIP 102	4	4	5	4	4.25
RoIP 302	3	4	4	3	3.5
ACU-1000	3	2	3	3	2.75
Gateway NXU-2A	3	2	3	4	3

Based on the results shown in the Table 2, it was determined that the RoIP 102 equipment with an average of 4.25 is the most optimal for carrying out the present investigation.

2.3 Phase 3: Test Environment Implementation

Once the design of the network based on ROIP technology was obtained, we proceeded to implement it in the communications network of the Portoviejo Fire Department and one of its stations, located in one of the rural parishes of the city, connected through the internet, integrating the new components, without interrupting the existing network or creating points of vulnerability. First of all, the fire station was connected to the communications network of the administrative office of the institution, using OpenVPN software installed on a Pfsense server and a TPLink router as a client.

Subsequently, the Grandstream UCM6204 IP PBX was configured, which will be accessed by the gateway devices connected to a radio base from the different stations through a VPN connection, where the creation of SIP extensions corresponding to each fire station and other areas necessary for testing was performed, for this, the IP and port assigned using the DTMF mode in RFC2833 was entered through a web browser.

In the selected fire station, we proceeded with the installation and configuration of a router ROIP 102 gateway, for this is accessed from a web browser to the IP assigned to the LAN port, where the LAN port configuration parameters were set to allow the connection to the telephone exchange through the virtual private network, the call configuration, specifying the SIP group number, username and password registered in the telephone exchange, and the IP of the SIP Proxy server.

Before connecting the ROIP 102 gateway to the base radio station, the accessory connector pins of the Motorola PRO 5100 radio were configured using a Motorola programming cable and the Professional Radio CPS Software. In the Accessory Pins option, setting pin 3 as External PTT Microphone Input, pin 8 as Output for detection of CSQ squelch carrier tones. In the Accessory Configuration option, the parameters of the audio transmission are configured, such as the type of transmission, PTT data sources, alarm times, bounce duration, among others.

Once the Motorola PRO 5100 pins were configured, the Channel port was connected to the load balancer through an RJ11 connector, which in turn was connected to the accessories port of the Motorola PRO5100 radio.

Finally, the 3CX Phone IP telephony software was installed on a computer in the administrative office network, and calls were made between portable radios connected to the base radio connected to the IP network, the softphone, and IP telephones, checking the correct operation of the ROIP system.

To obtain the values of the network parameters, the Packet Capture tool was used, included in the Pfsense software installed in the network of the administrative office of the institution. Once the captures of the packets transmitted during the communications carried out during the test period were obtained, they were loaded into the VoIP monitor software, to perform the corresponding analysis.

2.4 Phase 4: Analysis of Results

The analysis of the results obtained during the implementation phase of a test environment was carried out, which allowed us to deny or accept the hypothesis proposed and thus be able to conclude on the advantages and disadvantages of this technology.

In this phase, parameters such as bandwidth, delay, jitter, packet loss through the IP Internet protocol were measured by monitoring the components of the communications network with the Pfsense software. In addition, the VoIP monitor software was used to analyze the packet captures, using the ITU Y1540 and Y1541 recommendations as a reference for the evaluation [13].

Another parameter evaluated was the quality of experience (QoE) in communication, for which the mean opinion score (MOS) test tool was used, carried out on the institution's operational personnel who participated in the communication tests between the different devices, to verify the performance of the proposed system, using the ITU-T P.800 recommendation as a reference [12].

3 Results and Discussion

3.1 Design of a Network Implementing RoIP

The proposed design (Fig. 2) comprises the following elements: portable radios assigned to the institution's operational personnel, mobile radios located in the different rescue units, and base radios installed in the fire stations of the urban and rural parishes of Portoviejo city, with the IP-based communications network of the administrative office, through virtual private network (VPN) connections over the Internet, where the Grandstream UCM6204 equipment located in the demilitarized zone (DMZ) acts as a centralized solution for IP communications, allowing the different ROIP gateways to communicate with each other, or with the other voice over IP equipment already existing in the network, whether these are IP phones, as well as computers and smartphones through Softphones applications, including with the traditional circuit switched network (PSTN).

3.2 Performance Verification

Many studies in peer-reviewed publications [1,3] agree that the most important metrics to evaluate the actual performance of a network in a VoIP context are: delay, jitter, and packet loss. Other metrics that can be considered significant are the number of interruptions, which represents the number of cuts that the user experiences in the telephone call, and the duration of the interruption, which can be seen from the moment the routing protocol warns that the route has been lost until the availability of a new valid route. This represents the magnitude of the outage that the user experiences in his telephone call [8].

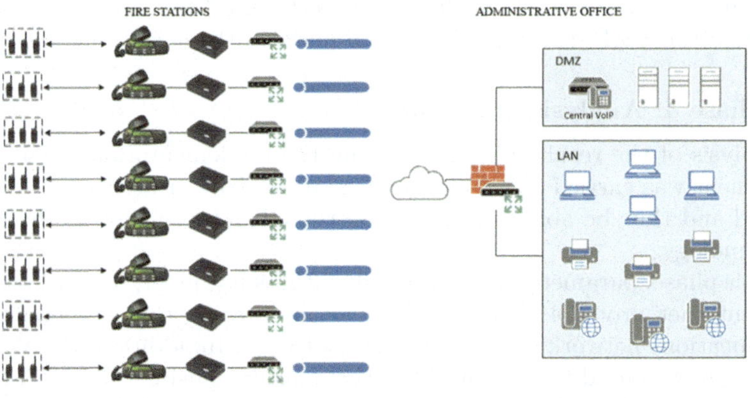

Fig. 2. Proposed topology implementing ROIP technology. **Source:** Own elaboration

The ITU, in its recommendation Y.1540, defines the parameters that can be used to assess performance. These parameters are applied end-to-end, point-to-point, and between the parts of the network that contribute to the provision of services. This recommendation defines several parameters, of which the four main ones are [3]: IP packet transfer delay or IPTD, IP packet loss ratio or IPLR, IP packet delay variation or IPDV, and IP packet error ratio or IPER [13]

In addition, recommendation Y.1541 specifies performance values for the above parameters (Table 3), where classes 0 and 1 are defined for real-time, delay-sensitive, and high iteration applications such as VoIP.

Table 3. Quality parameters. **Source:** ITU [14]

Parameter	QoS classes					
	Class 0	Class 1	Class 2	Class 3	Class 4	Class 5
IPTD	100 ms	400 ms	100 ms	400 ms	1 s	U
IPDV	50 ms	50 ms	U	U	U	U
IPLR	1×10^{-3}	1×10^{-3}	1×10^{-3}	1×10^{-3}	1×10^{-3}	U
IPER	1×10^{-4}					U

3.3 Verification of Network Parameters

Initially, the bandwidth available in the VPN connection between the administrative office of the institution, where the SIP telephone exchange is located, and the fire station with the radiofrequency equipment, connected to a RoIP Gateway, was evaluated using iPerf software, for this analysis the software was installed on two computers (client and server), and tests were performed during different hours. Figure 3 shows the histogram with the results, where an average bandwidth of 424 Kbps was determined.

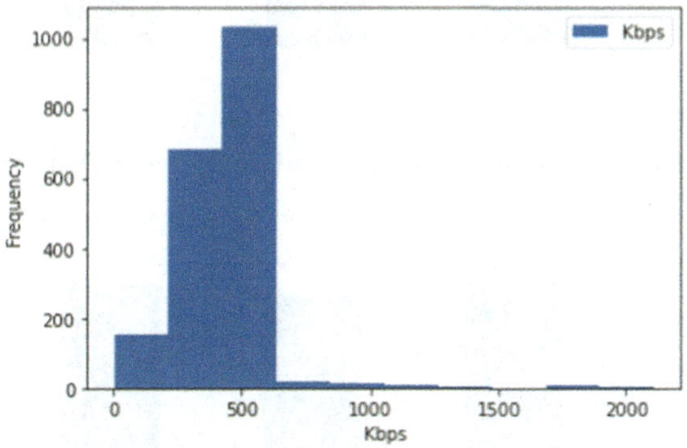

Fig. 3. Histogram of VPN bandwidth. **Source:** Own elaboration

Jitter Performance Verification

The delay that occurs in data flows is not constant, and this delay variation causes a new problem known as jitter. This can appear due to congestion in the network, but mainly due to incorrect bit synchronization between network elements, and appears as a narrowing and lengthening of the pulse width at the receiver [3]. Jitter performance verification was performed by a statistical check of the telephony network response and the RoIP network response.

Jitter verification of the institution's network was performed on a sample of 90 calls made with three different codecs, at different times of the day. Figure 4 shows the results of the jitter evaluation by codec, with the lowest average jitter being that of the G711a codec with 7.54 ms, complying with recommendations P.800 and Y.1541.

Delay Verification

This is a parameter that indicates the delay in the arrival of data flows at their destination in the network. Its effects are different according to whether the transport layer protocol used is TCP or UDP [2]. In all tests performed, the average delay value was 300 ms, complying with the Y.1541 recommendation, which indicates that the acceptable delay for a call must be less than or equal to 400 ms (Fig. 5).

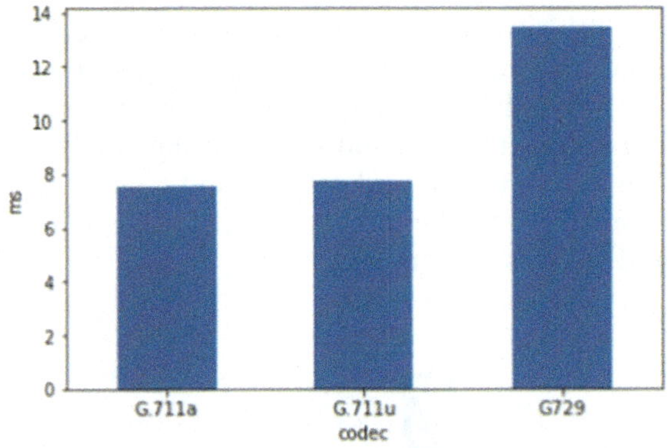

Fig. 4. Jitter evaluation results per codec. **Source:** Own elaboration

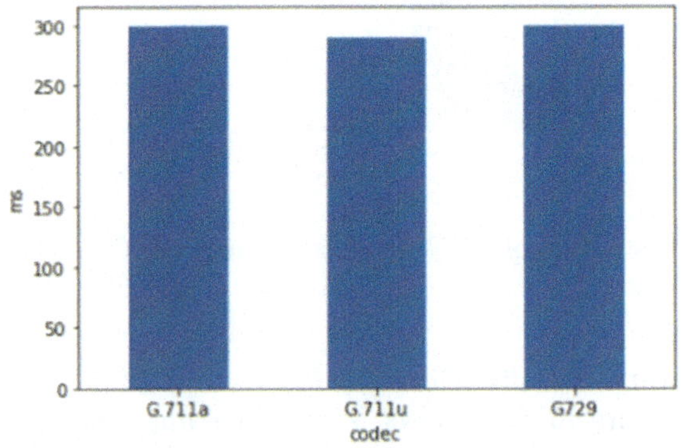

Fig. 5. Delay evaluation results. **Source:** Own elaboration

Checking the Packet Loss Rate

Packet loss occurs very frequently as a result of network congestion, one of the consequences in VoIP calls is the deterioration of the voice signal. The results represented in Fig. 6 show that the packet loss rate parameter is mostly 0, having a maximum average value of 0.0038% for the G.711a codec and an overall average of 0.001%, being the one recommended by ITU (International Telecommunication Union, 2011) for VoIP communications.

Verification of Bandwidth Usage Performance
The bandwidth is directly related to the codec and this is why when it does not have the appropriate level to support the codec used, the quality of service can be affected because it starts to present saturation in the channel [1].

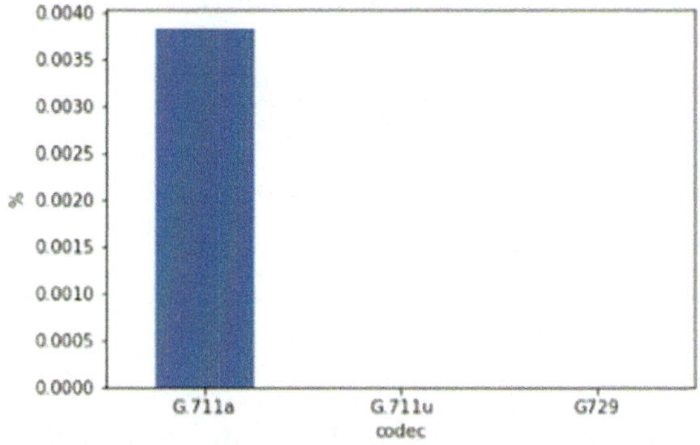

Fig. 6. Packet loss rate during per codec tests. **Source:** Own elaboration

In Fig. 7 it can be seen that the maximum bandwidth required for calls between RoIP equipment and the other VoIP devices in the network is 81.6 Kbps, while the average bandwidth available in the VPN connection is 424 Kbps. According to the design of the proposed topology, each VPN connection will have only one RoIP device, so it was verified that the current bandwidth is sufficient to guarantee the correct transmission of data.

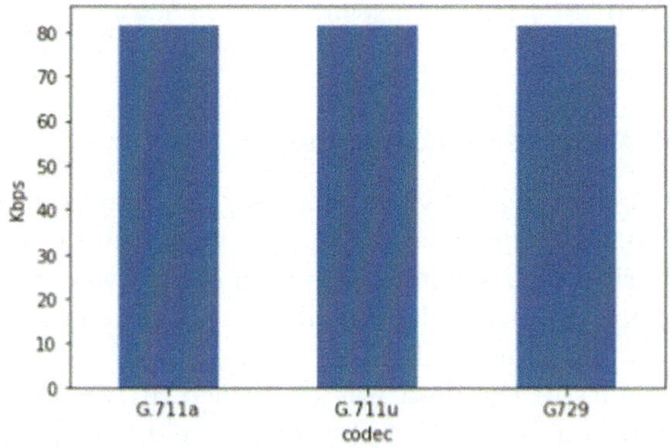

Fig. 7. Bandwidth used during testing per codec. **Source:** Own elaboration

MOS Verification

The verification of the bit error rate is a measure of the physical quality of the transmission, in VoIP and RoIP systems with a high error rate causes the voice at the receiver is not clear, to verify the performance is performed a subjective measure which consists of conducting a phone call quantifying the quality of the same. The MOS (Mean Opinion Score) technique was chosen, which tests a considerable volume of callers to evaluate the clarity of the communication [9]. Table 4 shows the opinion scales.

Table 4. MOS scale for call quality of experience. **Source:** ITU [12]

Score	Quality of experience
5	Excellent
4	Good
3	Regular
2	Mediocre
1	Bad

Table 5 shows the results of the MOS quality of experience. These values are obtained through automatic tests performed by the VoIP monitor software, which can establish voice quality estimates by analyzing the network parameters delay and packet loss, according to the ITU-T E model, which indicates that the MOS does not represent the audio signal but the network parameters.

Table 5. MOS evaluation results. **Source:** Own elaboration

codec	count	mean	std	min	25%	50%	75%	max
G.711a	30	4.47	0.164317	3.6	4.5	4.5	4.5	4.5
G.711u	30	4.486667	0.07303	4.1	4.5	4.5	4.5	4.5
G729	30	4.443333	0.20957	3.5	4.5	4.5	4.5	4.5

3.4 Statistical Analysis of Experiments

For the selection of the type of hypothesis test, the Shapiro test was performed for each of the variables, which poses the null hypothesis that a sample comes from a normal distribution. A significance level of 0.05 was chosen, and we have an alternative hypothesis that the distribution is not normal. Details of the results obtained are given in the Table 6.

Once the normality of the data was confirmed, it was determined that the samples did not follow a normal distribution, therefore, to perform the hypothesis test, the nonparametric Friedman test was used for each of the parameters analyzed, obtaining the results presented in the Table 7.

Table 6. Shapiro test results. **Source:** Own elaboration

Variable	Result	Observation
Jitter	6.66E−18	H1 The sample does not follow a normal distribution
Packet loss rate	stat = 0.081, p = 0.000	H1 The sample does not follow a normal distribution
Delay	stat = 0.389, p = 0.000	H1 The sample does not follow a normal distribution
Bandwidth used	stat = 1,000, p = 1,000	H0 The sample follows a normal distribution
MOS	stat = 0.217, p = 0.000	H1 The sample does not follow a normal distribution

Table 7. Friedman test results. **Source:** Own elaboration

Variable	Result	Observation
Jitter	stat = 8.588, p = 0.014	The mean of the samples are not equal
Packet loss rate	stat = 2.000, p = 0.368	The mean of the samples are equal
Delay	stat = 2.000, p = 0.368	The mean of the samples are equal
MOS	stat = 1.600, p = 0.449	The mean of the samples are equal

4 Conclusions

- Based on the literature review, the performance of RoIP technology was analyzed and it was determined that the objective evaluation of the performance of the proposed design should be based on parameters considered in other research oriented to evaluate VoIP systems since RoIP technology is essentially VoIP, which incorporates control signals for PTT (Push To Talk) to enable transmission and COR (Carrier Operated Relay) to indicate the reception mode.
- It can be concluded that within the Portoviejo Fire Department, in all the environments under study, there is adequate communication network infrastructure to implement RoIP, making it feasible and viable.
- From the performance evaluation of the proposed communication network design, through the analysis of bandwidth, delay, jitter, packet loss, and MOS parameters, it is concluded that the RoIP system complies with the ITU P.800, Y.1540, and Y.1541 recommendations, with a statistically significant sample.
- The Radio over IP system prototype presented a good performance, guaranteeing the quality of communication between radiofrequency equipment with VoIP devices, allowing to have a scalable communication system that efficiently integrates the radio communication systems of the Portoviejo Fire Department stations.

- According to the tests performed, it was determined that the G.711a codec is suitable for the proposed design, presenting a better performance concerning jitter, and similar values in the other network parameters, so it is proposed as the main alternative for audio encoder and decoder.

5 Recommendations

To give continuity to the present investigation it is recommended:

- Analyze new RoIP gateways with improved codec support, access to mobile networks such as 3G, and determine the performance of each for possible implementation in rescue vehicles.
- To use this study as a basis for the design and subsequent implementation of the improvement of the voice network of the Portoviejo Fire Department, given that it has contemplated a conscious analysis of the feasibility for the implementation of an alternative solution already proven in the management of networks with the same degree of complexity and responsibility, framing it in a standard that ensures the quality of service and safety.
- This research has been developed on Motorola PRO5100 and EP450 radios; however, it is open to the possibility of performing the corresponding tests on the different models and brands of radiofrequency equipment available in the market.

References

1. Blanco Almeida, M.F.: Evaluación de los Parámetros que Afectan la Calidad de Servicio en Telefonía IP. Evaluation of Parameters Affecting the Quality of Service (QoS) in IP Telephony (17):61–71 (2015)
2. Buñay, P., Pastor, D., Paguay, P., Moreno, S.: Vista de Análisis de la Arquitectura DIFFSERV sobre redes MPLS para la provisión de QoS en aplicaciones en tiempo real (VoIP) (2019)
3. Cuellar Quiñones, J., Olave, A., Valencia, L.: Análisis basado en parámetros QoS para el servicio VoIP sobre wireless en un campus universitario. Ventana Inform. **31**, 11–23 (2014)
4. Kester, Q.A.: An integrated RoIPg communication network for effective collaboration during emergency and disaster management. In: Bissyande, T.F., Sie, O. (eds.) AFRICOMM 2016. LNICST, vol. 208, pp. 174–184. Springer, Cham (2018). https://doi.org/10.1007/978-3-319-66742-3_17
5. Pimentel, R.E., Ortega, J.C.: Diseño para la evolución de la red de voz de control de tránsito aéreo de la Fuerza Aérea utilizando tecnología Radio sobre IP. Master's thesis, Pontificia Universidad Católica de Perú (2019)
6. Riofrío Córdova, A.A.: Solución basada en ROIP (radio over ip) como sistema alternativo o complementario para servicios troncalizados de misión crítica. Master's thesis, Pontificia Universidad Católica del Ecuador (2015)
7. Servicio Integrado de Seguridad ECU 911: ECU 911 se repotencia con ampliación de la Red Nacional Troncalizada - Servicio Integrado de Seguridad ECU 911 (2019)

8. Sing-borrajo, P.: MANET Performance evaluation of VoIP in MANET, November 2014
9. Suin Uyaguari, D.R.: Diseño e implemetación de una red RoIP de dos vías para la empresa MaxiGroup. Master's thesis, Universidad Politécnica Salesiana (2016)
10. Tabsombat, S., et al.: Push-to-talk control scheme for radio over IP communication. In: 2010 7th International Conference on Service Systems and Service Management. IEEE (2010)
11. Torres Chiluisa, C.A.: Investigación y diseño de una solución de comunicaciones electrónicas para los lugares sin cobertura de radiocomunicación de los cuerpos de bomberos en la provincia de Imbabura. Master's thesis, Pontificia Universidad Católica del Ecuador (2016)
12. Unión Internacional de Telecomunicaciones. P.800: Métodos de determinación subjetiva de la calidad de transmisión (1996)
13. Unión Internacional de Telecomunicaciones. Y.1540: Servicio de comunicación de datos con protocolo Internet - Parámetros de calidad de funcionamiento relativos a la disponibilidad y la transferencia de paquetes del protocolo Internet (2002)
14. Unión Internacional de Telecomunicaciones. Y.1541: Objetivos de calidad de funcionamiento de red para servicios basados en el protocolo Internet (2011)
15. Vélez Casanova, J.P.: Análisis de un sistema de radio sobre IP (ROIP RADIO OVER IP) para fortalecer la comunicación entre los Cuerpos Bomberos de la Provincia de. Master's thesis, Universidad Católica de Santiago de Guayaquil (2018)

Communication System Between People with Hearing Impairments and Their Patients

Marely del Rosario Cruz-Felipe(✉), Jenmer Maricela Pinargote-Ortega, and Gabriel Primitivo Demera Ureta

Facultad de Ciencias Informáticas, Universidad Técnica de Manabí, Portoviejo, Ecuador
{marely.cruz,maricela.pinargote,gabriel.demera}@utm.edu.ec

Abstract. At the Technical University of Manabi, there is a group of students with hearing disabilities, among them are students who are studying clinical laboratory, who need to interact with patients and be able to communicate correctly at the time of carrying out their practices. The present research shows a system in which various interfaces are evaluated to achieve communication between people with hearing disabilities and their patients. To develop the system, the Extreme Programming methodology was applied, using the Android system as the base platform using the Android Studio IDE, under the Java and XML language, Text to Speech extension as support for the reproduction and conversion of text into audio, and SQLite as a data store for phrases, words, and images. The application makes it easier for hearing impaired students to instruct their patients on the clinical examination they carry out. Tests were carried out on the system by calculating the precision with different interfaces, and it was obtained that the one with the lowest margin of error is the interface where images and keywords are applied, this being the one used in the construction of the proposed system.

Keywords: Hearing impairment · Android · Patients

1 Introduction

Technological advances have made it possible to improve the quality of life of people with disabilities, to fend for themselves [1]. Thanks to these advances, digital inclusion is achieved under similar conditions [2, 3].

People with hearing disabilities have an intellectual capacity similar to that of listeners; their only problems derive from the difficulty in acquiring and internalizing oral language [4–6].

The development of communication between two people is essential when carrying out an activity, but there are groups of people who cannot communicate correctly with others; this could be due to different circumstances, one of which is hearing impairment, which prevents fluid communication between two or more people. Different types of solutions have been developed, adhering to measures such as sign language [7, 8], and other learning methods that have undoubtedly been of utmost importance in their lives.

M. Botto-Tobar et al. (Eds.): CIT 2021, LNNS 405, pp. 192–204, 2022.
https://doi.org/10.1007/978-3-030-96043-8_15

Technology-supported applications have also been implemented, and as it progresses, numerous possibilities for improving the effectiveness of information exchange between two or more people with hearing impairment increase [1, 3, 7, 9].

In the Unit of Inclusion, Social Equity and Gender of the Technical University of Manabi there are three students with hearing impairment in the clinical laboratory career. The University does not have systems that provide students with dynamic communication between the student and the surrounding people.

Students with hearing disabilities have limited knowledge of sign language, not all are at the same level, when referring to handwritten words, they do not fully understand them, they only understand keywords, not phrases, which limits their communication with people who do not have this type of disability.

Taking into account the characteristics of the students, who do not correctly master the sign language, and find it difficult to adapt to the applications; it is necessary to design a system with characteristics adaptable to these students.

The objective of this paper is to obtain a communication system between people with hearing impairment and their patients. The paper is structured as follows: Sect. 2 shows the methodology applied, Sect. 3 presents the results and discussions, and Sect. 4 shows the conclusions.

2 Materials and Methods

2.1 Study Context and Participants

An investigation is carried out that uses the observation method to define the requirements of communication of students with hearing disabilities, starting from conducting different interviews with them to find how each one transmits their ideas to the patients. An experimental study is also carried out to validate the results obtained in the software.

The population for this study is a group of students and professors from the clinical laboratory career of the Technical University of Manabi (Ecuador), in which there were six students with hearing disabilities, and the remaining members participate in evaluating the understanding of the ideas transmitted by students with disabilities.

2.2 Data Initial

For the construction of the data, tests were carried out on the students through several interviews that allowed them to evaluate the most suitable way to express an idea. In Fig. 1, a test is shown where each student was asked to transmit an idea through text in their native language (Spanish).

In Fig. 1, the student tries to transmit the sentences: How old is she? Is she pregnant? Did you come fasting? As can be seen, the student writes sentences without writing rules, which is manifested in all the students; it was also observed that each response with different sentences to the same request, which makes it challenging to identify the idea they want to convey. For this, it is decided to use keywords that are common in the writing of all students.

Another test carried out through interviews with students with disabilities was to evaluate the use of images, which helped to gain clarity in the idea to be transmitted;

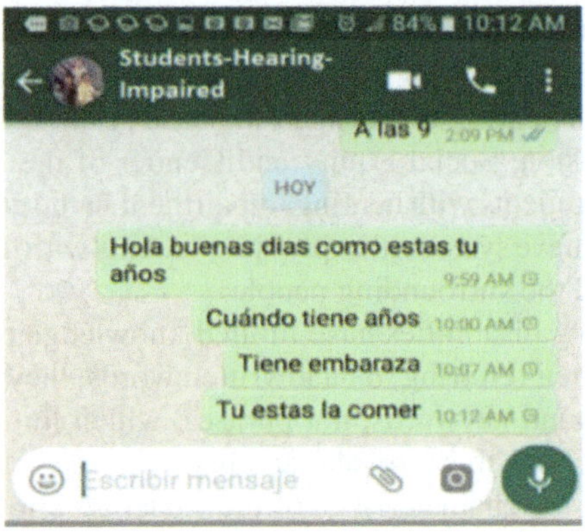

Fig. 1. Example of phrases transmitted by WhatsApp from students with hearing disabilities.

however, on some occasions, there are different interpretations by them. Therefore, representative images identified by the students were interrelated with keywords also identified by all; this allowed the initial data to be used in the construction of the software.

2.3 Procedure in the Construction of the Software

A database is built with keywords and representative images starting from the data obtained, which are shown in Fig. 2.

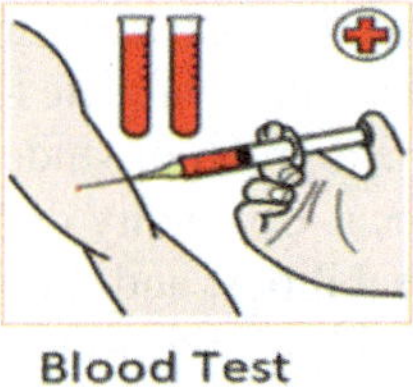

Fig. 2. Format of the data stored in the database.

The database is built with SQLite. The software was developed using the agile methodology eXtreme Programming (XP) by adopting an iterative development process [5, 6], and Android Studio, Java language, XML, Text To Speech were used to translate text to audio.

In Fig. 3, a flow chart with the steps in the construction of the software is shown.

2.4 Experimental Validation

Tests are carried out on the operation of the software, obtaining the margin of error from the calculation of efficiency and recall.

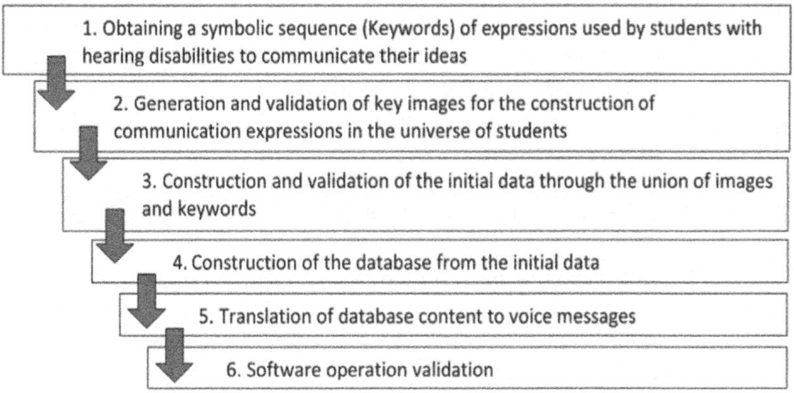

Fig. 3. Steps in building the software.

3 Results and Discussions

3.1 Requirement's Specification

Through observation sheets it was obtained inconveniences presented by students with hearing impairment of the clinical laboratory career of the Technical University of Manabi:

- Lack of an application that allows establishing possible communication between a person with a hearing impairment and their patient.
- Devices that exist in the market do not meet the characteristics of students with hearing impairment.

For the development of the application, requirements were identified, which were obtained through meetings with students with hearing impairment of the clinical laboratory career, in which, it was possible to analyze the friendliest interface to use, for this purpose an interface similar to the application of the social network WhatsApp was defined since it is the most frequently used by students.

Based on the needs the following requirements were specified:

1. **Functional Requirement's**

   - The interface will allow visualizing images and texts.
   - The application can display the main category interface which will also have images and would have the words used according to the chosen category.
   - The application will allow new categories to be added, with their respective image.
   - The interface will allow new words to be added, with their respective image and phrase.
   - The application will allow entering an empty category and within this, it will show the possibility of accessing the interface that allows the customer to add words.

- The interface would have a text playback button, which when pressed will emit a sound in the form of dictation, reading the phrase chosen by the user.
- The application would allow the user to filter words to speed up the use of it.

2. **Non-Functional Requirement's**

- The application needs the list of updated Text to Speech languages and downloaded in Spanish.
- The application needs to be installed on Android version 4.4 or later to work properly.
- The application requires a mobile device with a screen and its audio playback terminals to function normally.

Figure 4 shows the functionalities that were implemented in the application.

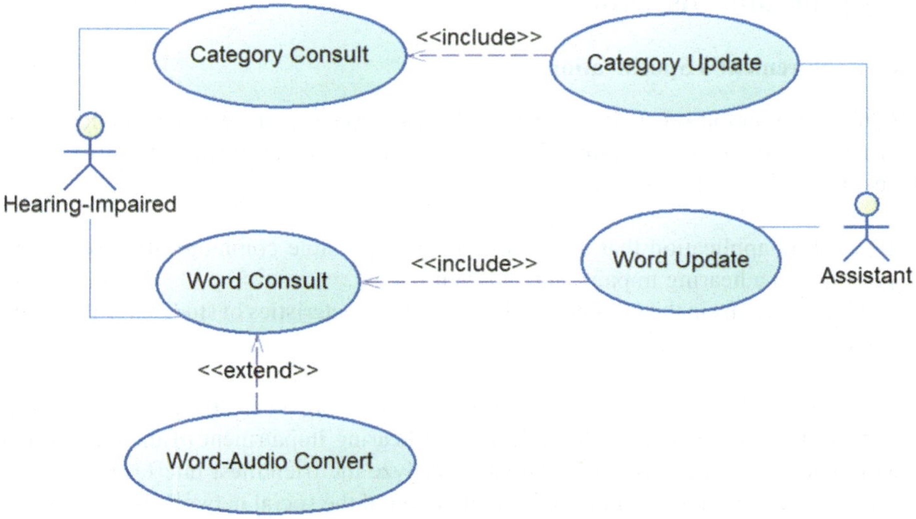

Fig. 4. Application functionalities.

Different interviews were carried out with the students with hearing disabilities, during the development of the application, to cover the requirements of each student and the interfaces to be used were evaluated:

- Use of the text: None of the tests carried out through the writing or comprehension of the text was passed, they only understood certain keywords.
- Use of images: We proceeded to present them on a computer with several images according to their area of study, in which they had to write or express that they understood the image. The results were a bit favorable, they understood certain images.
- Phrase construction: It was required that they structure sentences from the selection of several images, it was complex for the students, who failed to understand the different

types of tests performed with this structure, in any case, they chose a single image and its representation in the text below, thus discarding the rest of structured images as phrases.
- Keywords with images: The use of keywords under images were tested, with the restriction that in the application, they have a correctly structured phrase hidden, which is reproduced in audio so that the hearing people can hear it, and the people non-listeners understand its meaning.

It was concluded that the best option for communication interfaces is to use keywords with images, since it was the one that the students adapted satisfactorily, thus achieving that by selecting a single image, with their keywords below, the developed interface reproduce a correctly structured sentence, allowing them to improve communication between people with hearing impairment and their patients.

3.2 Implemented Application

For the development of the prototype, the IDE Android Studio development environment was considered [12], the language that operates within this IDE is JAVA as the main language and as XML modeling templates, in addition to its GRADLE compilation system that facilitates the implementation of the application in the devices in a simpler way, without needing to be creating several APPs, also considering a storage system or database known as SQLite, implemented in this IDE, this in order to develop quality interfaces. The vision of this methodology is that it allows the ease and effectiveness in the construction of the different interfaces, which are stable and adapt adequately to the needs of students with hearing impairment.

It was also considered mobile devices with an Android system. It should be noted that most mobile devices with intelligent systems currently work with the Android operating system, which is why it is chosen as the main platform for the operation of the application interfaces. Android also allows the use of native libraries, among which one is very useful for the study, such as Text to Speech, thus working fluidly and in several versions of it.

Within the IDE of Android Studio [7], the base template was used, in which they are defined as Activity, applying a blank interface, which allowed to design from the beginning, an interface that adapts to the end-users, which are the students with hearing impairment (Fig. 5).

In the same way, it was tried to resemble the design to that of a keyboard, which contains what is selected on the back, and what is shown on the top, that is to say, that an image is selected with its keyword, and this display a phrase in a text field at the top of the screen (Fig. 6).

Functional tests of the interface were carried out with the hearing-impaired students, verifying that they easily adapted to the application and were able to interact with it easily.

Through the search element and the implementation of existing functions, a word search method was defined, which filters by entering the desired text at the top of the application and in the same section where the results are displayed, you can select the desired word and reproduce it (Fig. 7).

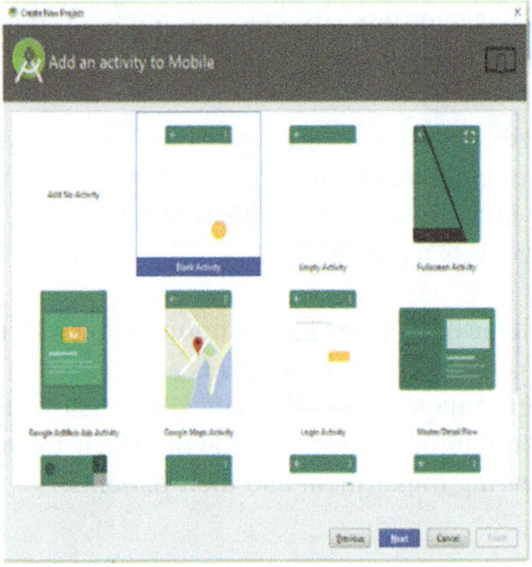

Fig. 5. Android studio templates.

Fig. 6. Word selection interface.

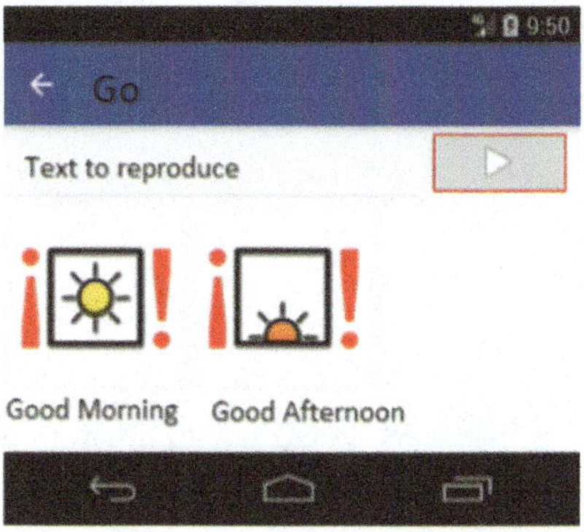

Fig. 7. Word search interface.

Subsequently, a view was implemented in the form of categories, which within each of them should show the words belonging to it, dividing the words like greetings, actions, exams, among others. This without affecting the result of the general search for elements, and allowing an additional user (who would be the Assistant), to be able to add new categories, as well as new words within any of them (Fig. 8 and Fig. 9).

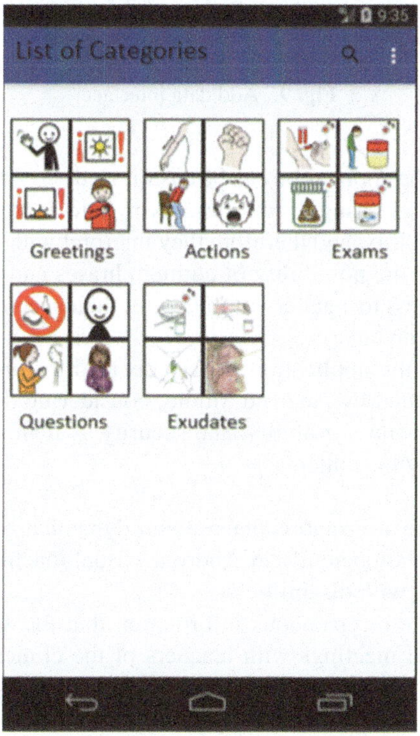

Fig. 8. List of categories interface.

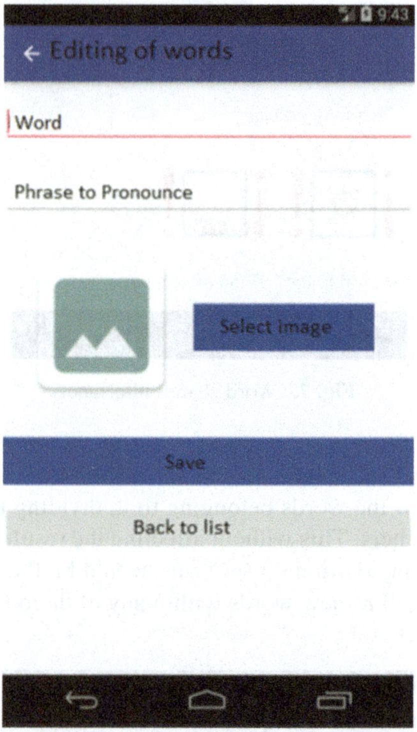

Fig. 9. Add data interface

As shown in Fig. 6, the application can be customized according to the characteristics of the types of users with hearing disabilities, taking into account that each of them has different degrees of disability, and therefore they interpret with greater or lesser ease the phrases to use. We offer the possibility of adding phrases and images that are selected by users, facilitating access to a gallery of images of their specialty, and select the image according to their requirements.

To perform tests of the application, it was decided to run the applications in the virtual environment granted by Android Studio, considering the different aspects such as content, function, structure, usability, and security. A review was made through the observation of the following contents:

- Dynamic: Refers to the information that remains dynamically in the application, these tests were performed competently in Android virtual machines, Android devices of teachers, and students with disabilities.
- Semantics: Incomplete or erroneous information that the application may have, in which through several meetings with teachers of the clinical laboratory career, the phrases contained in the application, images, and keywords have been correctly structured.

Students with hearing impairment used the application, demonstrated that they could manipulate it without any problem. Likewise, other people with hearing impairment and without disability managed to manipulate the application in a very comfortable and simple way.

The application of some security systems was not necessary, since the main purpose of the prototype is that anyone with a hearing impairment can use it. For performance tests and proper functioning and to verify its efficiency, we proceeded to use strategies that will help provide adequate results for proper evaluation.

- Tests were carried out in suitable environments to measure the effectiveness of the application. A clinical laboratory work environment was simulated with the help of teachers and a sign language interpreter. The hearing-impaired students used the application assuming the role of clinical laboratory staff providing patient care.
- The application was tested several times through the Android Studio test system. It started with 10 tests in the word interface, we managed to correct typing and design errors. Additionally, the content effectiveness and usability were tested on various virtual devices, and we replicated the fixes on other interfaces. The virtual testing interface used is shown in Fig. 10.

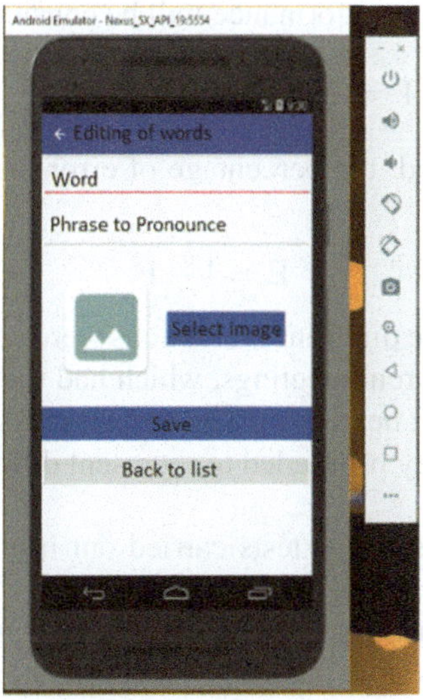

Fig. 10. Testing the application in the android Studio IDE interface.

In addition, multiple tests were run on physical devices with different characteristics. Precision indicators were used to evaluate the interfaces.

The application was tested periodically in meetings with teachers and students, who provided requirements that were implemented without any error. The final test yielded positive usability results since the hearing-impaired students were able to use the application without any problem and quickly became familiar with its functionalities.

3.3 Evaluations of the Implemented Interfaces

Based on the fact that there are no applications found in consulted bibliographies, that perform the same functionalities of the developed application, taking into account that this is a personalized application for special educational needs, it is not possible to make a comparison with others according to the percentage of error, but it is based on calculating this percentage and interpreting its results to validate the usefulness of the application.

For the evaluation of the interfaces, the Precision (precision) and Recall (recovery) indicators were used, widely used to evaluate the accuracy and efficiency of the systems through the collection of information. These are defined by the following equations [13, 14, 15]:

$$\text{Precision} = \frac{\text{CC}}{\text{CC} + \text{Fp}}$$

$$\text{Recall} = \frac{\text{CC}}{\text{CC} + \text{Fn}}$$

Where *CC* is the number of tests that were favorable when performed with people with hearing impairment, *Fp* (Positive False) tests that seemed favorable, but were not, and *Fn* (False negative) are all tests performed that may not be favorable, but they were.

There is also *F* which is a measure that combines *precision* and *recall*. The higher its value, the better the system performance will be considered.

$$\text{F} = 2 * \frac{\text{Precision} * \text{Recall}}{\text{Presicion} + \text{Recall}}$$

Once *F* has been obtained, the percentage of error that the system has (*E*) can be calculated.

$$\text{E} = 1 - \text{F}$$

The results obtained in the different evaluations were adjusted through the coordination and organization of different meetings, which had the participation of the students of the clinical laboratory with hearing disabilities, teachers, and other people who were part of the study, in which they proceeded to represent different types of content in each assessment.

Table 1 shows the results of the tests carried out in the first iteration, obtaining a margin of error of 18%.

Corrections were made to the application, and it was possible to obtain a lower margin of error (2%), as shown in Table 2.

Four possible forms of interfaces to be evaluated with the hearing-impaired students of the clinical laboratory career were presented. In each session, they had to manipulate different data that were represented in texts, sentences and images, and combinations of these. It was shown that the best results and with fewer errors are given through the use of keywords and images, achieving an error percentage of 2%, and that, compared to the other tests carried out, it is a minimum and acceptable percentage, since If there is an error, in a word, it is easy according to human behavior, to be able to interpret the complete sentence or through communication to request a repeat of what has been wanted to communicate.

Table 1. Interface's evaluation results.

Evaluation	CC	Fp	Fn	E
Text only	2	6	4	71%
Image only	10	5	6	35%
Phrase's construction	2	6	7	76%
Keywords and image	15	3	4	18%

Table 2. Final results for interface evaluation.

Evaluation	CC	Fp	Fn	E
Text only	4	7	5	60%
Image only	8	2	5	30%
Phrase's construction	2	6	6	75%
Keywords and image	20	0	1	2%

4 Conclusions

The mobile application, with all its functions such as adding words and categories, modifying them, also with the search function and its main function, which is to reproduce phrases in audios, was adequate, allowing the person with hearing impairment to improve the quality of communication that it has with its patient.

Through the application, students will be able to give instructions focused on complying with the professional protocols that correspond to the clinical laboratory career, facilitating communication between the laboratory worker and the patient.

The application can be replicable, providing the options to add new phrases and images, allowing its content to be adaptable to different needs and to be applied in various study areas, helping people with hearing disabilities to integrate more with society, as students and professionals.

From the tests carried out on the application, it was found that the degree of error is 2%, which is acceptable for this type of application.

As future work, it is planned to improve the system by implementing additional functionalities such as questionnaires, sheets, pens or others, that complement the application.

References

1. Anjankar, S., Ambhore, K.: Speech signal processing implementation for hearing impairment based on FPGA. In: 2015 Global Conference on Communication Technologies (GCCT), pp. 164–168 (2015)

2. García, P., Pinargote, J., Véliz, V., De la Herrán, A., Montiano, B.: Formación y Transformación para la Educación Inclusiva En la Universidad. Editorial Universidad de Granada, Granada (2017)
3. Goncharenko, A., Voronova, L., Artemov, M., Voronov, V., Bezumnov, D.: Development using wireless technologies for people with hearing impairments. In: 2019 24th Conference of Open Innovations Association (FRUCT), pp. 104–109 (2019)
4. Padilla, A.: Disability: context, concept and models. Rev. Colomb. Derecho Int. **16**, 381–414 (2010)
5. Hernández, M.: The concept of disability: from a disease to a rights approach. Revista CES Derecho **6**(2), 46–59 (2015)
6. Pinargote, J., Cruz, M., Bowen, L., Demera, G.: Discapacidad y Resiliencia, in Resiliencia experiencias investigativas y prospectivas, pp. 1–204 (2018)
7. Saleem, K., Alagha, I.: System for people with hearing impairment to solve their social integration. In: 2015 5th International Conference on Information and Communication Technology and Accessibility (ICTA), pp. 1–6 (2016)
8. Ojeda, J., Piedra, J., Bernal, C., Iribarne, L.: Sign communication for people with disabilities using kinect technology at home. In: 2016 8th International Conference on Games and Virtual Worlds for Serious Applications (VS-GAMES), Barcelona, Spain, pp. 1–2 (2016)
9. Montes Mora, J.F., Aldana Porras, J.M., Castiblanco Franco, D.F.: Promoting the self-efficacy of deaf people through the application of translator of signifiers. In: Botto-Tobar, M., Cruz, H., Díaz Cadena, A. (eds) Artificial Intelligence, Computer and Software Engineering Advances. CIT 2020. Advances in Intelligent Systems and Computing, vol. 1327. Springer, Cham (2021). https://doi.org/10.1007/978-3-030-68083-1_26
10. Pressman, R., Maxim, B.: Software Engineering-A Practitioner´s Approach, 8th edn. McGraw-Hill Education, New York (2015)
11. Sohaib, O., Solanki, H., Dhaliwa, N., Hussain, W., Asif, M.: Integrating design thinking into extreme programming. J. Ambient. Intell. Humaniz. Comput. **10**(6), 2485–2492 (2018). https://doi.org/10.1007/s12652-018-0932-y
12. Ribas, J.: Desarrollo de aplicaciones para-Android. 2018 edn Anaya Multimedia, España (2018)
13. Peraza, B., Pardo, Y.: Video sensor para la detección y reconocimiento de rostros. Revista Digital Sociedad de la Información **52**, 1–5 (2015)
14. Sun, Y., Robinson, M., Adams, R., Te Boekhorst, R., Rust, A., Davey, N.: Using sampling methods to improve binding site predictions. In: ESANN 2006 proceedings of the European Symposium on Artificial Neural Networks Bruges (Belgium), pp. 26–28 (2006)
15. Robinson, M., et al.: Improving computational predictions of cis regulatory binding sites. Pac. Symp. Biocomput. **11**, 391–402 (2006)

Computerized Tomography Images Processing Using Artificial Intelligence Techniques

Shirley Chuquín(✉) and Erick Cuenca

Yachay Tech University, Urcuquí, Ecuador
shirley.chuquin@yachaytech.edu.ec

Abstract. The study and analysis of images have been increased with time due to the extraction of different features. This research area is leadership by Image Processing because it contributes with interesting techniques and many advances. There are several applications in image processing, such as medicine, agriculture, sports, and object recognition. In medicine, images reveal their importance to improve the processes to treat patients. It also helps the doctors to make better diagnostic. Besides, the environment of medical images is vast and includes x-ray, magnetic resonance, and computerized tomography images. However, the usage of CT images needs much medical staff to detect the organ with disease or with any problem. Thus, it is necessary to apply segmentation methods to improve diagnostics or disease detection. Methods that involve image segmentation require an in-depth study of traditional and non-traditional methods. In that sense, Artificial Intelligence (AI) methods are crucial in the studies of CT image segmentation due to those methods automated all of the processes. The majority contributions of AI are the novel methods from its subfields of Deep Learning and Neural Networks. They facilitate the processes through the usage of algorithms with learning and prediction. It is essential to mention that Image processing techniques need to work with AI techniques to solve segmentation problems and improve the precision of the results. Therefore this paper aims to present state-of-the-art image segmentation using Computed Tomography images.

Keywords: Images processing · Artificial intelligence · State-of-the-art

1 Introduction

Nowadays, the study of images has been increased due to the necessities of different areas. For instance, image analyses are vital in medicine, agriculture, sports, and object recognition. In the medical domain, images reveal their importance to improve the processes to treat patients. Therefore, there is extensive research to contribute to medicine by segmentation and identifying areas of interest (e.g., organs). The pre-process phase is crucial to achieving better results. Several methods are proposed, such as reducing noise, improving illumination,

M. Botto-Tobar et al. (Eds.): CIT 2021, LNNS 405, pp. 205–218, 2022.
https://doi.org/10.1007/978-3-030-96043-8_16

or detecting the boundaries. The environment of medical images is vast and includes x-ray, magnetic resonance, and computerized tomography images.

Currently, researches has been focuses on Computerized Tomography (CT) images to apply image processing techniques due to its essential role in detecting diseases on the liver [30], lungs [11], brain [16], skull [26], and Covid-19 disease [39]. Consequently, to develop the usage of CT images, it is necessary to segment the organs applying image processing techniques and thus make a preventative treatment and diagnostic. However, the segmentation of CT images is a challenging task due to several factors such as the complex background, dataset size, dimensions, noise, illumination, and shadows.

Image processing needs to improve its techniques using novel methods. For instance, Artificial Intelligence (AI) has different ways to automated processes like Machine Learning (ML), Artificial Neural Networks (ANN), and Convolutional Neural Networks (CNN). Machine Learning algorithms work on methods based on supervised learning, unsupervised learning, and reinforcement learning. It aims to learn a model and make predictions or recognize patterns. By comparison, ANNs simulate the human brain to learn from data through neurons and finally extract the required information. Similarly, CNNs also learn from data, but they use deeper layers to recognize features, especially from images.

Methods that involve image segmentation require an in-depth study of traditional and non-traditional methods. Over the years, the fields of Artificial Intelligence have been revolutionizing, from basic Machine Learning techniques to Deep Learning models, increasing the precision and accuracy of each of the methods.

This paper aims to present state-of-the-art image segmentation using Computed Tomography images. It is organized as follows. The Sect. 2 presents definitions to contextualize the lecture on the topic. A discussion about used techniques from 2006 to 2021 is described in Sect. 3. Finally, Sect. 4 presents the conclusion of this work.

2 Background

This section explains concepts in terms of image processing and organ segmentation using artificial intelligence methods.

2.1 Artificial Intelligence

Artificial Intelligence (AI) is one of the Computer Science branches aiming to understand machine intelligence behavior [28]. The study of AI involves concepts and methods to solve problems through learning with data inputs/outputs to create relationships and extract information [27]. Two characteristics of AI are autonomy and adaptivity. The first one refers to the ability to perform tasks without constant guidance by the user. The second one involves the ability to improve performance by learning from data or experience. In this regard, AI has branches to work on like Deep Learning, Machine Learning, Natural Language Processing, and Computer Vision.

AI can be applied in different fields such as medicine [20], industry [34], data mining [7], robotics [9], and others. One of the most considerable advantages is on medicine by detecting anomalies or pathologies from digital images. For example, Tizhoosh et al. [33] investigated the challenges and opportunities to use artificial intelligence algorithms in digital pathology. They mention the importance of work with clinical imaging and AI. Similarly, Kim et al. [17] worked on studies to evaluate the performance of Artificial Intelligence algorithms for diagnostics using medical images.

2.2 Machine Learning

Machine Learning (ML) is a subfield of AI, and it allows machines to learn from data and predict information by recognizing patterns [12]. There are three subareas of ML used depending on the problem attacked: Supervised Learning, Unsupervised Learning, and Reinforcement Learning. A supervised model learns from a labeled dataset (i.e., data including output tags), and its task is to predict the correct output or label in a test dataset. These prediction models are suitable for *classification* problems, where a simple case can be a binary classification (i.e., when the answer is yes/no). The fields of application are diverse; for example, classify whether an email is a spam or not. Using traffic images, predict which traffic symbol is in the picture. On the other hand, an unsupervised model uses an unlabeled dataset, and its main task is to *discover* characteristics and data structure (e.g., patterns or associations). Since there are no labeled outputs, these descriptive algorithms make sense of the data by grouping similar items into clusters or associating them. Finally, the reinforcement algorithms use AI agents to choose the best actions to maximize a numerical reward signal [31]. The choice of the best actions is based on previous feedback. These algorithms are very suitable in video game or robot navigation environments.

2.3 Artificial Neural Networks

Artificial Neural Networks (ANNs) are algorithms that process information similar to biological neurons found in the human brains [1]. These networks have a large number of neurons that process and transmit information to each other. The principal purpose is learning from data that describes a phenomenon and then extracting information/knowledge [24]. A key feature of ANNs is processing a lot of information simultaneously; this behavior is suited when dealing with large datasets.

Deep Learning. Deep Learning (DL) is an advanced field of Machine Learning based on which computers learn through experiences and understand the world with a hierarchy of concepts. For this reason, it is not necessary that a human formally determine all of the knowledge needed by the computer [10]. Deep Learning methods structure the data in connected *layers* in a Neural Network.

The basic idea is to pass the input data of the system through each layer to processing information and thus predict or classify data. Deep Learning algorithms can use advanced techniques such as Convolutional Neural Networks, Recurrent Neural Networks, Deep Confidence Networks, or Auto Encoder methods.

Nowadays, Deep Learning techniques are considered robust in the field of Machine Learning. It has dramatically impacted research areas, from image analysis to automotive, finance, and medical industry [29]. Because the effect of DL on medicine has been helpful, researches worked on different studies like recognition of breast cancer [4,15,42], extraction of brain tumor [16,19,36], and organ segmentation [29,30,41].

2.4 Convolutional Neural Networks

Convolution Neural Networks (CNNs) is part of the Deep Learning field, and it maintains a linear operation between convolution matrices. The objective of CNNs is to solve problems that ANNs do not solve. One important aspect is to obtain features when input propagates toward deeper layers [2]. Due to their ability to adapt to different applications, their typical applications are: image classification [14,23], edge detection [38], decoding facial recognition [22,40], and analyzing documents [3,18].

A CNN has three fundamental layers: a convolutional layer, a pooling layer, and a fully connected layer. The first layer can be very complex to manage and compute. A CNN notably decreases that number because it works with a set of pixels instead of one by one. Figure 1 shows the CNN architecture for an image classification. The CNN input is an image, which typically is followed by a sequence of convolutional, pooling, and fully connected layers. The output layer is modified depending on the application. For instance, the outputs on object detection, the outputs represent the position, dimensions, and class of a particular object.

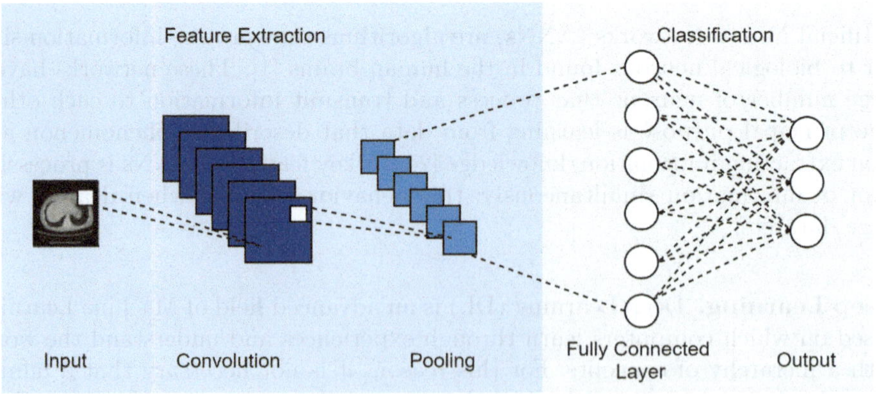

Fig. 1. A CNN architecture showing the layers: convolutional, pooling, and fully connected.

CNN's are multi-layer neural networks that mainly work to recognize features from any images. The CNN architecture determines how the layers will be set up to a particular model. It is significant to mention that researchers also develop new layers to add to their models. It aims to increase the accuracy or precision of their models. The Convolutional Layers form the most common architectures. It is the most important in CNN's architecture[2]. It consists of a group of filters to extract features from the input image. Mathematically, the convolution is represented by the following formula:

$$S(i,j) = (I * K)(i,j) = \sum_m \sum_n I(m,n)K(i-m, j-n) \quad (1)$$

Where I represents the input image, an image is represented by $i \times j \times r$ where r is the number of channels. K represents the kernel. Then, m and n represent the filter that can be of 3×3, 5×5, 7×7, or 1×11. The pixels of the image are represented by the subscripts i, j, where the convolution will be operated. This process is represented by Fig. 2.

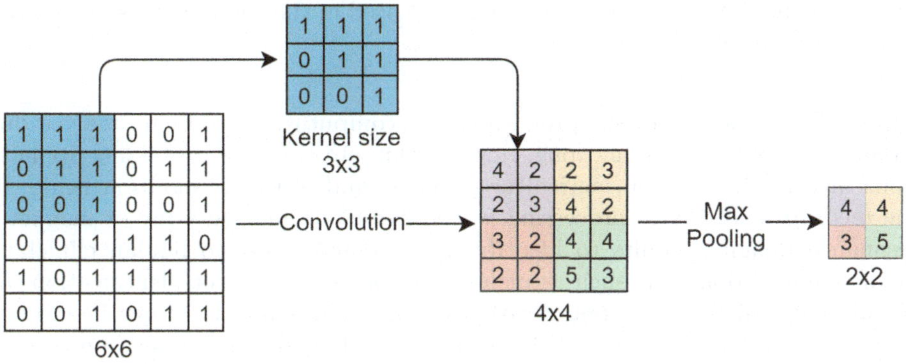

Fig. 2. Representation of convolution and pooling Layer

CNN's usually perform four phases to training the model: data preparation, feature extraction, classifier, and validation. Data Preparation is the process to normalize the input of the CNN. In other words, images are modified to attempt the CNN architecture. For example, in the AlexNet architecture [21], the input image must have dimensions of $227 \times 227 \times 3$ corresponding to height, width, and the number of channels of the input images, respectively. This phase is recommended to perform to obtain a better feature extraction and accuracy. The second phase is feature extraction. In this phase, convolutional operations are performed and the activation functions. The next phase is the classifier that makes predictions bases on the fully connected layers. The last phase is validation. In this phase, the model validates the predictions performed by the fully connected layers and applies the backpropagation rules to train the model.

Once the model is trained, the predictions can be obtained by providing an input image that will pass through all the stages.

The convolutional neural networks are applied in computer vision because of the significant advantages against traditional methods such as ANN and ML. For example, the benefits of the usage of CNNs can be seen on the fewer parameters, distortion tolerance, automatic feature extraction, work with a large amount of data, and high accuracy. All of the mentioned can see in the results of works that use images for detection [13] or classification.

2.5 Computer Vision

Due to the increasing amount of data in image format (e.g., medical, transport, social media), the computer vision domain increases its importance. This field aims to interpret and analyze images; however, this task is challenging due to the inverse problem of vision. It occurs when a lack of information exists. Thus, the application of new methods such as AI is necessary. There are many applications of computer vision such as machine inspection, retail, 3D model building, medical imaging, automotive safety, and biometrics [32]. As we mention, subfields of AI (e.g., Machine Learning, Deep Learning) are used in the computer vision domain to provide a solution.

Image Processing. Image processing is a computational task to analyze different types of images (e.g., .png, .jpg, .dicom, etc.). Image processing applications involve the medical, agriculture, sports, and object recognition fields. In the medical field, image processing is well suited to analyze x-ray or magnetic resonance imaging results to identify diseases and assist to human clinicians in the medical diagnosis such as melanoma and breast cancer identification. In the agricultural field, the optimization of plantations and the identification of missing vegetation are essential for production. For instance, image processing can help create models and determine the proper plant treatment depending on crops and regions. It can also help detect weeds' locations in images from cereal fields, classify health and unhealthy leaves plants, etc. The sports field is another domain where image processing has been applied. For instance, analyzing the vast amount of video games (video is a set of image frames) summarizes or detects events (e.g., scores, soccer goals, ball possession, etc.). Image processing techniques also detect anomalies in different scenarios or identify a specific object to analyze its behavior.

Computerized Tomography Images. A Computerized Tomography (CT) image refers to an x-ray image. Image slices can either be displayed individually or stacked together by the computer to generate a 3D image of the patient that shows the skeleton, organs, and tissues as well as any abnormalities the physician is trying to identify. In this case, Fig. 3 shows an example of a CT image, which shows the slide view of a liver taken from LiTS dataset [5].

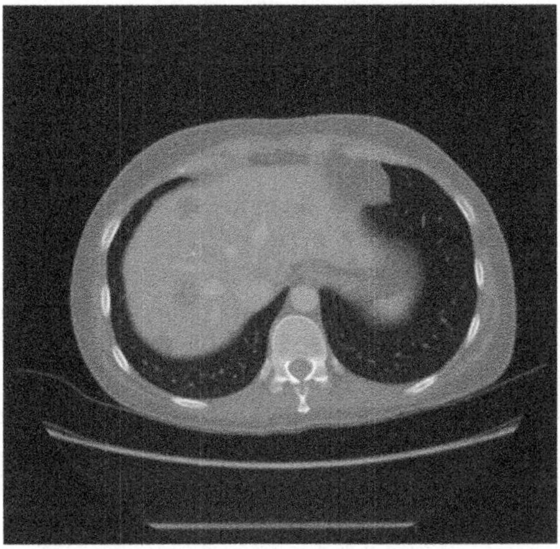

Fig. 3. Example of a computerized tomography image from liver [5]

Several challenges arise when processing images such as noise, complex background, dataset size, and shadows. A common problem is the digital image noise. Noise refers to the visual distortion or undesirable information added to images due to bad conditions during acquisition, digitization, or transferring. The noise can be seen as random speckles on a smooth surface. Noise plays a vital role in image classification because it can be misapprehended due to its presence. For example, Gigandet et al. [8] states that noise and shadows are the main factors of classification errors. Also, the shadow and illumination of an image are two important variables in image processing. Shadow refers to the variation in the intensity of light in a scene produced by an object interposed between the scene and the light source. Besides, lighting problems may appear indoors. During the day, outdoor environments can often undergo gradual changes. Finally, a critical aspect is the size of the dataset since the models need a lot of data to train and perform the right image classification; thus, get enough amount of data is an important issue in image processing.

3 Discussion

Several authors work on segmentation computerized images using different traditional and non-traditional techniques. For example, Susomboom et al. [30] worked on an approach based on a hybrid method to achieve a single-organ segmentation automatically. They used CT images of the liver in a DICOM format[1] and applied a binary classification model to obtain a probability image of interest. Then, the split-and-merge algorithm is applied to removing the noise and the region-growing algorithm to get the segmentation of boundaries.

[1] https://www.dicomstandard.org [last access April 17, 2021].

As we mention, Machine Learning methods have a considerable role in Computerized Tomography image segmentation. For instance, Seng et al. [26] designed a framework to extract objects from CT images using Reinforcement Learning (RL). The approach is based on obtaining several sub-images by the input image where each RL agent operates to identify the suitable value for each object in the image. Every state is associated with actions, while the reward function measures reward for each action of the RL agent. Subsequently, the information is saved in a Q-Matrix, and the results can be applied in future processes. The accuracy on cranial images segmentation is above 93%. Hybrid approaches also have been proposed. For instance, Guo et al. [11] combine thresholding and the Chan-Vense algorithm to segment many types of images. This approach improves the performance by calculating the similarity between each point and its neighborhood. They used lung images obtained from the Picture Archiving and Communication System (PACS), commonly used to secure store and transmit medical images.

Due to advances in the Artificial Intelligence domain, the usefulness of Artificial Neural Networks has increased over the years. Xu et al. [37] for instance, proposed a Membership Function Convolution Neural Network (MFCNN) approach. It starts the preprocessing phase by passing a Gaussian filter to the input images. Then, they use thresholding and the improved CNN segmentation method to achieve a higher dice coefficient compared with a traditional convolutional neural network. It is worth mentioning that image segmentation is a challenge that implies many studies and testing. In this regard, Wang et al. [35] propose a framework for multi-organ segmentation using Organ-Attention-Network with Reverse Connections (OAN-RCs). The OAN is a deep convolutional network that reduces the complex background. It aims to enhance the discriminative information for the target organs. Besides, in the first stage, RCs give the lower layers more semantic information. This approach is well adapted to the study of different organ sizes.

One of the critical studies for Computerized Tomography images is the detection of diseases like cancer. In that sense, the liver is one of the organs with special attention. For that reason, Sakboonyara et al. [25] propose an efficient method for liver segmentation. They mention the U-Net, which is a fully convolutional neural network, and the use of the mean-shift clustering algorithm to improve the precision of liver localization. Also, they introduce statistical thresholding to improve the accuracy.

Similarly, Zhou et al. [41] talk about the importance of CT segmentation. They made an implementation based on multi-organ segmentation. They use partially-labeled datasets of the pancreas where the problem is the background containing organs of interest. The authors applied the Prior-aware Neural Network (PaNN) to overcome the context background problem. PaNN assumes average organ size distributions in the abdomen should be close to their empirical distributions, based on prior statistics from the fully-labeled dataset. They used 30 abdominal datasets to train and compare with traditional algorithms like ResNet50, ResNet101, and 3D-UNet, obtaining better results by the influence of PaNN. Dabiri [6] proposed a study of medical images to find the middle axial slice at the L3 level automatically from a full or partial body CT image. The algorithm uses a fully convolutional classifier with 1,200 images in a CNN with encoder and decoder architecture.

Science advances through research, even during the COVID-19 pandemic. One example is the study of image segmentation in patients with COVID-19 by Yan et al. [39]. The authors suggest using a new deep convolutional neural network called COVIDSegNet to segment infected regions and the entire lung from chest CT images. The proposed network focuses on Feature Variation (FV) block and Progressive Atrous Spatial Pyramid Pooling (PASPP). The dataset consists of 165,667 images from 867 patients with COVID-19. The importance of the study is the comparison between traditional architectures such as FCN, UNet, VNet, UNet++, and the proposed method. The experiments also were compared between the lung and COVID-19 CT images, obtaining higher results on precision.

Table 1 shows an overview of related works based on Artificial Intelligence techniques to image processing. As we can see, Artificial Neural Networks (ANN) are the most used methods for image segmentation with Computerized Tomography images in recent years. There are several architectures and models for ANN; they can be modified or customized according to the necessities. For instance, Zhou et al. [41] performed a PaNN technique to overcome the context background problem. The loss function based on Dice Coefficient has a vital role in image segmentation techniques because many works compare the samples to shows if the algorithm proposed is better. Xu et al. [37] compared a Membership Function Convolution Neural Network with traditional CNN obtaining a higher dice coefficient which means the proposed method is nearest to perfect overlap. Although the Artificial intelligence methods have advanced, image processing techniques are combined with the current methods because the input images, in many cases, need pre-processing. For instance, Xu et al. [37] combined Gaussian Filter with a CNN model to reduce the noise before the processing. It worth mentioning that Deep Learning methods influence image segmentation due to the immerse fields of that.

Several works [11, 26, 30] used image processing methods like thresholding and split-and-merge algorithms. Nevertheless, using Machine Learning algorithms, the process of segmentation had an advantage due to the automation and the learning stage to improve the accuracy of the methods.

Table 1. Resume of methods to segment CT images comparing the usage of Neural Networks, Deep Learning, Machine Learning and other methods

Approaches	Year	Method	Neural networks	Deep learning	Machine learning	Other methods
Yan et al. [39]	2021	Architecture with feature variation and progressive atrous spatial pyramid pooling	x			
Dabiri et al. [6]	2020	Algorithm of L3		x		
Wang et al. [35]	2019	OAN - RCs and statistical fusion	x			
Sakboonyara et al. [25]	2019	U-Net and Mean-Shift Histogram	x			
Zhou et al. [41]	2019	PaNN	x			
Xu et al. [37]	2017	Membership function convolution neural network	x			
Guo et al. [11]	2015	Iterative Chan-Vese				x
Seng et al. [26]	2011	Reinforcement Learning			x	
Susomboom et al. [30]	2006	Pixel-based, split-and-merge, and region growing algorithms				x

4 Conclusions

Image processing is crucial in many domains such as computer sciences, industry, security, and medicine. In recent years, the study of Computerized Tomography (CT) images reveals its importance in the medical field by helping to detect diseases like tumors, cancer, or control organs with problems. Models based on Artificial Intelligence and Machine Learning algorithms allow the specialist to manipulate and obtain information from images. The automation of algorithms using advanced techniques such as Convolutional Neural Networks can improve the accuracy and performance of results. A common challenge in image processing, especially in the medical field, is obtaining a vast and proper dataset. However, thanks to new methods such as data augmentation, this limitation is improved.

Previous work reveals that it is necessary to use traditional image processing techniques in the data preparation stage of the algorithms. For that reason, most of the time, image processing techniques combine Artificial Intelligence algorithms. It is essential to mention that the most accurate and used methods are Artificial Neural Networks due to training and the establishment of hyperparameters.

References

1. Abraham, A.: Artificial neural networks. In: Handbook of Measuring System Design (2005). https://doi.org/10.1002/0471497398.mm421
2. Albawi, S., Mohammed, T.A.M., Alzawi, S.: Understanding of a convolutional neural network. IEEE (2017)
3. Chen, L., Wang, S., Fan, W., Sun, J., Satoshi, N.: Deep learning based language and orientation recognition in document analysis. In: 2015 13th International Conference on Document Analysis and Recognition (ICDAR), pp. 436–440 (2015). https://doi.org/10.1109/ICDAR.2015.7333799
4. Chorianopoulos, A., Daramouskas, I., Perikos, I., Grivokostopoulou, F., Hatzilygeroudis, I.: Deep learning methods in medical imaging for the recognition of breast cancer. In: 2020 11th International Conference on Information, Intelligence, Systems and Applications (IISA), pp. 1–8. IEEE Computer Society, Los Alamitos, CA, USA, July 2020. https://doi.org/10.1109/IISA50023.2020.9284373. https://doi.ieeecomputersociety.org/10.1109/IISA50023.2020.9284373
5. Codalab: Lits - liver tumor segmentation challenge. https://competitions.codalab.org/competitions/17094#learn_the_details
6. Dabiri, S., et al.: Deep learning method for localization and segmentation of abdominal CT. Comput. Med. Imaging Graph. **85**, 101776 (2020)
7. Durand, T., Hattingh, M.: Data mining and artificial intelligence techniques used to extract big data patterns. In: 2020 2nd International Multidisciplinary Information Technology and Engineering Conference (IMITEC), pp. 1–8. IEEE Computer Society, Los Alamitos, CA, USA, November 2020. https://doi.org/10.1109/IMITEC50163.2020.9334069. https://doi.ieeecomputersociety.org/10.1109/IMITEC50163.2020.9334069
8. Gigandet, X., Cuadra, M.B., Pointet, A., Cammoun, L., Caloz, R., Thiran, J.: Region-based satellite image classification: method and validation. In: IEEE International Conference on Image Processing 2005, vol. 3, pp. III–832 (2005). https://doi.org/10.1109/ICIP.2005.1530521
9. Gomez-Espinosa, A.: Developing robot manipulators in mechatronics and robotics. In: 2018 International Conference on Mechatronics, Electronics and Automotive Engineering (ICMEAE), pp. 44–49. IEEE Computer Society, Los Alamitos, CA, USA, November 2018. https://doi.org/10.1109/ICMEAE.2018.00015. https://doi.ieeecomputersociety.org/10.1109/ICMEAE.2018.00015
10. Goodfellow, I., Bengio, Y., Courville, A., Bengio, Y.: Deep Learning, vol. 1. MIT Press, Cambridge (2016)
11. Guo, S., Wang, L.: Automatic CT image segmentation of the lungs with an iterative Chan-Vese algorithm. In: 2015 International Conference on Informatics, Electronics and Vision (ICIEV), pp. 1–4. IEEE Computer Society, Los Alamitos, CA, USA, June 2015. https://doi.org/10.1109/ICIEV.2015.7334070. https://doi.ieeecomputersociety.org/10.1109/ICIEV.2015.7334070
12. Hashimoto, D.A., Rosman, G., Rus, D., Meireles, O.R.: Artificial intelligence in surgery: promises and perils. Ann. Surg. **268**(1), 70–76 (2018). https://doi.org/10.1097/SLA.0000000000002693
13. Iorga, C., Neagoe, V.: A deep CNN approach with transfer learning for image recognition. In: 2019 11th International Conference on Electronics, Computers and Artificial Intelligence (ECAI), pp. 1–6 (2019). https://doi.org/10.1109/ECAI46879.2019.9042173

14. Islam, M.T., Karim Siddique, B.M.N., Rahman, S., Jabid, T.: Food image classification with convolutional neural network. In: 2018 International Conference on Intelligent Informatics and Biomedical Sciences (ICIIBMS), vol. 3, pp. 257–262 (2018). https://doi.org/10.1109/ICIIBMS.2018.8550005
15. Jannesari, M., et al.: Breast cancer histopathological image classification: a deep learning approach. In: 2018 IEEE International Conference on Bioinformatics and Biomedicine (BIBM). pp. 2405–2412. IEEE Computer Society, Los Alamitos, CA, USA, December 2018. https://doi.org/10.1109/BIBM.2018.8621307. https://doi.ieeecomputersociety.org/10.1109/BIBM.2018.8621307
16. kaitouni, S.E., Tairi, H.: Segmentation of medical images for the extraction of brain tumors: a comparative study between the hidden Markov and deep learning approaches. In: 2020 International Conference on Intelligent Systems and Computer Vision (ISCV), pp. 1–5. IEEE Computer Society, Los Alamitos, CA, USA, June 2020. https://doi.org/10.1109/ISCV49265.2020.9204319. https://doi.ieeecomputersociety.org/10.1109/ISCV49265.2020.9204319
17. Kim, D.W., Jang, H.Y., Kim, K.W., Shin, Y., Park, S.H.: Design characteristics of studies reporting the performance of artificial intelligence algorithms for diagnostic analysis of medical images: results from recently published papers. Korean J. Radiol. **20**(3), 405–410 (2019). https://doi.org/10.3348/kjr.2019.0025
18. Kosaraju, S.C., et al.: DoT-Net: document layout classification using texture-based CNN. In: 2019 International Conference on Document Analysis and Recognition (ICDAR), pp. 1029–1034 (2019). https://doi.org/10.1109/ICDAR.2019.00168
19. Li, Y., Xie, X., Liu, S., Li, X., Shen, L.: GT-Net: a deep learning network for gastric tumor diagnosis. In: 2018 IEEE 30th International Conference on Tools with Artificial Intelligence (ICTAI), pp. 20–24. IEEE Computer Society, Los Alamitos, CA, USA, November 2018. https://doi.org/10.1109/ICTAI.2018.00014. https://doi.ieeecomputersociety.org/10.1109/ICTAI.2018.00014
20. Liu, J., Zhai, X., Han, A., Ding, S., Wang, M., Yin, P.: The future development of traditional Chinese medicine from the perspective of artificial intelligence with big data. In: 2018 IEEE 4th International Conference on Big Data Security on Cloud (BigDataSecurity), IEEE International Conference on High Performance and Smart Computing, (HPSC) and IEEE International Conference on Intelligent Data and Security (IDS), pp. 204–209. IEEE Computer Society, Los Alamitos, CA, USA, May 2018. https://doi.org/10.1109/BDS/HPSC/IDS18.2018.00051. https://doi.ieeecomputersociety.org/10.1109/BDS/HPSC/IDS18.2018.00051
21. Pal, R., Saraswat, M.: Enhanced bag of features using Alexnet and improved biogeography-based optimization for histopathological image analysis. In: 2018 Eleventh International Conference on Contemporary Computing (IC3), pp. 1–6. IEEE Computer Society, Los Alamitos, CA, USA, August 2018. https://doi.org/10.1109/IC3.2018.8530540. https://doi.ieeecomputersociety.org/10.1109/IC3.2018.8530540
22. Qu, D., Huang, Z., Gao, Z., Zhao, Y., Zhao, X., Song, G.: An automatic system for smile recognition based on CNN and face detection. In: 2018 IEEE International Conference on Robotics and Biomimetics (ROBIO), pp. 243–247 (2018). https://doi.org/10.1109/ROBIO.2018.8665310
23. Ramanath, A., Muthusrinivasan, S., Xie, Y., Shekhar, S., Ramachandra, B.: NDVI versus CNN features in deep learning for land cover classification of aerial images. In: IGARSS 2019, 2019 IEEE International Geoscience and Remote Sensing Symposium, pp. 6483–6486 (2019). https://doi.org/10.1109/IGARSS.2019.8900165

24. Sadiq, R., Rodriguez, M.J., Mian, H.R.: Empirical Models to Predict Disinfection by-Products (DBPs) in Drinking Water: An Updated Review. 2 edn., Elsevier Inc., Amsterdam (2019). https://doi.org/10.1016/B978-0-12-409548-9.11193-5. http://dx.doi.org/10.1016/B978-0-12-409548-9.11193-5
25. Sakboonyara, B., Taeprasartsit, P.: U-net and mean-shift histogram for efficient liver segmentation from CT images. In: 2019 11th International Conference on Knowledge and Smart Technology, KST 2019, pp. 51–56 (2019). https://doi.org/10.1109/KST.2019.8687816
26. Seng, W., Chitsaz, M.: Medical image segmentation by using reinforcement learning agent. In: International Conference on Digital Image Processing, pp. 216–219. IEEE Computer Society, Los Alamitos, CA, USA, March 2009. https://doi.org/10.1109/ICDIP.2009.14. https://doi.ieeecomputersociety.org/10.1109/ICDIP.2009.14
27. Shahin, M.A.: State-of-the-art review of some artificial intelligence applications in pile foundations. Geosci. Front. **7**(1), 33–44 (2016)
28. Shapiro, S.C.: Artificial Intelligence (AI), pp. 89–93. Wiley, Hoboken (2003)
29. Stefaniga, S., Gaianu, M.: An approach of segmentation method using deep learning for CT medical images. In: 2019 21st International Symposium on Symbolic and Numeric Algorithms for Scientific Computing (SYNASC), pp. 273–279. IEEE Computer Society, Los Alamitos, CA, USA, September 2019. https://doi.org/10.1109/SYNASC49474.2019.00045. https://doi.ieeecomputersociety.org/10.1109/SYNASC49474.2019.00045
30. Susomboon, R., Raicu, D., Furst, J., Channin, D.: Automatic single-organ segmentation in computed tomography images. In: Sixth International Conference on Data Mining (ICDM 2006), pp. 1081–1086. IEEE Computer Society, Los Alamitos, CA, USA, December 2006. https://doi.org/10.1109/ICDM.2006.24. https://doi.ieeecomputersociety.org/10.1109/ICDM.2006.24
31. Sutton, R.S., Barto, A.G.: Reinforcement Learning: An Introduction. MIT Press, Cambridge (2018)
32. Szeliski, R.: Computer Vision: Algorithms and Applications. Springer, Cham (2010). https://doi.org/10.1007/978-1-84882-935-0
33. Tizhoosh, H.R., Panta nowitz, L.: Artificial intelligence and digital pathology: challenges and opportunities (2018)
34. Wang, Y., et al.: On autonomous systems: from reflexive, imperative and adaptive intelligence to autonomous and cognitive intelligence. In: 2019 IEEE 18th International Conference on Cognitive Informatics & Cognitive Computing (ICCI*CC), pp. 7–12. IEEE Computer Society, Los Alamitos, CA, USA, July 2019. https://doi.org/10.1109/ICCICC46617.2019.9146038. https://doi.ieeecomputersociety.org/10.1109/ICCICC46617.2019.9146038
35. Wang, Y., Zhou, Y., Shen, W., Park, S., Fishman, E.K., Yuille, A.L.: Abdominal multi-organ segmentation with organ-attention networks and statistical fusion. Med. Image Anal. **55**, 88–102 (2019)
36. Xiao, Z., et al.: A deep learning-based segmentation method for brain tumor in MR images. In: 2016 IEEE 6th International Conference on Computational Advances in Bio and Medical Sciences (ICCABS), pp. 1–6. IEEE Computer Society, Los Alamitos, CA, USA, October 2016. https://doi.org/10.1109/ICCABS.2016.7802771. https://doi.ieeecomputersociety.org/10.1109/ICCABS.2016.7802771

37. Xu, J., Liu, H.: Segmentation of pulmonary CT image by using convolutional neural network based on membership function. In: 2017 IEEE International Conference on Computational Science and Engineering (CSE) and IEEE International Conference on Embedded and Ubiquitous Computing (EUC), vol. 1, pp. 198–203. IEEE Computer Society, Los Alamitos, CA, USA, July 2017. https://doi.org/10.1109/CSE-EUC.2017.42. https://doi.ieeecomputersociety.org/10.1109/CSE-EUC.2017.42
38. Xue, W., Wenxia, X., Guodong, L.: Image edge detection algorithm research based on the CNN's neighborhood radius equals 2. In: 2016 International Conference on Smart Grid and Electrical Automation (ICSGEA), pp. 115–119 (2016). https://doi.org/10.1109/ICSGEA.2016.38
39. Yan, Q., et al.: Covid-19 chest CT image segmentation network by multi-scale fusion and enhancement operations. IEEE Trans. Big Data **7**(01), 13–24 (2021). https://doi.org/10.1109/TBDATA.2021.3056564
40. Zhang, Y., Zhao, M., Yan, L., Gao, T., Chen, J.: CNN-based anomaly detection for face presentation attack detection with multi-channel images. In: 2020 IEEE International Conference on Visual Communications and Image Processing (VCIP), pp. 189–192 (2020). https://doi.org/10.1109/VCIP49819.2020.9301818
41. Zhou, Y., Li, Z., Bai, S., Chen, X., Han, M., Wang, C., Fishman, E., Yuille, A.: Prior-aware neural network for partially-supervised multi-organ segmentation. In: 2019 IEEE/CVF International Conference on Computer Vision (ICCV), pp. 10671–10680. IEEE Computer Society, Los Alamitos, CA, USA, November 2019. https://doi.org/10.1109/ICCV.2019.01077. https://doi.ieeecomputersociety.org/10.1109/ICCV.2019.01077
42. Zou, W., Lu, H., Yan, K., Ye, M.: Breast cancer histopathological image classification using deep learning. In: 2019 10th International Conference on Information Technology in Medicine and Education (ITME), pp. 53–57. IEEE Computer Society, Los Alamitos, CA, USA, August 2019. https://doi.org/10.1109/ITME.2019.00023. https://doi.ieeecomputersociety.org/10.1109/ITME.2019.00023

Digital Model to Predict Failures of Porous Structures in DLP-Based Additive Manufacturing

Ivannova Jumbo-Jaramillo and Hernan Lara-Padilla(✉)

Universidad de Las Fuerzas Armadas ESPE, Av. General Rumiñahui s/n, 171-5-231B Sangolqui, Ecuador
hvlara@espe.edu.ec

Abstract. Additive manufacturing (AM) technologies benefit innovation and development as they enable designers and scientists to prototype models to evaluate their initial performance. A novel digital model to study the mechanical behavior of porous structures during their fabrication is presented. The digital light projection (DLP), a type of additive manufacturing, is still non-reliable for constructing porous structures because of the phenomena known as separation force. This separation force produces high mechanical stress on the structures during the building process. If the structure being constructed is porous, it usually is fragile, then the action of separation force results in breaking or distortion of the construct. The digital model consists of three modules: Computer-Aided Design (CAD), Slicing and Analysis/Report. As proof of concept, a TPMS-based structure was evaluated. The performance of the digital model enables possibilities to detect early failures avoiding waste of time and resources. The focus of the study is to report some practical instructions to design porous structures to obtain the maximum advantages when these constructs will build with DLP-based additive manufacturing.

Keywords: Digital model · Additive manufacturing · Digital light processing

1 Introduction

Additive manufacturing (AM) methods have great potential for transforming research in many fields of engineering and materials science. Because additive manufacturing is expected to impact different industries significantly, several difficulties are necessary to master. Some of these difficulties are the variation of material properties, lower part accuracy, reduced repeatability and consistency, and the need for qualification and certification standards [1]. Initiatives and plans to overcome the mentioned challenges have been created in recent years. The National Institute of Standards and Technology (NIST) demands the necessity to develop traceability systems and closed-loop controls, which are essential for material characterization, the performance of process and equipment, assurance of the part and the capability to qualify and certify components and functions [2].

M. Botto-Tobar et al. (Eds.): CIT 2021, LNNS 405, pp. 219–228, 2022.
https://doi.org/10.1007/978-3-030-96043-8_17

Vat photopolymerization, a category of additive manufacturing (AM) process, is defined by ISO/ASTM 52900:2015 as an "additive manufacturing process in which liquid photopolymer in a vat is selectively cured by light-activated polymerization." Stereolithography (SLA), the first AM process to be patented and commercialized, is a vat photopolymerization technique [3]. In SLA, layers or masks of a liquid photoresin are consecutive photo-crosslinked into a solid or porous 3D construct. SLA is considered the most functional AM technology with high accuracy and precision [4]. SLA is performed in two different manners: using a laser beam or digital light projection [5]. The method that uses digital light projection is named digital light processing (DLP). In DLP, ultraviolet or blue light is projected from a source through the transparent and non-adhering bottom surface of the resin basement (Fig. 1). The motion of the build platform is in a stepwise method in the ascending direction out of the resin volume. The main problem with the DLP-based AM process is that after the projection of each mask the constructs need to be separated from the bottom surface. This separation generates a high mechanical separation force performing on the structures during the manufacturing process. If the structure being constructed is porous, it usually is fragile, then the action of separation force results in breaking the construct.

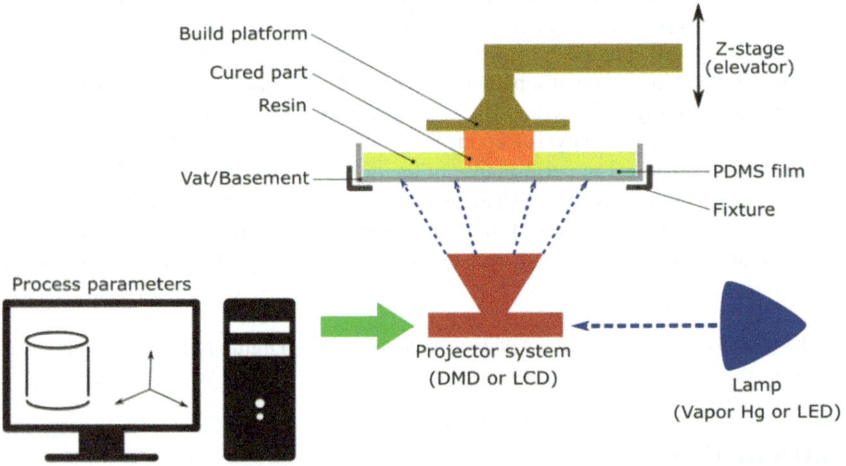

Fig. 1. DLP-based additive manufacturing process

Separation force restricts the manufacturing process reliability, print velocity, printable volume, and the lifespan of the constrained surface (usually a PDMS film). Also, high separation force induces fabrication defects and continued failures during the printing process. The commercial DLP systems do not have instrumentation or control devices to detect these shortcomings during the structures' fabrication. The users find out the defects in constructs only when the overall time process ends (several hours in many cases).

The digitalization of manufacturing process is the basis of smart manufacturing [6]. A digital model can be defined as the reproduction of an actual physical process or object without automatic data exchange between the physical and digital objects [7]. This comprehensive digital representation of an AM process contains essential information such as geometric and design features, material properties and process parameters. As the data is interlinked, a digital model can be automated to a high degree [8].

Here we present a digital model to represent the design and build process of solid and porous structures using DLP. This digital model intent to optimize the design of the structures and increase the productivity of parts manufactured with DLP, reducing the variations and costs during the printing process. This study is focused on theoretical and computational approaches to the design of solid and porous structures, separation force, green strength, and complemented with areas such as simulation-based engineering and computational materials engineering applied to the DLP process.

2 Methodology

2.1 Design Factors and the Capabilities of DLP

The DLP-based AM process is completely digital. Volumetric geometries are defined as 3D digital representations of the solid or porous geometry intended to be fabricated, frequently using algorithms to define complex models based only on digital data representations. These model data is then digitally processed to produce 2D layers/masks information, including AM tool-paths and correspondent process parameters. The processed information of the model is then digitally sent to command a DLP-based AM process.

2.2 Resolution

DLP is capable to obtain superior resolutions than some other AM processes. DLP resolution in the XY-direction is determined by the mask projector system (DMD or LCD) and can vary from 50 to 200 μm. Adjust this parameter in a DLP machine could be extremely difficult. Z-direction resolution usually ranges from 25 to 200 μm, and it is a function of the layer thickness let by the machine. The layer thickness depends on the movement characteristics of the Z-direction linear stage actuator and the curing process produced by the machine. When a very fine vertical (Z) resolution is set, it is a compromise between velocity and quality (anisotropy and surface finish).

2.3 Print Orientation

DLP is a bottom-up SLA approach. This kind of machine photocures the part from the bottom and pulls it out of the vat of photoresin. The separation force affected by the suction between the build platform with a print sticking to the bottom of the vat, is the main drawback in this process. The separation force is related directly to the vertical cross-sectional area and its perimeter [9]. For this reason, the reduction of the cross-sectional area along the Z-axis is the best practice to orient solid or porous structures for DLP prints. The structure design should be adapted to reduce large cross-sectional areas and avoid possible failures during the manufacturing process.

2.4 Support Material

DLP requires support material for holding overhanging features during the printing process. The main reason is that the uncured photoresin is not capable of supporting these features on its own. The support material is eliminated in the post-processing stage. In most DLP machines, the process of place support material to the overhanging features of the structure is automatic. However, in cases where the surface finish is critical, the operators can manually edit the supports.

2.5 Isotropy

Using DLP processes is possible to achieve printed parts with comparatively isotropic properties (especially, mechanical properties). This is because the join between layers is produced by a chemically bond to one another as they cure, resulting in near-identical mechanical properties in all directions.

2.6 Separation Force in Printing Irregular Solid Geometries

For irregular solid geometries, the separation force can be computed by the equation [9]:

$$F = \frac{8\mu V}{h^3}\frac{A}{L}\int_0^{2\pi} r(\theta)^3 d\theta \tag{1}$$

Equation (1) shows that when the printing geometry is a solid form (that is the mask is totally filled), the separation force is a function of the dynamic viscosity μ, gap size h, separation speed V and printing geometry. In the case of irregular solid geometries, the impact of printing geometry on separation force could be mathematically modeled considering the area to perimeter ratio A/L and the radius r of each mask/layer. Keeping constant some process parameters such as gap size and separation speed, and using the same material (photoresin), a constant printing area with a smaller perimeter will support a higher separation force. This situation implies the probability of cause defects or failures during the printing process.

2.7 Separation Force in Printing Porous Structures

Porous structures are made up of an interconnected network of plates, solid struts, or small unit cells (periodic or random) [10]. The porosity (P) is the most critical characteristic of porous structures and it is defined as

$$P = 1 - \frac{\rho_P}{\rho_M} \tag{2}$$

where ρ_P is the density of the porous structure and ρ_P is the material density from which the porous structure is made.

For porous structures, the influence of print geometry on separation forces becomes more difficult to understand, as the liquid filling and constrained surface deformation are more complicated than the circumstances of printing solid geometries. Pan et al. [9]

studied the relation of porosity with separation force. This study reports the measurement of the separation force of four structures with different degrees of porousness (φ). The porousness degree is defined as the radio of bounding box area to printing area of each mask. The experiment shows that the separation force intensifies with the degree of the porousness of the print geometry. For example, set the same printing area, a higher degree of porousness increases the "effective" separation force (F_e). However, the relationship between porousness degree and separation force is very complicated.

$$F_e = \varphi F \tag{3}$$

2.8 Green Strength

The green strength (GS) can be defined as the resistance of the cured polymer (during printing) to breaking under tension. The tensile stress is produced by the separation force. The stacked photo-cured polymer layers not only need to be enough strong and stable to preserve adherence to each other during the manufacturing process but also the pho-to-cured layers need to separate from the bottom surface of the vat without failures in each stage of photo-curing of the masks.

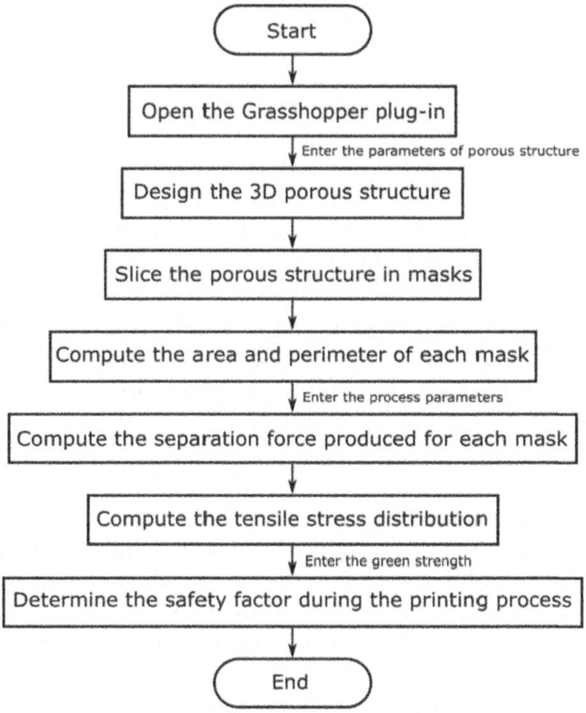

Fig. 2. Flow chart for generating the proposed digital model for porous structures.

The digital model was implemented using the CAD Software Grasshopper™ (https://www.grasshopper3d.com/). The parametric design features of Grasshopper allow controlling the structural parameters and elements of the model precisely. Besides, it is possible to measure the geometric properties of the porous structures such as porosity, and surface area and perimeter of each mask.

2.9 Modules

CAD Module
The CAD module is capable to design regular and irregular porous structures. The regular porous structures design was based on Triply Periodic Minimal Surfaces (TPMS) theory. Irregular porous structures design can be implemented using algorithms such as the Voronoi tessellation method. The digital model was implemented using the CAD Software Grasshopper™ (https://www.grasshopper3d.com/). The parametric design features of Grasshopper allow controlling the structural parameters and elements of the model precisely. Besides, it is possible to measure the geometric properties of the porous structures such as porosity, and surface area and perimeter of each mask.

Triply periodic minimal surfaces (TPMS) are of special interest to many scientists in the field of design and fabrication for bone tissue engineering applications due to the high area-volume ratio, high porosity, modulated structure, among others. These properties improve characteristics such as mechanical behavior, permeability, cell adhesion, degradation, and resorption. TPMS are defined by the isosurface function,

$$f(x, y, z) = \sum_{hkl} |F(h, k, l)| cos(hX + kY + lZ - \alpha_{hkl}) = t \tag{4}$$

where $X = 2\pi x/a$, $Y = 2\pi y/a$, $Z = 2\pi z/a$, are the positions of the structure in the space, a denotes the length of the unit cell, α_{hkl} denotes the phase angle, and $|F(h, k, l)|$ denotes the symmetry of the structure [11].

The TPMS porous structures were constructed through the combination of plug-in Millipede and a C# Script. The isosurface function corresponding to each TPMS can be edited by setting different offset values of the constants of Eq. 4. Finally, the structures were softened using the command Catmull-Clark Subdivision of the plug-in Weaverbird. The result is a porous structure in mesh format.

Slicing Module
Standard Tessellation Language (STL) is considered as the facto file for additive manufacturing. However, the STL file must be sliced into 2D-layers or masks of contours. These masks are projected through the resin vat to form a printed structure. The traditional methods to obtain the masks are performed before the projection stage, often affecting the printed parts' quality and it is a time-consuming computer job.

The aim of this module is to slice the porous structures in real-time directly from the mesh file. The term real-time means the slicing algorithm is able to obtain the projection masks immediately after the CAD model with low computational time. The slicing algorithm was implemented using the Silkworm Slicer command and complemented by a C# Script.

Analysis and Report Module
The aim of this module is to evaluate if the geometry of porous structure is capable to be manufactured using a DLP machine. Firstly, a user interface was incorporated in order to load the photopolymer (photoresin) properties and process parameters discussed in Sect. 2.6. Usually, this information is proportioned by the materials producer for DLP machines, except the green strength, which has to be tested. The study presented by [12] offers an experimental setup to obtain this property.

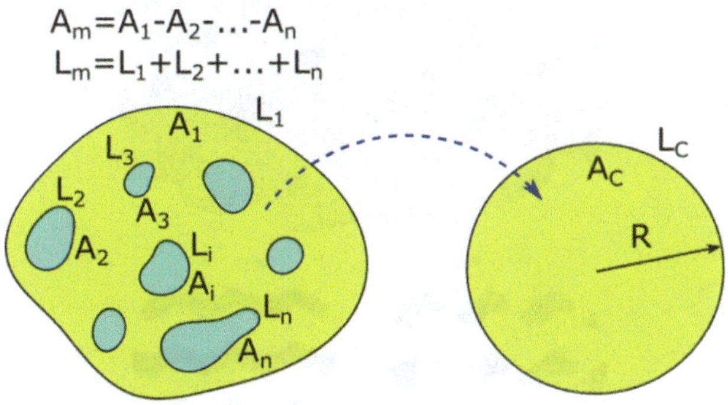

Fig. 3. Relation between a porous structure mask and its equivalent circle.

This module also obtains the surface area and perimeter of each mask of the 3D structure. Applying the equivalence between the area of the $i-th$ mask and a circle of radius R (Fig. 3) and using Eqs. 1 and 3, the separation force can be estimated from:

$$F_m = 32\sqrt{\pi}\varphi_m \frac{\mu V A_m^{3/2}}{h^3 L_m} \tag{5}$$

where the subscript m correspond to the $i-th$ mask, and the porousness of each mask is

$$\varphi_m = \frac{A_1}{A_2 + A_3 + \cdots A_n} \tag{6}$$

The separation force for each mass was computed using a C# Script, recorded with the commands FillCell Excel and reported utilizing the command Pterodatyl. Finally, applying an iterative method with the plug-in Anemone, the maximum tensile stress for each mask was computed. The maximum tensile stress of a mask is calculated by the quotient between the separation force produced for the consecutive masks and the current area. The tensile stress is represented by the symbol S_{ij}. For example, S_{25} means the tensile stress on the mask 2 (already cured) produced by the separation force created when the 5th mask is being cured. Then, the maximum tensile stress for the $i-th$ mask is calculated by

$$S_i^{max} = max\left(S_{i,i+1}, S_{i,i+2}, \cdots, S_{i,j-1}, S_{i,j}\right) \tag{7}$$

Finally, relating the maximum tensile stress (S_i^{max}) and the green strength (GS), the factor of safety for each mask is computed using the equation

$$SF_i = \frac{S_i^{max}}{GS} \tag{8}$$

Using the Eq. 7 is possible determine the possible masks where the porous structure will fail during the construction using a DLP system.

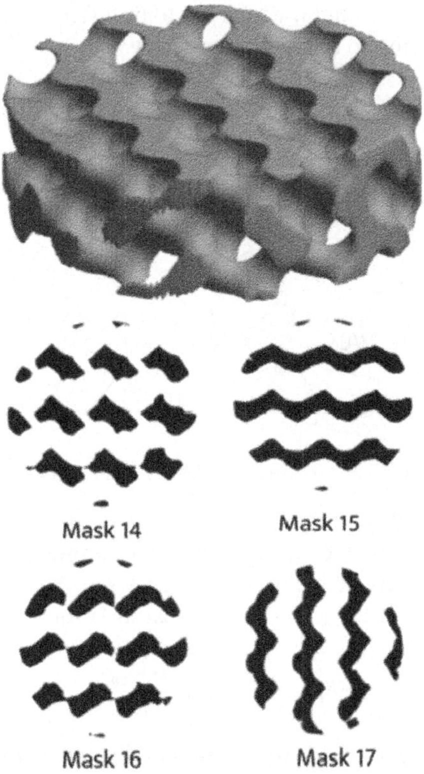

Fig. 4. TPMS structure used to test the digital module.

3 Results

As a proof of concept, a cylinder with a porous structure was obtained applying the Eq. 4. The values used for the Eq. 1 were: dynamic viscosity $\mu = 1$ cP, gap size $h = 200\ \mu$m, separation speed $V = 1$ mm/s, and $GS = 0.8$ MPa.

The TPMS structure named gyroid was designed using a pore size and strut size of 1400 μm and 1000 μm, respectively. The estimated porosity was 62% (Fig. 4).

According to Eq. 8, the factor of safety was computed for each layer, assuming the complete porous structure is printed. Using this approach, was possible to estimate the layers where the part will fail during the construction. The results are shown in the Fig. 5.

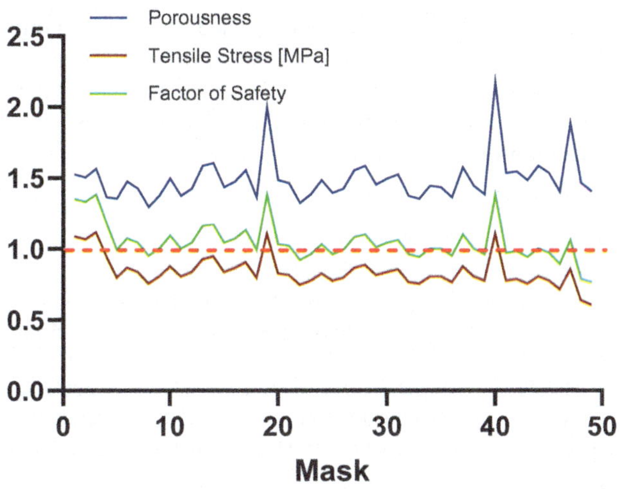

Fig. 5. Factor of safety analysis of TPMS porous structure

4 Discussion

The digital design of a porous structure is the earliest stage during the additive manufacturing process. It is essential to predict possible failures during the printing process to avoid the loss of time and resources. According to Fig. 5, understand why part orientation affects the quality of a printed porous structure. An excessive separation force can cause distortions or failures in the printed part. It is necessary to orient structures to keep the Z-axis cross-sectional area approximately constant to avoid extreme stress in specific masks. In some cases, a design may need to be modified to obtain a factor of safety greater than one.

The digital model and its modules permit evaluating the mechanical performance of a porous structure during the DLP-based printing process. This performance is related to several process parameters, material properties and geometric design. In this study, the relations between cross-sectional area, perimeter, porousness and part orientation were studied and used in order to minimize the failures of a porous part during its construction.

It is necessary to continue exploring other kinds of porous structures such as irregular (e.g. Voronoi and foams) and regular (e.g. lattices and gradient). More evaluations will allow obtaining more information to improve this novel digital model. Also, it is important to consider an additional dynamic module to study the influence of the cyclic load of the separation force on the masks of the structure. Finally, it is critical to test the designed models in a real DLP machine in order to evaluate their precision.

5 Conclusion

The article presents a new methodology to design a porous structure in order to predict possible failures during the DLP-based additive manufacturing. This methodology was implemented in three modules of a digital model. The mechanical behavior of porous structures was evaluated, and a factor of safety is used as a result to improve the design process. As a proof of concept, a TPMS porous structure was designed and evaluated. In the case study design process, the digital model demonstrated the validation of the adaptability of the approach and the capacity to produce design options of porous structures that are based on valid information.

References

1. Chua, C.K., Wong, C.H., Yeong, W.Y.: Roadmap on additive manufacturing standards. In: Standards, Quality Control, and Measurement Sciences in 3D Printing and Additive Manufacturing, pp. 31–55. Elsevier (2017)
2. Pellegrino, J., Makila, T., McQueen, S., Taylor, E.: Measurement science roadmap for polymer-based additive manufacturing. National Institute of Standards and Technology, Gaithersburg, MD, NIST AMS, pp. 100–105, December 2016. https://doi.org/10.6028/NIST.AMS.100-5
3. Kodama, H.: Automatic method for fabricating a three-dimensional plastic model with photo-hardening polymer. Rev. Sci. Instrum. **52**(11), 1770–1773 (1981). https://doi.org/10.1063/1.1136492
4. Dean, D., et al.: Continuous digital light processing (cDLP): highly accurate additive manufacturing of tissue engineered bone scaffolds: this paper highlights the main issues regarding the application of continuous digital light processing (cDLP) for the production of highly accurate PPF scaffolds with layers as thin as 60 μm for bone tissue engineering. Virt. Phys. Prototyp. **7**(1), 13–24 (212). https://doi.org/10.1080/17452759.2012.673152
5. van Bochove, B., Hannink, G., Buma, P., Grijpma, D.W.: Preparation of designed Poly(trimethylene carbonate) meniscus implants by stereolithography: challenges in stereolithography. Macromol. Biosci. **16**(12), 1853–1863 (2016). https://doi.org/10.1002/mabi.201600290
6. Shao, G., Jain, S., Laroque, C., Lee, L.H., Lendermann, P., Rose, O.: Digital twin for smart manufacturing: the simulation aspect. In: 2019 Winter Simulation Conference (WSC), National Harbor, MD, USA, pp. 2085–2098, December 2019. https://doi.org/10.1109/WSC40007.2019.9004659
7. Kritzinger, W., Karner, M., Traar, G., Henjes, J., Sihn, W.: Digital twin in manufacturing: a categorical literature review and classification. IFAC-PapersOnLine **51**(11), 1016–1022 (2018). https://doi.org/10.1016/j.ifacol.2018.08.474
8. Tao, F.: Digital Twin Driven Smart Design. Elsevier, Waltham (2020)
9. Pan, Y., He, H., Xu, J., Feinerman, A.: Study of separation force in constrained surface projection stereolithography. Rapid Prototyp. J. **23**(2), 353–361 (2017). https://doi.org/10.1108/RPJ-12-2015-0188
10. Hollister, S.J.: Porous scaffold design for tissue engineering. Nat. Mater **4**(7), 518–524 (2005). https://doi.org/10.1038/nmat1421
11. Michielsen, K., Kole, J.S.: Photonic band gaps in materials with triply periodic surfaces and related tubular structures. Phys. Rev. B **68**(11), 115107 (2003). https://doi.org/10.1103/PhysRevB.68.115107
12. Lammel-Lindemann, J., Dourado, I.A., Shanklin, J., Rodriguez, C.A., Catalani, L.H., Dean, D.: Photocrosslinking-based 3D printing of unsaturated polyesters from isosorbide: a new material for resorbable medical devices. Bioprinting **18**, e00062 (2020). https://doi.org/10.1016/j.bprint.2019.e00062

Optimization of an Autonomous Learning Model for Detection COVID-19 Using Medical Images

Rodrigo Bastidas-Chalán[1,2](✉) and Paul Medina[3]

[1] Instituto de Postgrado, Universidad Técnica de Manabí, Portoviejo, Ecuador
rbastidas8382@utm.edu.ec
[2] Departamento de Ciencias Exactas, Universidad de Las Fuerzas Armadas ESPE, Sede Santo Domingo, Vía Santo Domingo-Quevedo km 24, Santo Domingo de los Tsáchilas, Ecuador
rvbastidas@espe.edu.ec
[3] Departamento de Ciencias Exactas, Universidad de Las Fuerzas Armadas Espe, 171103 Sangolquí, Ecuador
plmedina@espe.edu.ec

Abstract. Since COVID-19 appeared in 2019, detection methods based on medical images depend on the criteria of the specialist. This paper shows the optimization process of an autonomous image-based learning system. To improve the classification of positive images for COVID-19, obtained from a simple chest tomography, expert judgment and singular value decomposition are used. A computed tomography scan generates images in dicom format that must be converted to jpg or bmp format, so it is necessary to determine which format has better resolution. With the new processed images, matrix of size 145×145, a dataset is formed to train the neural network. When obtaining the neural network model, a reduction in training times between processed and unprocessed images is observed. Likewise, the predictions of the neural network, developed in python, for the detection of COVID-19 with processed images indicate results over 98%. Another important result is that singular value decomposition can be used to determine that images in bmp format provide more detail than images in jpg format, working with 90% of the original information in the image. Finally, the image classification is improved and the neural network is optimized.

Keywords: Optimization · COVID-19 · Neural network

1 Introduction

From December 2019, the pandemic caused by the COVID-19 changed the world, causing a governmental, social and health impact [1]. At first, patients with unexplained cases of pneumonia were found in Wuhan-China, caused by a type of coronavirus that had never been found before, it was called Severe Acute Respiratory Syndrome (SARS-CoV2), while the disease was called COVID-19. On March 12, 2020, the World Health Organization declared COVID-19 a pandemic [2]. In some countries, no matter if they are first world, like United States of America [3] or third world countries like Ecuador,

M. Botto-Tobar et al. (Eds.): CIT 2021, LNNS 405, pp. 229–240, 2022.
https://doi.org/10.1007/978-3-030-96043-8_18

where the reality exceeded the statistics, due to the large number of deaths that could not be quantified [4].

For the detection of Covid-19 there are countless tests, including: the polymerase chain test known as PCR and the so-called rapid tests [5], which have advantages in terms of sensitivity, but disadvantages in terms of costs, infrastructure, staff training and among others. On the other hand, chest computed tomography (CT) scans allow immediate diagnosis, avoiding inappropriate triage and late diagnosis of patients, also reducing possible infections to other people [6].

However, the drawback of CT is that its diagnosis depends on the professional, since it is he who is capable of detecting the affectation in the lungs and this criterion has a high degree of subjectivity, since it depends on the expertise and experience of the professional for proper diagnosis. Based on this premise, it is intended to use machine learning techniques, including neural networks, which provide support to the medical professional, for a better diagnosis of the patient [7].

Based on the antecedents exposed above, in which the importance of a correct diagnosis is manifested, it is imperative to ask the question: ¿Why improve the diagnosis based on medical images of patients with COVID-19?

In particular, for these autonomous learning systems, based on images, to achieve an adequate classification, it is necessary that they have adequate training, as optimization is necessary, for this reason, we will seek to develop robust models in order to achieve an adequate classification and therefore a better diagnosis.

2 Materials and Methods

The following stages has been realized in the present investigation.

2.1 Image Collection

A clinical dataset was used in which patients are anonymized. It must be verified that the people from whom the medical images were obtained, with a positive diagnosis of COVID-19, have a clinical examination that validates their diagnosis.

2.2 Medical Imaging Dataset

A simple chest CT generates around 70 medical images (dicom format), that is, for each patient we have 70 images that show the cross-section of the lungs in sequence. Due to the number of images, it is essential to be able to classify them according to the greater detail of the affectation in the lungs caused by SARS-CoV2 (COVID19).

In the present study, we work with simple chest tomographies of patients who present high affectation in the lungs, this is due to the fact that they present greater detail. It should be noted that the CT scans were obtained from a multidetector computed tomography.

A first classification of the images is carried out using the criteria of experts [8], that is, consulting a specialist radiologist with experience in the detection of COVID-19 using CT.

Because CT are generated in dicom format, they can only be viewed by medical programs such as Agnosco Dicom viewer, therefore it is necessary to convert them to a format that is capable of being processed by the programs that will be used. You can work in jpg or bmp format, however it is necessary to determine which format can be stored with the smallest size and best possible resolution.

2.3 Sample Selection

Because 119 patients are managed and from each one there are 70 images with a diagnosis of severe COVID-19 involvement, it is necessary to take a sample with 95% confidence. For this, the standard formula was used [9]:

$$n = \frac{Z_{\alpha}^{2} Npq}{d^{2}(N-1) + Z_{\alpha}^{2} pq} \quad (1)$$

where
n: sample size
N: population size
Z: corresponding value of the Normal distribution ($Z_{\alpha=0.05} = 1.96$)
p: expected prevalence of the value to be evaluated.
q: complement of p, q = 1–p
d: standard error

So:

$$n = \frac{1.96^{2}(119)(0.9)(0.1)}{0.05^{2}(119-1) + 1.96^{2}(0.9)(0.1)} \quad (2)$$

$$n = 64{,}21 \quad (3)$$

The sample was taken from 64 people, including patients with a positive and negative diagnosis for COVID-19 and patients with any other pathology.

2.4 Singular Value Decomposition (SVD)

For improved reconstruction and noise reduction in medical imaging the singular value decomposition can be used [10]. For SVD the Matlab commands can be used to obtain the singular value "S" [11].

Given a matrix A, of size mxn, its decomposition in singular values is defined by:

$$A = U \Sigma V^{T} \quad (4)$$

where:
U is a matrix of size m × m
Σ is a diagonal mxn matrix
V is a transpose matrix of size n × n.

2.5 Selecting the BMP Format

Medical images in both jpg and bmp formats are interpreted as a matrix of size 512 × 512. To compare which one has the best resolution and definition, the SVD will be used, the one that generates at least 90% of original information and that has the smallest possible size matrix will be considered.

In this first procedure, SVD is used to compare some images of a patient (see Fig. 1), both in jpg and bmp formats.

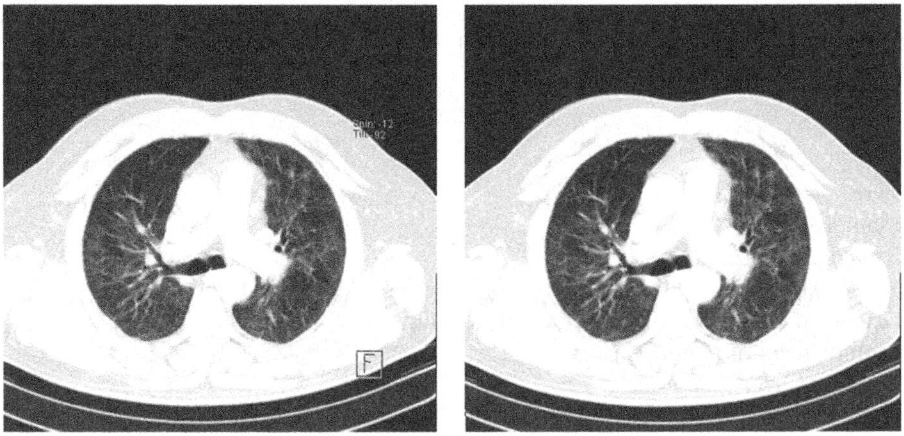

Fig. 1. Comparison of images with positive COVID19. Left (jpg format), Right (bmp format).

As you can see in the results of Table 1, with smaller matrices, the bmp format provides more information than the jpg format. For example, for image 1, with a matrix of size 146 × 146 in jpg format, 90% is reached, while for the same image, with a matrix of size 140 × 140 in bmp format, 90% is reached.

Likewise, for image 2, the size of the matrix that is needed, in bmp format is 132 × 132, while in jpg format it is 138 × 138. The same occurs for the other 8 images analyzed, in bmp format a smaller matrix is needed to reach 90% of the information. The remaining 10% of the information that is omitted corresponds to the noise of the image, of values that do not contribute the characteristics that are needed [12].

To determine the optimal size of the matrix of the bmp images, it is necessary to evaluate the dataset with 1127 positive images for COVID-19 and verify which matrix size provides at least 90% of the information.

Table 1. Comparison of a patient image in jpg and bmp format

Image	Size jpg	Information jpg	Size bmp	Information bmp
1	146 × 146	90.05	140 × 140	90.05
2	138 × 138	90.01	132 × 132	90.07
3	137 × 137	90.04	131 × 131	90.09
4	139 × 139	90.06	133 × 133	90.13
5	144 × 144	90.01	138 × 138	90.07
6	147 × 147	90.02	142 × 142	90.08
7	144 × 144	90.05	139 × 139	90.11
8	140 × 140	90.08	134 × 134	90.09
9	136 × 136	90.00	130 × 130	90.02
10	140 × 140	90.02	134 × 134	90.08

Likewise, it is important to mention that there are images with sizes of 90 × 90 or smaller, which manage to provide 90% of the information. However, most matrices achieve that percentage with a size of 145 × 145, as indicated in Table 2:

Table 2. Number of images that provide more than 90% of information

Size jpg	Quantity of images > 90%	Quantity of images < 90%
90 × 90	148	979
95 × 95	176	951
100 × 100	233	894
105 × 105	304	823
110 × 110	422	705
115 × 115	549	578
120 × 120	634	493
125 × 125	687	440
130 × 130	736	391
135 × 135	778	349
140 × 140	829	298
145 × 145	865	262
150 × 150	893	234

With a matrix of size 90 × 90, there are 148 of 1127 images with 90% of the information, while with a matrix of size 145 × 145, 865 of 1127 images have more than 90% of the information. That is, 76.75% of the dataset is reached, ensuring greater specificity and greater number of images.

It is very important to mention that the sample mode criterion is used to determine the size of the matrix of the images in bmp format that are needed when calculating the singular value "S". In this case, the size of the matrix of the medical image that provides at least 90% of the information, in most images, is 145 × 145.

It should be noted that larger matrices, such as those 150 × 150 or larger, also provide information. However, for this study, they are not considered an appropriate size, since the objective is to work with a matrix that is the smallest possible size.

2.6 Improved Medical Image Dataset

The images processed from matrices of size 145 × 145 (see Fig. 2) form the dataset of images positive for COVID-19, which in turn, serves to train the neural network. Due to the number of files that a CT generates, it is necessary to determine a range of the eigenvalue "S" that allows the images with better details to be automatically classified by programming, as indicated in Table 3.

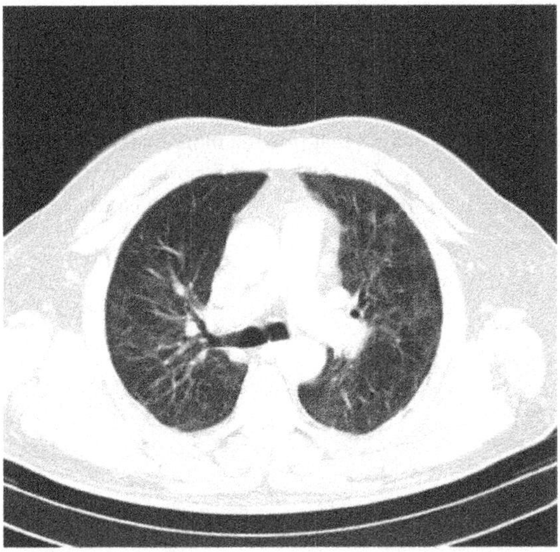

Fig. 2. Positive images for COVID-19 with a matrix of size 145 × 145.

It is necessary to state that the new images were created from a matrix of size 145 × 145, but their size is still 512 × 512. Also, since noise is removed, the generated file size in Kilobytes is smaller than the original images. Table 3 indicates the singular values "S" of all the images that make up the dataset.

The range of singular values obtained for Covid-19 Images corresponds to all the processed images that make up the dataset for the neural network to be trained and is between 1.417e+05 and 3.056e+05. With respect to the other raw images there is no major difference.

Table 3. Singular value range

Type of images	Size	Minimum	Mean	Maximum
COVID-19 images	145 × 145	1.417e+05	2.529e+05	3.056e+05
COVID-19 images	512 × 512	1.417e+05	2.539e+05	3.099e+05
No COVID-19 images	145 × 145	2.173e+05	2.573e+05	2.888e+05
No COVID-19 images	512 × 512	2.183e+05	2.586e+05	2.902e+05

The range of singular values obtained for no COVID-19 Images correspond to all the processed images that make up the dataset for the neural network to be trained and is between 2,173e+05 and 2,888e+05. With respect to the other raw images there is no major difference.

The range found allows the images to be classified for a correct diagnosis of the disease, fulfilling the criteria of experts. However, since there is a considerable dataset, with all types of images, these ranges can vary significantly, so it is necessary to use a neural network to properly classify the images.

2.7 Neural Network

The last stage consists of developing a neural network, with Resnet34 architecture, using the fit one cycle method [13]. It is necessary that the medical images are in 3 different sets: training, validation and testing, starting with a percentage of 70%, 20% and 10% of the total images respectively, until reaching 80%, 10% and 10%, as shown in Table 4 (matrix of size 512 × 512) and Table 5 (matrix of size 145 × 145).

For the first training of the model (unprocessed images), with a distribution of the total of train images (70%), validation (20%), and test (10%), an error of 3.11% was obtained with a training time 48:08 min total. For the train (80%), validation (10%), and test (10%) distribution, an error of 0% is obtained with a total training time of 35:22 min. This is because the network neuronal is over-fit.

For the second training of the model (raw images), with a distribution of the total of train images (70%), validation (20%), and test (10%), an error of 2.01% was obtained with a training time 47:07 min total. For the train (80%), validation (10%), and test (10%) distribution, an error of 0% is obtained with a total training time of 51:50 min. This is because the network neuronal is over-fit.

Comparing the two previous models, the error obtained is less than in comparison to the first model, likewise, the training times are less in the second model, optimizing not only research times but also computer resources. Therefore, this second model is used for subsequent analysis in the detection of COVID-19.

In each of these datasets, there must be images with a positive and negative diagnosis, which must be classified in their respective dataset. In this way, the neural network model is trained and validated with images of positive and negative cases of COVID-19.

It is important to consider that when executing the programming codes, as well as the image processing algorithms and neural networks, there is an enormous expenditure

Table 4. Neural network result (unprocessed images)

% Images to dataset	Epoch	Train loss	Validation loss	Error rate	Time [min]
70%, 20%, 10%	0	0.322402	0.241869	0.068889	12:14
	1	0.196146	0.192080	0.057778	11:55
	2	0.135109	0.127875	0.046667	12:00
	3	0.106955	0.081959	0.031111	11:59
80%, 10%, 10%	0	0.314988	0.088856	0.053571	09:46
	1	0.181351	0.010972	0.008929	08:31
	2	0.145491	0.027080	0.000000	08:32
	3	0.097659	0.013131	0.000000	08:33

Table 5. Neural network result (processed images)

% Images to dataset	Epoch	Train loss	Validation loss	Error rate	Time [min]
70%, 20%, 10%	0	0.3301317	0.068671	0.020089	12:05
	1	0.243110	0.068689	0.024554	11:45
	2	0.165036	0.115972	0.035714	11:39
	3	0.144975	0.050863	0.020089	11:38
80%, 10%, 10%	0	0.088287	0.047014	0.017857	12:47
	1	0.087869	0.070983	0.022321	13:07
	2	0.070411	0.010523	0.004464	12:58
	3	0.080234	0.010267	0.000000	13:01

of computing resources, so a robust computer or a virtual computer can be used [14], for example, Google Colab.

3 Results

The singular values allow handling ranges in which there is a pre-classification for images of patients with COVID-19. Also, the images can be reconstructed using 90% of the original information. The prediction results when analyzing some processed images (70%, 20% and 10%) are indicated in Table 6.

In the first 4 images (see Fig. 3), the model predicts detection over 99% in positive images for COVID-19. Likewise, for the following 4 images (see Fig. 4), the model predicts detection over 99% in negative images for COVID-19.

Table 6. Neural network prediction with COVID-19 and no COVID-19 images

Image	Probability of COVID-19	Probability of no COVID-19
1	0.9899	0.0101
2	0.996	0.004
3	0.9955	0.0045
4	0.9981	0.0019
5	0.0112	0.9888
6	2.4336e−05	9.9998e−01
7	6.7694e−05	9.9993e−01
8	0.0032	0.9968

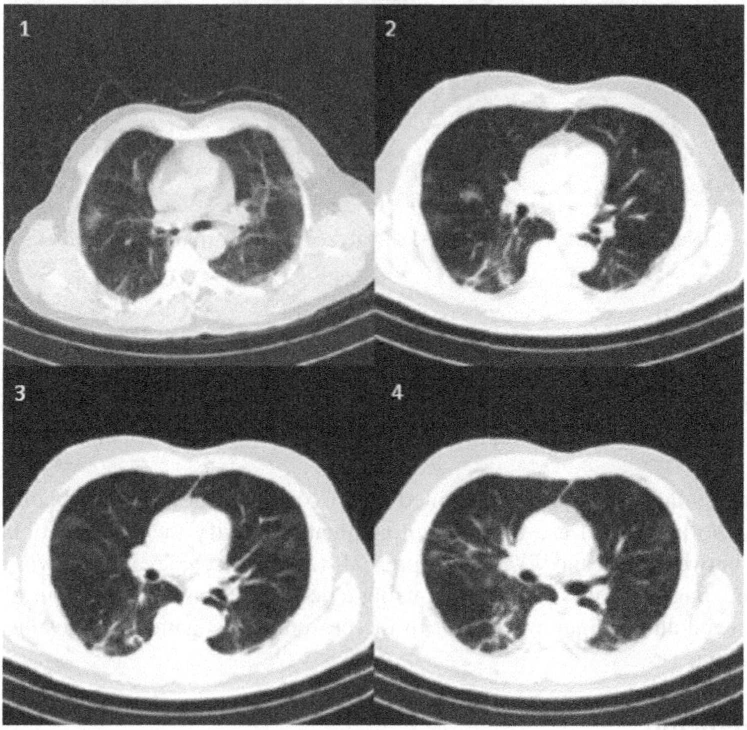

Fig. 3. Positive images for COVID-19 analyzed with distribution 70%, 20%, 10%.

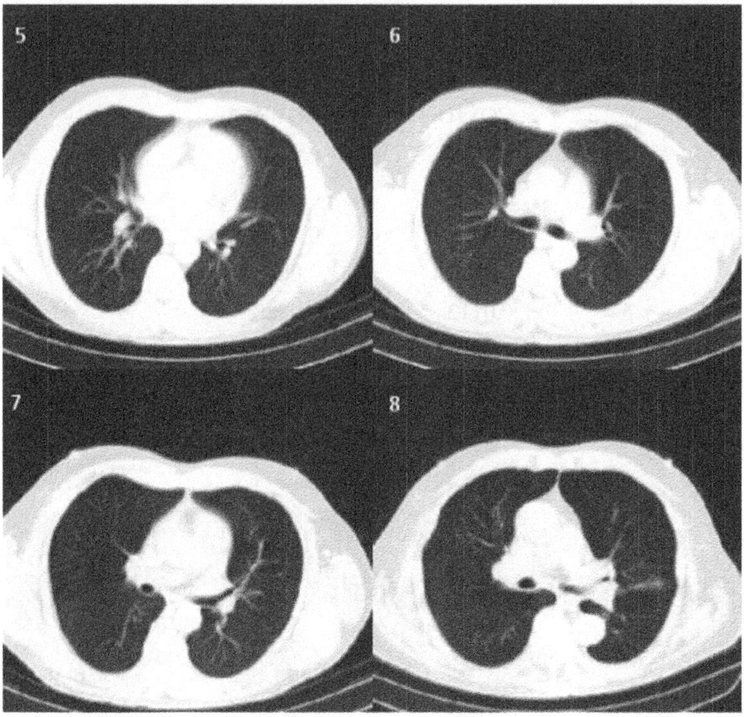

Fig. 4. Negative images for COVID-19 analyzed with distribution 70%, 20%, 10%.

4 Discussion

In an investigation [15] for the classification of medical images of CT, using multi-objective differential evolution–based convolutional neural networks, results with errors greater than 1% were obtained. However, in the present study errors less than 1% were obtained.

Medical images can also be obtained from X-rays, in the study [16] convolutional neural networks (CNN) are used to automatically classify the images, obtaining an accuracy between 80% and 90%. In the present study, an accuracy of 99% was obtained.

For the diagnosis of X-ray images, using a deep CNN based architecture CovXNet [17], achieved an accuracy of 97.4%. In the present investigation, the results improve that precision.

5 Conclusion

When working with processed images (matrix of size 145 × 145) the training time of the neural network is reduced from 48:08 to 47:07. In addition, the storage space of the processed images is less than that of the raw images.

An image dataset can be improved not only by changing from jpg to bmp format, but also by eliminating noise, that is, working with a matrix of size 145 × 145 but containing 90% of the original information.

The neural network was optimized both in training times, image storage, lower cost of computer resources and in prediction probability for the classification of images, with which the proposed objective is reached.

6 Recommendations

This research provides a tool to help specialist doctors. Under no circumstances does it eliminate the criteria of an expert, rather it complements it and should be used to improve the evaluation of a patient. Research can be complemented for cases of COVID-19 with mild or moderate diagnosis, to improve the study of the range of singular values found.

Also, the dataset could continue to be modified, increasing the images to the training base and achieving, for example, a model with a percentage of 96%, 2% 2%. However, for this investigation the maximum percentage of 70% of the total images in the "train" dataset, since the results indicate that the network is over-fit.

This investigation can be complemented with a study to discriminate the set of 70 images normally delivered by a CT, and pre-classify the images normally delivered by a CT.

Data Availability. Matlab and Python code used in this study is available in: https://drive.google.com/drive/folders/1rF5z-F22Gz0Vn_g3Y6uwq_bStaVFNTdq?usp=sharing.

References

1. Proaño, C.: On the macroeconomic and social impact of the coronavirus pandemic in Latin America and the developing world. Intereconomics **55**, 159–162 (2020)
2. Sun, J., Aghemo, A., Forner, A., Valenti, L.: COVID-19 and liver disease. Liver Int. **40**(6), 1278–1281 (2020)
3. Karaye, I., Horney, J.: The impact of social vulnerability on COVID-19 in the US: an analysis of spatially varying relationships. Am. J. Prev. Med. **59**(3), 317–325 (2020)
4. Molina, N., Mejias, M.: Social impact of COVID-19 in Brazil and Ecuador: where reality surpasses the statistics. Edumecentro **12**(3), 277–283 (2020)
5. Calvache, J., Rodríguez, A., Martínez, C., Paucar, V.: Usefulness of polymerase chain tests, rapid tests and CT scans in patients with Covid-19. J. Am. Health **3**(2), 32–39 (2020)
6. Pareja, J., Anicama, S., Perez, P., Pecho, S., Amado, J.: Importance of the implementation of the chest tomography to contribute to the early diagnosis and timely triage of patients with COVID-19 in Peruvian hospitals. Acta Méd. Peruana **37**(2), 239–241 (2020)
7. Pham, T.: A comprehensive study on classification of COVID-19 on computed tomography with pretrained convolutional neural networks. Sci. Rep. **10**(1), 1–8 (2020)
8. Gorina, A., Berenguer, I., Salgado, A., Álvarez, J.: The management of scientific information provided by the criteria of experts. Ciencias de la Informaci6n **45**(2), 39–47 (2014)
9. Badii, M., Castillo, J., Guillen, A.: Optimal sample size. Innovaciones de negocios **5**(9), 53–65 (2017)
10. Lyra, D., Carvalho, J., Azevedo, J.: Enhanced reconstruction of magnetic resonance data using singular value decomposition approximation. Ingeniería **17**(2), 35–45 (2012)
11. MathWorks homepage. https://la.mathworks.com/help/matlab/ref/double.svd.html;jsessionid=7c0d111aa3913e4333866507c1e9. Accessed 21 Nov 2016

12. Li, Y., Ma, W., Zhao, Y.: Application of digital image processing technology based on artificial intelligence in the analysis of medical images. Invest. Clin. **60**(6), 1548–1561 (2019)
13. Smith, L.N.: A disciplined approach to neural network hyper-parameters: Part 1-learning rate, batch size, momentum, and weight decay, pp. 1–21. ArXiv, arXiv:1803.09820 (2018)
14. Colab. https://colab.research.google.com/. Accessed 23 Mar 2021
15. Singh, D., Kumar, V., Kaur, M.: Classification of COVID-19 patients from chest CT images using multi-objective differential evolution–based convolutional neural networks. Eur. J. Clin. Microbiol. Infect. Dis. **39**(7), 1379–1389 (2020)
16. Varela, S., Melin, P.: A new approach for classifying coronavirus COVID-19 based on its manifestation on chest X-rays using texture features and neural networks. Inf. Sci. **545**, 403–414 (2021)
17. Mahmud, T., Rahman, M., Fattah, S.: CovXNet: a multi-dilation convolutional neural network for automatic COVID-19 and other pneumonia detection from chest X-ray images with transferable multi-receptive feature optimization. Comput. Biol. Med. **122**, 103869 (2020)

An Implementation of an Algorithm for Information Theft Using Deep Learning Techniques: An Ethical Hacking Event

Carlos Andrés Estrada, Walter Fuertes(✉), Joyce Denisse Castro, and Daniel Nuñez-Agurto

Department of Computer Science, Universidad de las Fuerzas Armadas – ESPE, Sangolquí 171103, Ecuador
{caestrada4,wmfuertes,jdcastro,adnunez1}@espe.edu.ec

Abstract. Ethical hacking is modern data protection and prevention technique proper to prevent information losses. Under this precept, this study aims to design and implement an algorithm capable of circumventing preventive security to perpetuate an attack to steal information from the victim. The algorithm programmed through Python, MatLab, and R, automates the export of a data repository saved in a .CSV file, which contains all the methods to prevent an attack. In this way, when the algorithm is executed, it is trained to disable all detection mechanisms. The algorithm uses the structure of a neural network based on Deep Learning, Bayesian networks, a mathematical model, and a Rectified Linear Activation function or ReLU. It was configured to execute the algorithm on each interpreted output data. The proofs of concept were developed in a controlled virtual network environment, using a Linux-based server. The results were validated based on resource management in both processing and memory, using monitoring standards and averaging curves. Thus, we demonstrate the successful unauthorized access to a remote server to steal all the victim's information and that the algorithm was undetectable and efficient.

Keywords: Deep learning · Ethical hacking · ReLU · Neural network

1 Introduction

Information security plays an essential role in everyday life since the evolution of technology increases the risk factors of vulnerability. In this aspect, electronic devices have an essential role in computer science by sending and receiving vital information from users. According to [7], the security of information transfer, messages, and user data depends on the guarantees provided by authentication protocols and encryption techniques [14].

This research presents an unpublished algorithm that seeks to change the current insecurity in computer systems. This algorithm can overcome any antivirus

M. Botto-Tobar et al. (Eds.): CIT 2021, LNNS 405, pp. 241–254, 2022.
https://doi.org/10.1007/978-3-030-96043-8_19

and also steal the victim's information. Specifically, we designed an algorithm that combined a mathematical model and a Rectified Linear Unit (ReLU) as part of Deep learning, employing the gradient algorithm; this can pass all the information to a distributed server in relevant phases of standard and corrupted information. We propose ReLU due to the speed up the learning convergence. Every time the proposed algorithm detects a failure, it is responsible for linking each difficulty to steal the information and store it on a server configured in Ubuntu. In this way, all the information is stolen. Our neural network works in such a way that it can eliminate any existing antivirus, being undetectable by any user.

In addition, data sciences techniques have been used for the analysis of relevant information. The algorithm can classify all the data into three different groups as partial information, damage information, and relevant information. In each group, there will be data that can serve an attacker, although it would be a new information entry strategy.

We first focus on simulating a process with a mathematical model that complies with the Turing test's statistical principles and Bayes' theorem and is represented through design, implementation, processing speed, and RAM efficiency. The main contribution of this study is the creation of an unpublished algorithm for information theft, which requires no other functionality than an authentic connection.

The remainder of this article is organized as follows: Sect. 2 explains the materials and methods used to develop the algorithm and the information theft process. Section 3 presents the evaluation of the results. Section 4 clarifies the conclusions and lines of future work.

2 Materials and Methods

2.1 Deep Learning Applied to Cybersecurity, A Slight Overview

There are different studies, which deal with deep learning and cybersecurity [12]. It is worth noting that since the development of technology and the Internet has been growing unlimitedly, people's lifestyles have been changing [3], namely, studies, work, entertainment, etc.etera. Because of this, cyber-attacks are becoming more dangerous and effective [20]. However, cybersecurity is designed to protect computers, networks, and data from attacks and unauthorized alteration or access [18].

One of the Deep learning goals is to solve the problem of finding a representation by learning it from the data. Another goal is to assume that depth, allowing learning a sequence of parallel instructions that transforms an initial input vector. Deep Learning often involves learning representations using a series of layers, which operate by processing an input generating a series of representations given as input to the next layer [13].

The exponential increase of users on the Internet, cloud services, Internet of Things devices are related to the number of attempts to find a way to break into these systems. Cybersecurity can have many methods; one of them is deep learning, a branch of Machine Learning that tries to learn many characteristics from data to use them later for decision-making [16].

It is necessary to improve the detection of security issues under IoT since its intrusion security detection determines if the IoT is in a dangerous environment by catching its data characteristics and analyzing the attack behavior. When it gets attacked, a response can be made to avoid loss [11].

Deep learning techniques are inspired by the capacity of the human brain when learning from previous experiences instantaneously. Cybersecurity insight nowadays is because of the increased internet usage and a wide variety of network applications [6,15].

Depp learning also is a powerful method of data exploration to learn about behaviors according to how IoT components and devices interact with one another within the IoT environment [2]. It has made end-to-end learning possible, even handling vast amounts of data without getting humans involved. Its algorithms require little time of running while testing compared to machine learning's [5]. When talking about artificial intelligence, we can relate it to deep learning since it has become a very relevant tool for applications, starting self-driving cars to surveillance and security [1].

Deep learning models are vulnerable to adversarial examples maliciously crafted inputs to trigger target Deep Learning models to misbehave [19], which significantly hinders the application of Deep Learning in security-sensitive domains [18]. Intensive research on adversarial learning has led to an arms race between adversaries and defenders [10]. However, in the last five years, Deep learning techniques have gained much interest due to the increased amount of data available and its algorithmic innovations and significant improvements in computing capabilities enabled by GPUs, which have made the fast training [16].

2.2 Design and Implementation of the Algorithm

The proposed algorithm uses an artificial intelligence technique called deep learning. According to [8], Deep learning is a form of machine learning that enables computers to learn from experience and understand the world in terms of a hierarchy of concepts. Because the computer gathers knowledge from experience, there is no need for a human-computer operator formally to specify all of the knowledge needed by the computer.

We programmed the algorithm with Python language by exporting a data store saved in a .CSV dataset. This repository contains all the methods to avoid an attack. In this way, when the algorithm is executed, it is trained to deactivate all the methods.

In addition, we use NumPy for data science, which allows us to perform numerical calculations based on the designed mathematical model, and it was configured to execute the algorithm on each interpreted output data. NumPy is a Python library that supports creating vectors and multidimensional arrays, along with an extensive collection of high-level mathematical functions to operate on them.

Afterward, we used a metaheuristic technique [17] that explains its use through learning or experience. The neural network that it was designed for would learn and deliver results according to the required classification.

The algorithm can classify all the data into three different groups, partial information, damage information, and relevant information [11]. In each of them, data can serve an attacker, but on the other hand, it is a new information entry strategy.

Then we used a direct connection to a remote server, which has been virtualized in Linux, which will direct cloning of the existing data and information. So in this way, it can be considered that the hack is effective when attacking. We have developed our customized algorithm that has not been published.

An experimental research proposal has been carried out, applying software development methodologies, and controlled virtualized environments, which allow emulating any event that can be generated during networks with this type of technology [4]. Also, we use a Bayesian network with several learning layers and helps to diversify the computational mathematics for the continuous analysis of data that will enter this input function to give a resultant functional output [9].

We have added different techniques concerning the control the insecurity in information systems. The algorithm uses the structure of a different neural network based on a ReLU. We developed in Python, MATLAB, and R programming languages, which were tested on Windows-based operating systems, using a server-based on Linux, through Ubuntu; this will determine the efficiency of the same. The mathematical modeling process is explained below (see Eq. 1):

$$L1 = ReLU(W_1X! + a_1 + b_1 - W_2Y!) \tag{1}$$

The neural network was concise, assembled in the following factorial form, where its learning definition works with the respective derivatives in each recursive equation (see Eq. 2).

$$W_1 = W_1X! - lr \times \frac{\partial L}{\partial W_2Y} \tag{2}$$

The equations comply with the respective loss rule. We used five layers to allow the algorithm to be efficient in the analysis of the data. Therefore, the output function generates the following result according to the linear algebra computer (see Eq. 3).

$$Y = w_{11} \times x_{11} \times a_{11} + w_{21} \times x_{12} \times a_{12} + w_{31} \times x_{13} \times a_{13} + w_{41} \tag{3}$$

The previous mathematical model expresses the amount of data by creating an intelligent algorithm based on the ReLU activation function. However, it should be emphasized that the present work consists of a modification in its structure, based on the tests of Godel and Turing for recursive computability. As can be seen, it contains a factorial point that must be analyzed in each entry. For its analysis is the number of insecurities and vulnerabilities, which, when executed in real-time, the mathematical model runs through each perspective. It finally deactivates the antivirus, and thus the algorithm is in charge of copying all the information in the virtualized server environment.

Figure 1 illustrates that the process for accessing information has several stages. Detecting all the data in this way can also distribute the information according to its relevance. The controlled environment that was created was through various logical aspects. The algorithm was developed in various programming languages in the first instance, including direct access by Python. Inferential statistics were used in R language to evaluate the mathematical model, which allowed to measure the access probability curves, the speed at which the algorithm was executed, and the propagation through all the files. So in this way, through training for different data, the neural network learns from the layers' data and generates an output based on the configuration parameters.

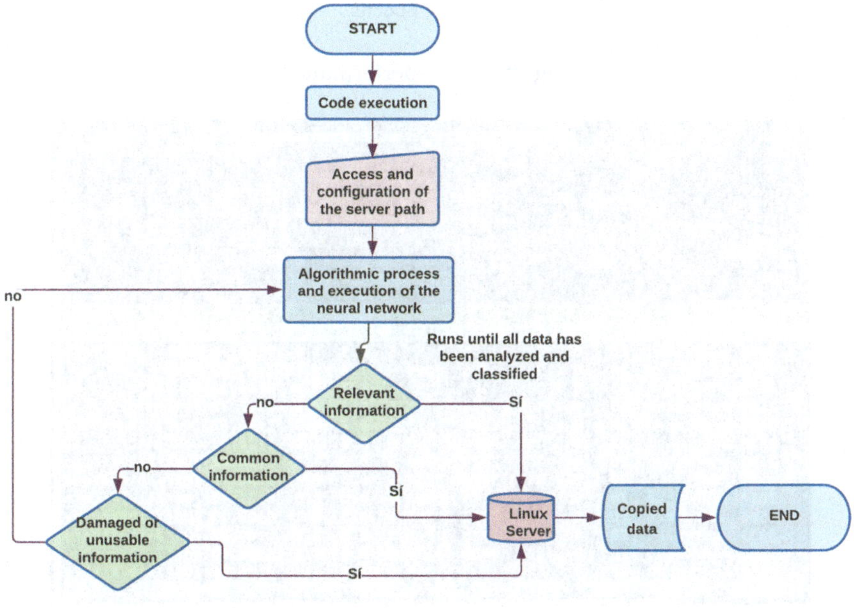

Fig. 1. Flowchart that explains how the algorithm works

The behavior of the mathematical model, which was designed for the study and development of this algorithm, has the following function, which shows its efficiency in the iteration of entered data (see Fig. 2).

The connection with the server must be established and created. Figure 3 shows the execution of the server and the performance statistics of the linked computer. We can observe in real-time during the execution of the algorithm.

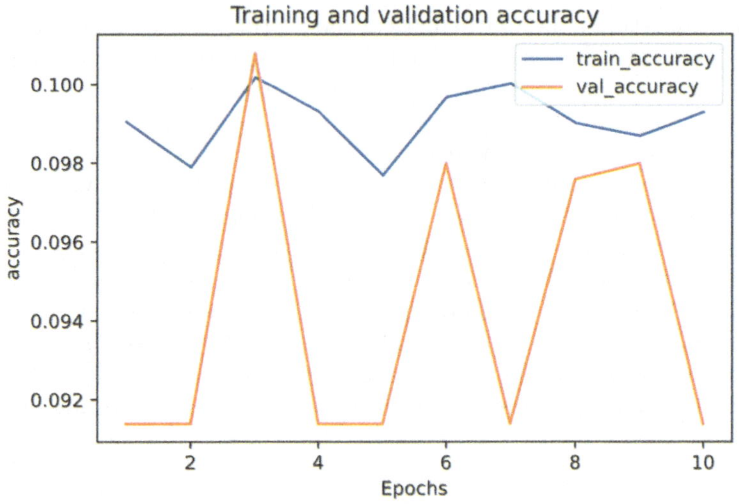

Fig. 2. ReLU performance

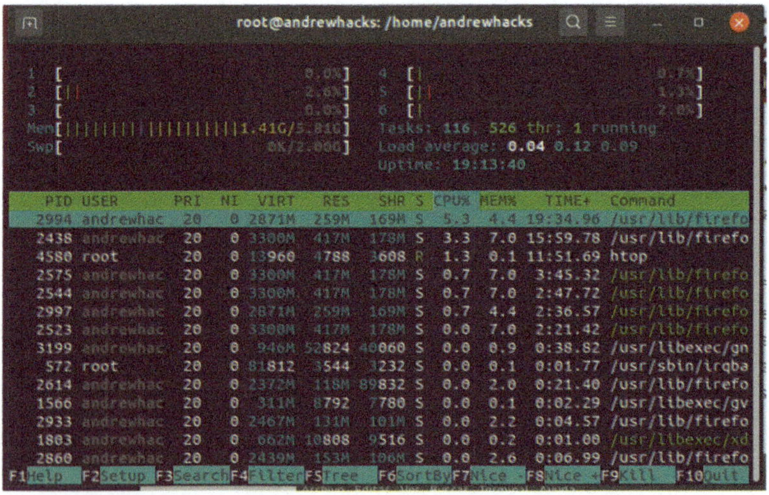

Fig. 3. Server connected and the performance statistics

On another side, we create a readData algorithm in charge of distributing and analyzing the information according to its neural network, stored in a Linux server (see Algorithm 1). We use Python, and other existing libraries for algorithm programming, with modifications in the ReLU, e.g., TensorFlow and Matplotlib. For the mathematical statistics, we employ a thread-based programming model through the Pandas library.

Algorithm 1. readData

```
Establish connection to the server
Run the algorithm
Read all data
if Antivirus exist then
   Disable the antivirus
else
   while Data Exists do
      if Data is relevant then
         Copy relevant information in the section "Relevant" created in the server
         database
      else
         if Data is common then
            Copy relevant information in the section "Common" created in the server
            database
         else
            Copy relevant information in the section "DoI" created in the server
            database
         end if
      end if
   end while
end if
Review the information
```

The design of the neural network is about five layers. As we can see in Fig. 4, it has a data input layer, three layers of classification and data reading, and an output layer, which can decode the information as required, then works recursively as long as it complies with the Turing model.

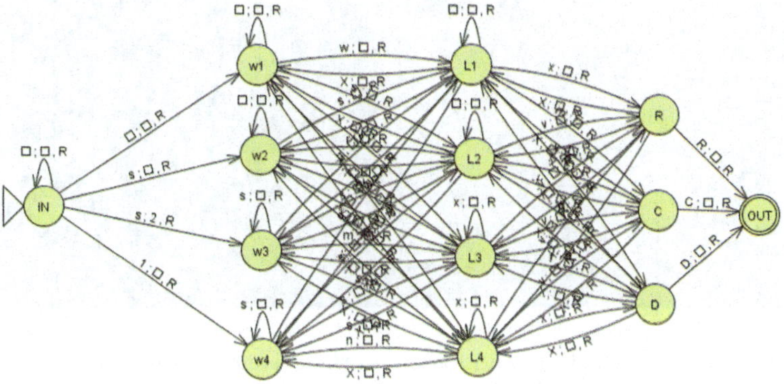

Fig. 4. Diagram of the neural network

The neural network is activated by the ReLU activation function applied at the input or IN layer. Subsequently, the network decrypts the security of the computer and all the information. However, it depends on its difficulty level since the network is trained by converting all the data to machine language, i.e., 0 and 1.

The learning network is activated by the automatic connection command when connecting with the computer and establishing a connection with the server. Also, during this process, it analyzes and deciphers if the information is encrypted or blocked. This analysis is commissioned by layers 2 and 3, so the information is legible and without security. At the end, the output is already a database with all the information copied and divided according to its relevance. Likewise, the fourth layer is in charge of its classification.

Each layer and connection is recursive until the information is completely copied; this is how its process is interconnected between layer and layer until it learns and becomes more efficient in applying another theft of information.

The connection is established, and the algorithm is executed. The first instance is the windows service, the existing files will be shown, and the antivirus status. Once the algorithm is executed in the results section, we will observe the copy made on the server.

Figure 5 shows the leading directory of the virtualized system, and the Windows 10 system was installed in the VirtualBox virtual machine. In contrast, Fig. 6 shows that the antivirus is activated before the connection to the server and the execution of the algorithm.

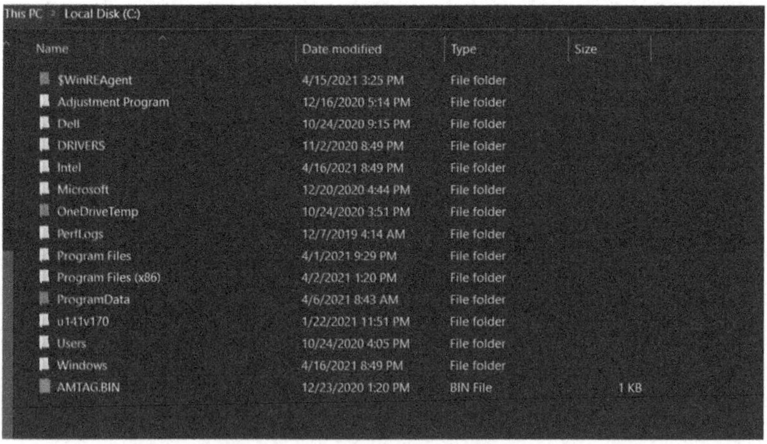

Fig. 5. Files and folders in the system directory

The proof of concept was analyzed using the R statistics library, which expresses the performance in real-time execution. The results were validated based on resource management in both processing and memory, using monitoring

Fig. 6. Status of the antivirus before the algorithm

standards and averaging curves. Thus, we determined that the algorithm was undetectable and efficient.

3 Results Evaluation and Discussion

We tested the algorithm using the neural Turing phases to evaluate computability, which was executed several times. Its evaluation results were exported in a .dat file, analyzed using R Studio's statistical software. Also, it allowed us to interpret the processing consumption in a system and its maximum number of data computable as the inputs enter the network. Table 1 shows the behavior generated during the execution of the neural network.

Table 1. Analysis of the neural network when it runs

	RAM		PROCESSING	
	Before test	After test	Before test	After test
Speed	30%	50%	25%	55%
Efficiency	15%	100%	35%	50%

The algorithm was working correctly during its execution once the server is copied with the information. As a consequence, the data statistics are increasing. Figure 7 shows the learning and decoding capacity of the data. Thus, the existing network distributes the data according to the propagation curve. Once the algorithm is executed, Fig. 8 shows in real-time the antivirus status when the same is running. In addition, the information is displayed and copied automatically. In Fig. 9, we can see that the algorithm is executed (i.e., and the success when the system has been deactivated), connecting to the remote server. Once it is executed and the system has an antivirus, the neural network is responsible for deactivating it. There are no problems when copying all the information from the system. Figure 9 shows the execution of the algorithm.

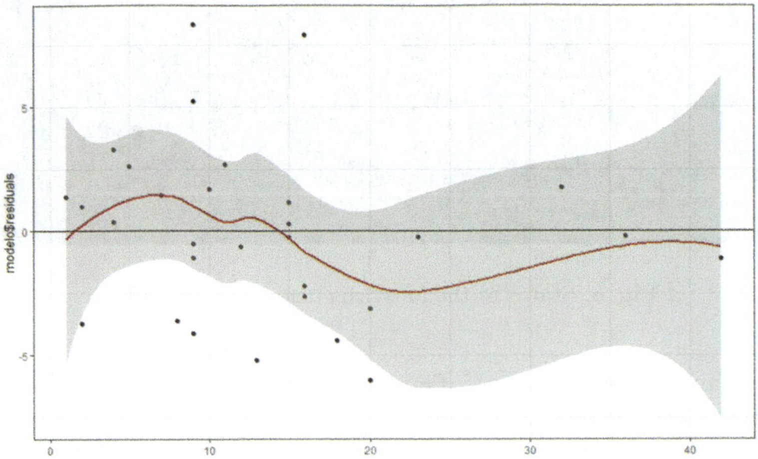

Fig. 7. Data propagation during the algorithm works

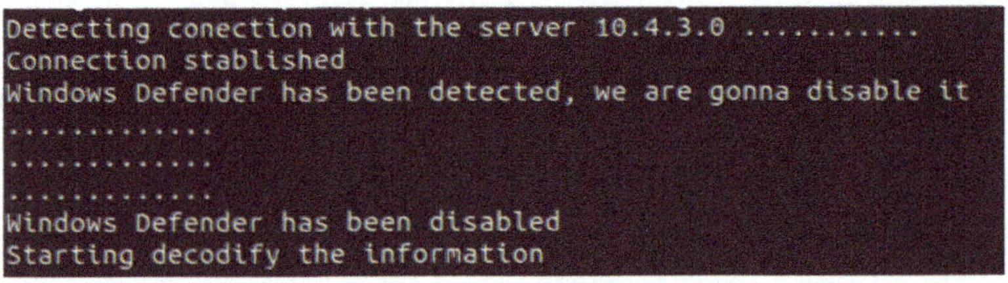

Fig. 8. Execution of the algorithm and deactivation of the antivirus

After the algorithm has deactivated the antivirus, it starts copying the information until it finally displays the server's Linux directory, reporting the information and its classification. Figure 10 shows the copying process, while Fig. 11 generates the results of the stolen information.

The results are as expected according to the computability test. Since the neural network was measured with the Turing test, the efficiency and speed of execution generated are accurate. Also, as shown in the illustrations above, the algorithm works correctly, copying the respective information. Since the statistics obtained in this test show that the mathematical model is a magnificent invention, since it is an unpublished algorithm, it is considered that it would revolutionize security in the field of computational sciences and have been designed with exhaustive programming.

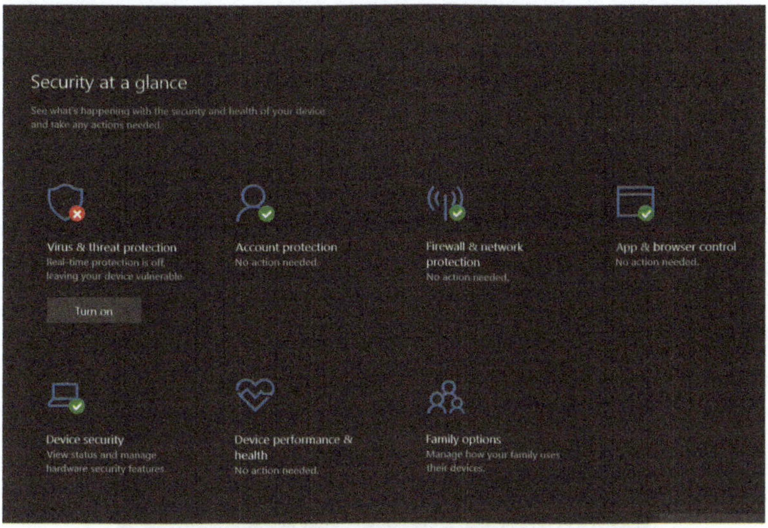

Fig. 9. Antivirus disabled

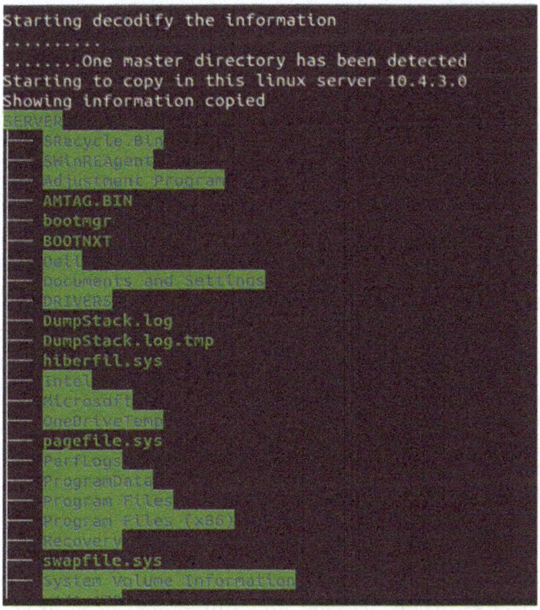

Fig. 10. Copying process

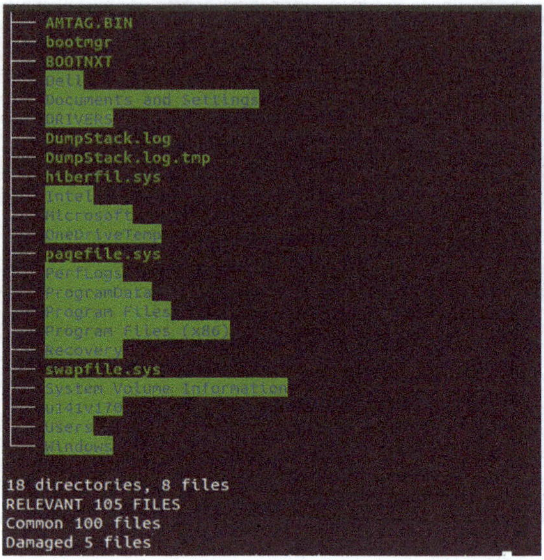

Fig. 11. Results of the information and their classification

4 Conclusions and Future Work

This study focused on designing and implementing an unprecedented algorithm to perpetuate the theft of the victim's information, capable of not being detected by an antivirus. The algorithm implemented with Python, Matlab, and R assembled a mathematical model and an activation function of a Rectified Linear Unit (ReLU) as part of Deep Learning. The algorithm achieved the training of a neural network to eliminate any existing antivirus, being undetectable by any user. With this, we demonstrate the theft and the transmission of all the victim's information to a distributed server. The results demonstrated the functionality of this training as part of an ethical hacking event. In this way, the algorithm could become a valid solution to reduce cyber-attacks perpetuated on Windows platforms.

As future work, we plan to strengthen this algorithm using machine learning and a mathematical model that uses convolutional networks.

Acknowledgments. The authors would like to thank the Universidad de las Fuerzas Armadas – ESPE of Sangolquí, Ecuador, for the resources granted for the development of the research project entitled: "Design and Implementation of the IT infrastructure and service management system for the ESPE Academic CERT", Code: PIC-ESPE-2020-ESPE-CERT.

References

1. Akhtar, N., Mian, A.: Threat of adversarial attacks on deep learning in computer vision: a survey. IEEE Access **6**, 14410–14430 (2018)
2. Al-Garadi, M.A., Mohamed, A., Al-Ali, A.K., Du, X., Ali, I., Guizani, M.: A survey of machine and deep learning methods for internet of things (IoT) security. IEEE Commun. Surv. Tutorials **22**(3), 1646–1685 (2020)
3. Benavides, E., Fuertes, W., Sanchez, S., Sanchez, M.: Classification of phishing attack solutions by employing deep learning techniques: a systematic literature review. Dev. Adv. Defense Secur. 51–64 (2020)
4. Chalapathy, R., Chawla, S.: Deep learning for anomaly detection: A survey. arXiv preprint arXiv:1901.03407 (2019)
5. Ferrag, M.A., Maglaras, L., Moschoyiannis, S., Janicke, H.: Deep learning for cyber security intrusion detection: approaches, datasets, and comparative study. J. Inf. Secur. Appl. **50**, 102419 (2020)
6. Geetha, R., Thilagam, T.: A review on the effectiveness of machine learning and deep learning algorithms for cyber security. Arch. Comput. Meth. Eng. 1–19 (2020)
7. González, C.: Desafíos de seguridad en redes 5g. Technol. Inside CPIC **3**, 36–45 (2019)
8. Goodfellow, I., Bengio, Y., Courville, A., Bengio, Y.: Deep Learning, vol. 1. MIT Press, Cambridge (2016)
9. Grigorescu, S., Trasnea, B., Cocias, T., Macesanu, G.: A survey of deep learning techniques for autonomous driving. J. Field Robot. **37**(3), 362–386 (2020)
10. Ling, X., Ji, S., Zou, J., Wang, J., Wu, C., Li, B., Wang, T.: Deepsec: a uniform platform for security analysis of deep learning model. In: 2019 IEEE Symposium on Security and Privacy (SP), pp. 673–690. IEEE (2019)
11. Lv, Z., Qiao, L., Li, J., Song, H.: Deep learning enabled security issues in the Internet of Things. IEEE Internet Things J. **8**(12), 9531–9538 (2020)
12. Mahdavifar, S., Ghorbani, A.A.: Application of deep learning to cybersecurity: a survey. Neurocomputing **347**, 149–176 (2019)
13. Ponti, M.A., Ribeiro, L.S.F., Nazare, T.S., Bui, T., Collomosse, J.: Everything you wanted to know about deep learning for computer vision but were afraid to ask. In: 2017 30th SIBGRAPI Conference on Graphics, Patterns and Images Tutorials (SIBGRAPI-T), pp. 17–41. IEEE (2017)
14. Salguero Dorokhin, É., Fuertes, W., Lascano, E.: On the development of an optimal structure of tree parity machine for the establishment of a cryptographic key. Secur. Commun. Netw. **2019** (2019)
15. Salloum, S.A., Alshurideh, M., Elnagar, A., Shaalan, K.: Machine learning and deep learning techniques for cybersecurity: a review. In: Hassanien, A.E., Azar, A., Gaber, T., Oliva, D., Tolba, F. (eds.) AICV 2020. AISC, vol. 1153, pp. 50–57. Springer, Cham (2020). https://doi.org/10.1007/978-3-030-44289-7_5
16. Singla, A., Bertino, E.: How deep learning is making information security more intelligent. IEEE Secur. Priv. **17**(3), 56–65 (2019)
17. Truong, T.C., Huynh, T.P., Zelinka, I.: Applications of swarm intelligence algorithms countering the cyber threats. In: Proceedings of the 2020 Genetic and Evolutionary Computation Conference Companion, pp. 1476–1485 (2020)
18. Xin, Y., et al.: Machine learning and deep learning methods for cybersecurity. IEEE Access **6**, 35365–35381 (2018)

19. Yashiro, R., Machida, T., Iwamoto, M., Sakiyama, K.: Deep-learning-based security evaluation on authentication systems using arbiter PUF and its variants. In: Ogawa, K., Yoshioka, K. (eds.) Advances in Information and Computer Security. IWSEC 2016. LNCS, vol. 9836, pp. 267–285. Springer, Cham (2016). https://doi.org/10.1007/978-3-319-44524-3_16
20. Zambrano, P., et al.: Technical mapping of the grooming anatomy using machine learning paradigms: an information security approach. IEEE Access **7**, 142129–142146 (2019)

MESH Networks to Optimize the Quality of Internet Service via WiFi in University Institutions

Rafael Sánchez-Pinargote[1](✉), Miguel Joseph Rodríguez Véliz[2], and Emilio Cedeño-Palma[2]

[1] Instituto de Postgrado, Universidad Técnica de Manabí, Portoviejo, Ecuador
rafael.sanchez@utm.edu.ec
[2] Facultad de Ciencias Informáticas, Universidad Técnica de Manabí, Portoviejo, Ecuador
{miguel.rodriguez,emilio.cedeno}@utm.edu.ec

Abstract. Nowadays the need to be connected through the internet in a more reliable and faster way is indispensable, Mesh networks are a great connectivity solution because with its configurable and self-regenerating mesh technology it can be implemented in urban and rural areas and university campuses offering favorable results to thousands of users. In this article we propose the design of a Mesh network to optimize the quality of internet service via WIFI in university educational facilities, for this we took as a model the Faculty of Computer Science of the Technical University of Manabi, taking as initial data the operation of the current network, in order to perform the design of the Mesh network and then the simulation of the same, for this we used the simulation tool Riverbed Modeler Academic Edition 17.5, the same one that allows to show viable results by means of modules and graphs for a future implementation benefiting teaching and administrative students of this faculty.

Keywords: Red Mesh · Riverbed Modeler Academic Edition 17.5 · Router

1 Introduction

1.1 Mesh Networking Approach

Wireless networks have undergone an important evolution in terms of mobility, more and more mobile devices are demanding this type of resources and tools, without the need for Ethernet cabling and physical connections between nodes, the user has the possibility to stay connected and avoid annoying network hopping.

Mesh networks are considered a promising solution to provide ubiquitous wireless Internet access, also an optimal response to the evolution of WIFI networks, because its wide wireless coverage, without signal loss and its technology composed of a Router and its satellites or access points that communicate with each other, ensure good resource management in the network and that the signal is not lost, even if a node stops working or if the user moves at any point or distance.

M. Botto-Tobar et al. (Eds.): CIT 2021, LNNS 405, pp. 255–267, 2022.
https://doi.org/10.1007/978-3-030-96043-8_20

A Mesh network is able to redirect the traffic through the network always in the optimal way to have the best possible signal. Wifi Mesh networks calculate which node/satellite is best to connect to at any given moment according to the status of other nodes, the connected devices, the distance to each of the satellites, signal strength and many other factors, in a completely transparent way to the user, who does not have to worry about which node he/she is connected to.

This article proposes to design a wireless network with Mesh technology, which guarantees the connectivity of different mobile devices, regardless of where they are located, with this is intended to provide a quality connectivity service at any point of the facilities of the Faculty of Computer Science: classrooms, laboratories, teachers' offices and administrative areas, without losing connectivity in case of displacement.

For this purpose, using the simulation tool Riverbed Modeler Academic Edition 17.5, I will compare the efficiency of the star network that currently exists in the Faculty of Computer Science with the proposed Mesh Network and determine the efficiency of the network through parameters such as delay, performance and load.

2 Materials and Methods

2.1 Mesh Wireless Network

Mesh networks, coupled networks, or infrastructure wireless mesh networks, unite the two topologies of wireless networks: the ad-hoc topology and the infrastructure topology. These networks self-organize and self-configure dynamically with the network nodes, automatically establishing an ad-hoc network and maintaining the connection. WMNs (Wireless Mesh Networks) consist of two types of nodes: mesh routers and Mesh clients.

In addition to the functions of a conventional wireless router, the mesh router contains additional functions to support the mesh infrastructure. Thanks to the multi-hop communications system, the same coverage can be achieved with less transmission power. Mesh routers make it possible to link devices to the network that are outside the coverage range of access points (APs) but are within the coverage range of any network card that is directly or indirectly within the coverage range of an access point in the network.

To enhance the flexibility of the mesh network, a mesh router contains multiple wireless interfaces based on the wireless technology itself. Despite these small differences, mesh routers are built on hardware similar to that of any router. Mesh routers have limited mobility and form the backbone of the network [4] (Fig. 1).

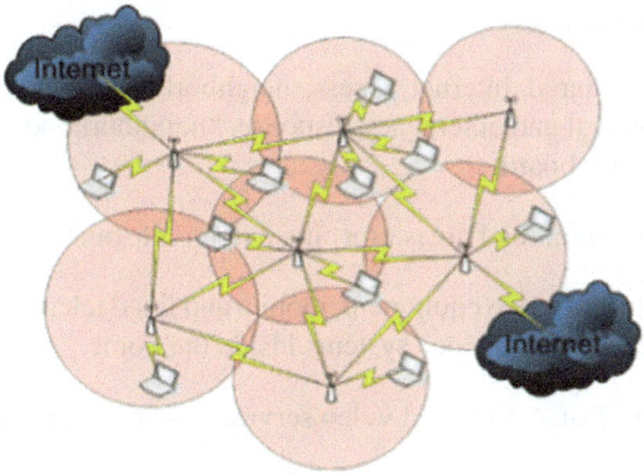

Fig. 1. Mesh network architecture. **Source.** Own development.

2.2 Features

Low Cost. Due to the use of a mature technology and the existence of the WiFi standard.

Decentralized Structure. In isolated rural areas, where much of the time can be spent on travel.

Power Supply and Integration. Mesh nodes need only minimal power, providing them with autonomy with natural energy such as solar power.

Flexibility. A node can join the network if it can see one of the neighboring nodes.

Redundancy. The devices that form the Mesh networks communicate with their peers through several routes, which allow connection to any destination by means of different operating nodes implemented in the network.

Installation. To install a Mesh Network it is only necessary to install the node with the mesh software pre-installed, as the network route is dynamic, when this node automatically locates a neighboring node, it puts it at the service of the network [6].

WMN Security. Security management in a WMN is one of the most important factors from the administration point of view. Some of the services to be protected are:

Confidentiality. Data are transmitted only to the entities or persons concerned.

Legitimacy. An entity has in fact the identity it claims to have, i.e. recognition of the users who own the service. Access control: ensures that only authorized actions can be performed by users registered in the system.

Non-repudiation. Protects the entities participating in an information exchange.

Availability. Ensures that network resources are accessible at all times to authorized users or entities [3].

2.3 Application Scenarios

Community Mesh. Shared Internet access, neighborhood surveillance and security (video cameras), medical and emergency response, "neighborhood eBay" (sale and flea market), virtual bulletin board.

Municipal Mesh. Commercial areas of a city, citizen information points and tourism.

Home Mesh. Audio and video equipment, mobile and fixed telephones, laptops, palms, smart switches, ambient intelligence systems inside the home.

Spontaneous Mesh. Data, voice and video services, peer-to-peer calls.

Campus Mesh. Universities or educational institutions [2].

2.4 Architecture

Infrastructure WMN. Connects a wired network to wireless devices through a base station, called an access point. The access point links the wireless network to the wired network and is used as the central controller of the wireless network, and coordinates the interconnection of wireless devices;

Client WMN. These are made up of client mesh devices, which behave as Ad-hoc or peer to peer connections with other client equipment in the mesh network, configured as an independent local area network that is not connected to a wired infrastructure; all stations are linked to each other, i.e. the network is independent.

Hybrid WMN. This topology combines the Ad Hoc network and the infrastructure topology. A hybrid WMN consists of a mesh router that forms the network structure. In addition, wireless clients can dynamically connect to the mesh network providing functions such as data packet routing [1].

In order to obtain the expected results, a study of the current network was conducted to verify its operation and shortcomings. Regarding the design and testing of the Mesh network, the design and simulation software Riverbed Modeler Academic Edition 17.5 was used, which allowed modeling and verifying the operation of the network in its future implementation.

3 Mesh Network Implementation

3.1 Current Network

The Faculty of Computer Science is formed by a star topology, with a single-mode fiber optic laying with the capacity to transmit at Gigabit 1000BaseT speeds, through its 11 servers that are located in the Data Center is responsible for technologically interconnecting all computer services to the faculty and various departments of the Technical University of Manabi.

To offer wireless internet service the Faculty has implemented 18 wireless access points (AP) in star topology, which allow data transmission between the devices connected to the wired network and wireless devices (Laptop, Tablets, Smartphones, etc.) (Fig. 2).

These AP devices are strategically located throughout the building, 8 on the first floor, 6 on the 2nd floor, and 4 on the 3rd floor; they have a coverage that reaches the entire building, but with the limitation that only allows connecting up to 20 users per AP.

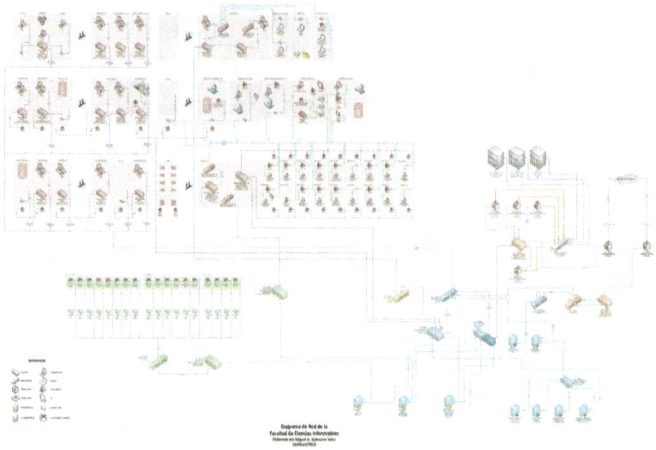

Fig. 2. Computer Science Faculty (FCI) network diagram. **Source.** Own development.

3.2 Mesh Network Design Proposal

The design of the Mesh network in the Faculty of Informatics consists of deploying a wireless mesh network, which is composed of access points (wireless network devices) to provide Internet access to all network users.

At the time of the elaboration of the design, the work is based on the documents and research previously done. Parallel work is done to identify the area in which the network will be designed along with the familiarization of the tools and modules offered by the Riverbed Modeler Academic Edition 17.5 program for the simulation of the design (Fig. 3).

Fig. 3. Area where the Mesh design will be developed. **Source.** Own development.

The physical and logical design of the network is carried out on the chosen tool. Figure 4 shows the distribution of the nodes in the building of the Faculty of Computer Science, which are distributed as follows:

- Router A and Router B: Floor 3.
- Router C and Router D: Floor 2
- Router E and Router F: Floor 1
- Router G and Router H: Courtyard.

Each node is located at a distance of 100 m.

Fig. 4. Physical design of the Mesh network at FCI. **Source.** Own development.

Similarly, Fig. 5 shows the logical design of the Mesh network, which is based on the space where it will be implemented in the future.

It consists of 3 parts:

- **Data Center.** It is constituted at the head of the network, where all the network administration and support equipment is located.
- **Nodes.** The nodes are the routers that are responsible for providing interconnectivity to all network users.
- **Users.** Finally, there are the users of the network, which in the simulation are represented under the subnetwork tool.

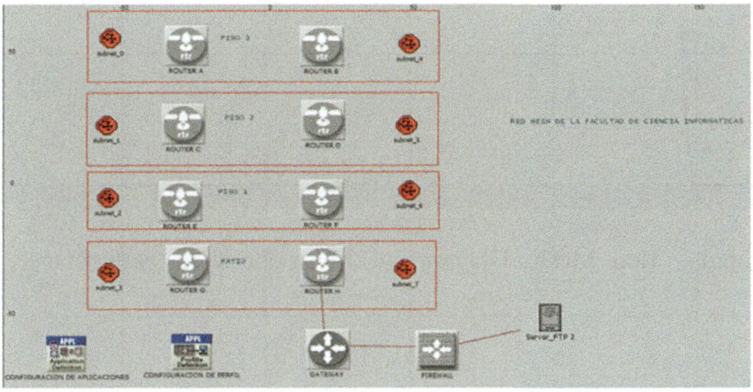

Fig. 5. Logical model - Riverbed Modeler Academic 17.5. **Source.** Own development.

3.3 Testing

Once the design is done, we proceed to load several parameters to the routers, Gateway, servers and application modules and profiles, it is also necessary to give parameters to the equipment that will access the environment in this case the PCs. All this will allow the tool to provide at the end graphs in which the behavior of the network can be evidenced.

In this case the project will evaluate and focus on measuring the behavior of the network in terms of performance, load and packets sent and received, as well as evaluating these 3 parameters with the current network of the Faculty of Computer Science.

4 Results

The results of the simulated scenarios were performed in the Riverbed Modeler Academic Edition 17.5 tool, which measured the parameters of performance, load and packets sent and received, these results were obtained in graphical form with data provided by the simulation tool.

4.1 Mesh Network Scenario (Throughput)

Figure 6 shows the throughput in bit/s in which the X axis shows the time in seconds and the Y axis shows the throughput, the graph shows how it grows to 3800 bits/s and after a few seconds the graph drops and stabilizes at 3400 bit/s.

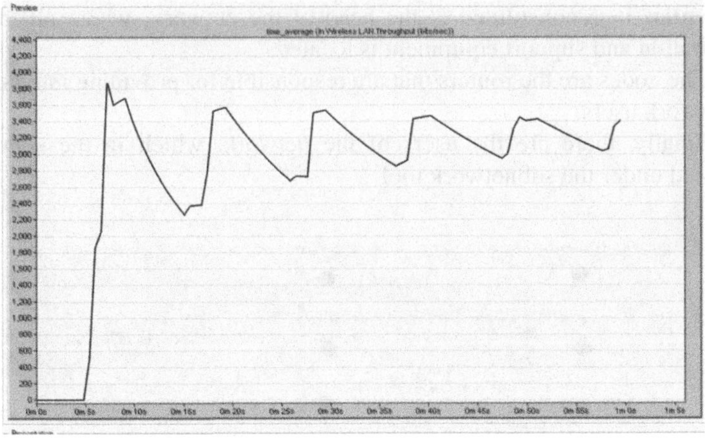

Fig. 6. Throughput - Red Mesh. Diagram of Mesh Network performance. **Source.** Own development.

4.1.1 Delay

Figure 7 indicates the network delay where the X-axis indicates the time in seconds and the Y-axis the delay, the graph reveals that the minimum value 0.00007 ms and the maximum value is 0.023 ms.

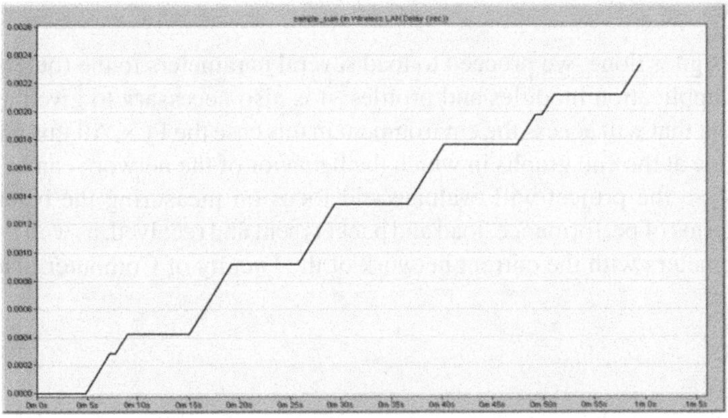

Fig. 7. Delay - Red Mesh. Mesh Network delay diagram. **Source.** Own development.

4.1.2 Packets Sent and Received

Figure 8 shows the transfer of files sent and received on the network where X provides the time in seconds and the Y axis the packets per second, the graph indicates a value of packets sent of 0.19 packets/sec and packets received of 0.09 packets/sec.

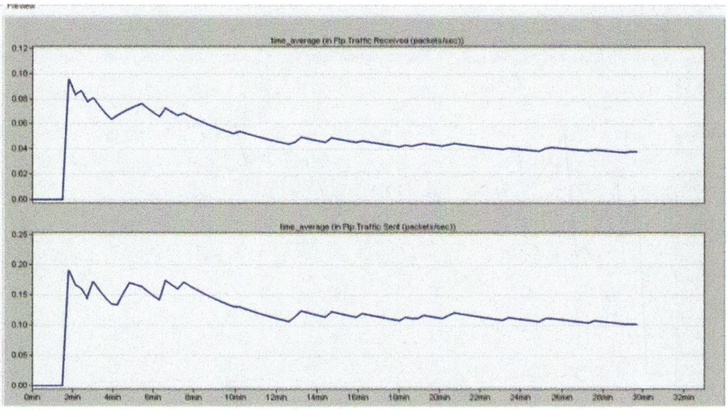

Fig. 8. Packets sent and received - Mesh Network. The upper part shows the packets sent and the lower part shows the packets received. **Source.** Own development.

4.2 Mesh Network vs. Current Network Scenario

To show the difference between the current network of the Faculty of Computer Science and the Mesh network to be designed, the simulation tool was used to measure the performance, load and delay parameters of the two networks.

4.2.1 Throughput

In Fig. 9, the graph shows the Mesh network throughput of 24,000 bit/s, while the actual network shows a maximum value of 15,000 bit/s.

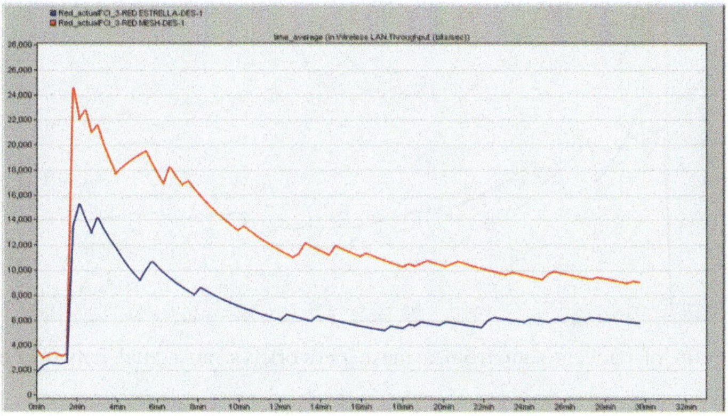

Fig. 9. Throughput-Red Mesh vs Red Actual. Diagram showing the performance of the mesh network vs. the current network. **Source.** Own development.

4.2.2 Delay

Figure 10 indicates in the graph an increase in the delay of the current network of 0.0084 ms as opposed to the Mesh network which shows a maximum value of 0.0007 ms.

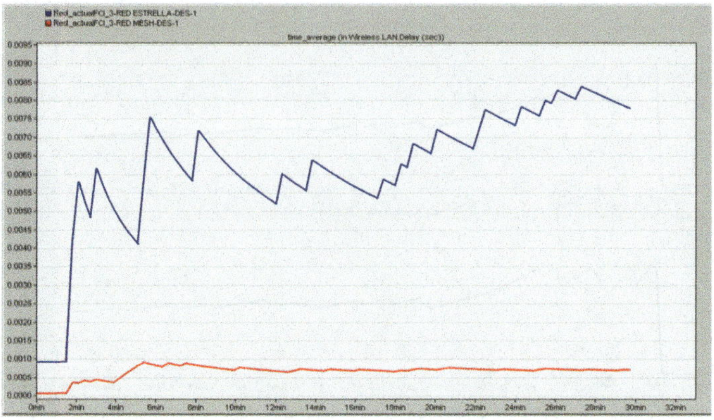

Fig. 10. Delay- Red Mesh vs Red Actual. Diagram detailing the delay of the mesh network vs. the current network. **Source.** Own development.

4.2.3 Packets Sent

Figure 11 shows an increase in file transfer sent from the Mesh network of 0.19 packets/sec as opposed to the current network which shows a maximum value of 0.11 packets/sec.

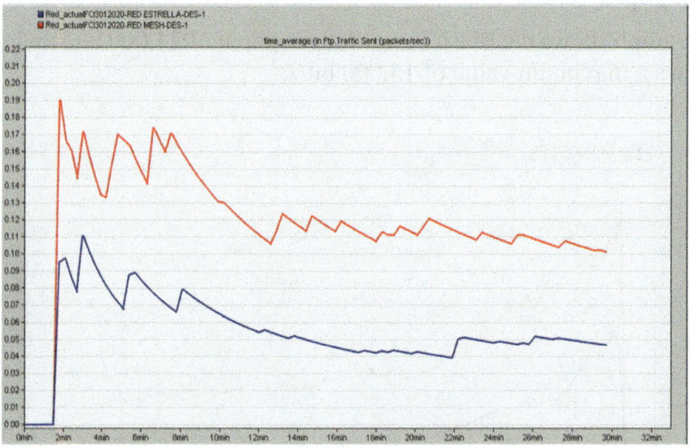

Fig. 11. Diagram of packets sent from a mesh network vs. an actual network. **Source.** Own development.

4.2.4 Packets Received

Figure 12 evidences an increase in file transfer received from the Mesh network of 0.10 packets/sec while the actual network shows a value of 0.05 packets/sec (Table 1).

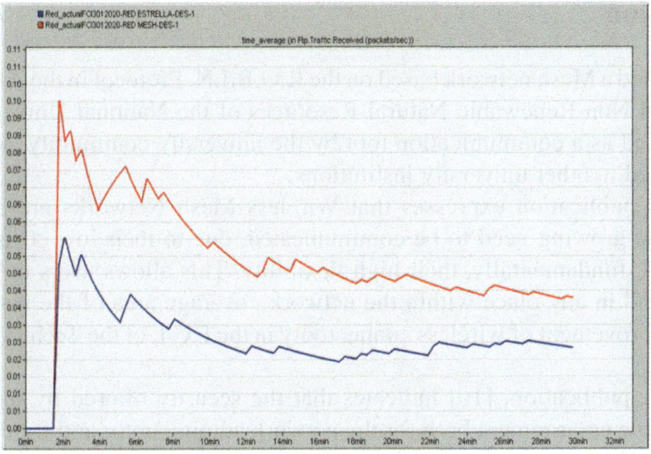

Fig. 12. Plot of packets received from the mesh network vs. the actual network. **Source.** Own development.

Table 1. Table of simulation parameters. **Source.** Own development.

Parámetros	Red Mesh		Comparación Red Mesh vs Red Actual	
	Valor mínimo	Valor máximo	Red Mesh	Red Actual
Throughput	3400 bit/s	3800 bit/s	24.000 bit/s	15.000 bit/s
Delay	0.00007 ms	0.023 ms	0.0084 ms	0.0007 ms
Paquetes enviados	0.19 packets/sec		0.19 packets/sec	0.11 packets/sec
Paquetes recibidos	0.09 packets/sec		0.10 packets/sec	0.05 packets/sec

4.3 Analysis of Results

From the results, it was observed in the simulation graphs acceptable values for the performance, delay sent and received packets of the Mesh network.

On the other hand, when comparing the results of the current network of the Faculty of Computer Science, it was possible to recognize remarkable values in the graphs; in what has to do with the performance of the current network, lower values are shown, unlike the Mesh network that indicates higher performance values; similarly in the delay, higher delay values are seen in the current network and lower delay values in the Mesh network, in what concerns the file transfers sent and received, an increase in the Mesh network is evident in the sent packets and a minimum loss of packets in relation to the current network.

5 Discussion

[5] implemented a Mesh network based on the R.O.B.I.N. Protocol in the Area of Energy, Industries and Non-Renewable Natural Resources of the National University of Loja, which was used as a communication tool by the university community, so this type of network is used in other university institutions.

[8] in his publication expresses that Wireless Mesh Networks are an innovative solution to the growing need to be communicated, due to their low costs, robustness, reliability and, fundamentally, their high flexibility. This allows users to be connected at any time and in any place within the network coverage area of the institution. This allows the improvement of wireless connectivity in the F.C.I. of the Technical University of Manabi.

In another publication, [10] indicates that the security offered by mesh networks allows a reliable environment both for the network administrator and for each user who intends to use the infrastructure. On the other hand [9] concludes in his research that with the implementation a greater control of the network is obtained since it is possible to manage the communication devices both the mikrotik routerboard and the ubiquiti APs which did not exist before, through these investigations it is possible to achieve that the implementation of a mesh network is proactive for the university institutions.

In addition [7], indicates that the wireless network equipment should be installed in strategic locations, achieving greater coverage throughout the career of Computer Engineering and Networking.

6 Conclusion

Mesh technology is modern, robust and flexible, because they are self-healing and self-configurable. In this way, Mesh wireless networks can develop greater potential in terms of services offered and performance.

With the design and simulation of the Mesh network in the Faculty of Computer Science, it was found that each node can be located 100 m away and that if a node stops working the network will continue to operate smoothly.

In addition, it was verified that device redundancy can be eliminated.

The results obtained showed that the performance of the network can be improved by using Mesh technology, as well as the transfer of files.

References

1. Buettrich, S., Escudero, A.: Topology and basic infrastructure of wireless networks. Tricalcar, Peru (2017). http://www.itrainonline.org/itrainonline/mmtk/wireless_en/04_Infrastructure_Topology/04_en_mmtk_wireless_basic-infrastructure-topology_slides.pdf
2. Cañarte, M., Parra, D.: Study and design of an access node to serve as a pilot for the implementation of a wireless mesh network in the faculty of electrical and computer engineering of the Espol. Espol, 1–7 (2010). https://www.dspace.espol.edu.ec/handle/123456789/10589

3. Cruz, V., Collazos, J.: Formulation of a Methodology to Design and Implement MESH networks as an alternative solution for community or rural networks. Support Project; Construction of a technological scheme for routing protocols in MESH networks (2016). https://repository.unilibre.edu.co/bitstream/handle/10901/10815/Tesis-Final.pdf?sequence=1&isAllowed=y
4. Gómez, B., Maimó, J., Merideño, J.: Wireless Mesh Networks. Enginyeria Informática Universitat de les Illes Balear, 9–12 (2010). https://edicions.uib.cat/ojs/index.php/enginy/article/view/48/30
5. Ochoa, J.: Design and Implementation of a Wireless Network Pilot for the center of the city of Loja, based on MESH Technology, Cuenca, Ecuador (2013). http://dspace.ucuenca.edu.ec/bitstream/123456789/3385/1/tesis.pdf
6. Sánchez, J.: Implementation of a wireless network for academic reinforcement in the classrooms of the computer and network engineering career, Jipijapa, Manabí, Ecuador (2020)
7. Simó, F., Osuna, P., Seoane, J., Martínez, A.: Development of autonomous Wi-Fi Mesh nodes to support rural telemedicine networks in isolated areas. BURJC DIGITAL, 1–4 (2009) https://burjcdigital.urjc.es/bitstream/handle/10115/2348/Desarrollo%20de%20nodos%20Mesh%20Wi-Fi%20aut%EF%BF%BDnomos%20como%20soporte%20a%20redes%20de%20telemedicina%20rural%20en%20zonas%20aisladas.pdf;jsessionid=B4E68DC24DC2CC96A0362408D4D1EC87?sequence=1
8. Vásconez, D., Manzano, V.: Wireless Mesh Network (WNM) standard 802.11 transmission and coverage optimization in schools in the province of Tungurahua, Ambato, Ecuador (2014). https://repositorio.uta.edu.ec/bitstream/123456789/6989/1/Tesis_t873mrt.pdf
9. Vásquez, D.: Design and implementation of a wireless network with the IEEE 802.11ac standard with quality of service and security for the administration of the communication service of a car sales company, based on mesh technology with ubiquiti and m equipment, Guayaquil, Ecuador (2019). http://repositorio.ug.edu.ec/bitstream/redug/40125/1/B-CINT-PTG-N.423%20V%C3%A1squez%20Flores%20Danny%20Rodolfo.pdf
10. Velasco, M.: Design of a WISP on the campus of the Universidad Técnica del Norte to provide wireless internet services using a wireless mesh scheme, Quito, Ecuador (2009). https://bibdigital.epn.edu.ec/handle/15000/4186?mode=full

Throughput and Latency Evaluation of a 4G LTE Network Driven by SDR and Open-Source LTE Software

Fernando Lara[1,3](✉), Daniel Altamirano[1,2], Jorge Arellano[1,3], and Miguel Castillo[1,3]

[1] WiCOM-Energy Research Group, Universidad de las Fuerzas Armadas ESPE, 171103 Sanqolquí, Ecuador
{mflara3,cdaltamirano,jfarellano,macastillo8}@espe.edu.ec

[2] Department of Electrical, Electronics and Telecommunications, Universidad de las Fuerzas Armadas ESPE, 171103 Sanqolquí, Ecuador

[3] Maestría de Investigación en Electrónica mención Telecomunicaciones, Universidad de las Fuerzas Armadas ESPE, 171103 Sanqolquí, Ecuador

Abstract. The use of software defined radio (SDR) on the deployment of wireless communication standards is widely investigated for its low cost in the implementation by using the same hardware and fourth-generation (4G) long-term-evolution (LTE) networks are also part of this new communications paradigm. Furthermore, 4G networks enabled by SDR allows practical learning in a real environments that facilitate the development of technical skills without the inherent difficulty and cost associated with radio frequency components and regulation. In this paper, a 4G LTE network driven by SDR hardware and software-radio-system-LTE (srsLTE) that is an open-source LTE software platform is evaluated in terms of the throughput and latency. Besides, three different scenarios are implemented in order to evaluate the 4G LTE network by traffic injection through Iperf3 application and calculating internet throughput using the Fast web tool. The results show that the reached measurements are near to the theoretical parameters of 4G LTE networks. Thus, the implemented network is can be used as a testbed and is useful for teaching.

Keywords: 4G LTE · Software Defined Radio (SDR) · LTE open-source network · Traffic injector

1 Introduction

Mobile communications have been and will be the center of attention for innovation and digital transformation, its great development allows access to information in a massive way and in real time. This means that there is a continuous research for innovation of new services and architectures that support mobile networks.

M. Botto-Tobar et al. (Eds.): CIT 2021, LNNS 405, pp. 268–280, 2022.
https://doi.org/10.1007/978-3-030-96043-8_21

Fourth-generation (4G) long-term-evolution (LTE) networks are currently expanded worldwide. Therefore, professionals with technical skills of this standard are necessary at the industry. However, there is an inherent difficulty to access the 4G LTE networks associated with radio frequency components, the system model and regulations. At this point, the use of software defined radio (SDR) hardware that provides greater flexibility appears as an option to implement 4G LTE mobile networks and their possible applications [8].

Today, several mobile wireless networks are implemented by using SDR hardware, which allows them to make configurations based on requirements, while maintaining the same hardware [12]. Indeed, the use of SDR technology for 4G LTE networks has given satisfactory results for the implementation of the radio frequency layer in the eNodeB (eNB) [1,15,20].

The literature establishes major developments of 4G LTE networks based on open-source LTE software platforms, such as OpenAirlnterface (OAI), OpenLTE and srsLTE [16,22]. The latter is not finished yet, so it is in constant innovation and development, but the positive thing about this software is that there is a lot of information for its implementation, the code structure is easy to modify and customise [9,21].

In [10], the use of SDR hardware is proposed to measure the main radio frequency parameters in 4G LTE networks; in [19], SDR is used to determine how vulnerable LTE networks are to interference as a denial of service (DoS) technique. Finally, in [3], the use of parallel programming and the use of SDR are proposed to use a personal computer as the core of a 4G LTE network.

Once 4G LTE networks are based on IPv4, evaluates the performance through measurements of metrics such as throughput and latency are necessary in order to understand the theoretical parameters of the standard [2,13]. There are several tools to perform this performance measures such as Iperf, Iperf3, distributed internet traffic generator (D-ITG), among others. In [18], the authors introduces the constant rate ultra short probing (CRUSP), that is a method to estimate available throughput in 4G LTE networks in the controlled LTE test cell, which consumes only a fraction of the data volume (less than 2 MB) and time (less than 50 ms), in comparison with conventional applications. Moreover, in [17] a test with CRUSP in live 4G LTE cells is performed and compared with Iperf3 traffic injection using the Kolmogorov-Smirnov test, finding similar results in performance measures.

This paper propound the implementation of a 4G LTE network driven by software-radio-system-LTE (srsLTE) that is an open-source LTE software platform [21] and SDR hardware of Ettus Research. The SDR hardware is the universal-software-radio-peripherical (USRP) X310 and the daughterboard CBX-120. By using this platform, the 4G LTE network is evaluated in terms of the throughput and latency. To this aim, three different scenarios are proposed and traffic injection through Iperf3 application [6] is used for calculating the throughput and latency. Additionally, Fast web tool [7] is used and comparisons with the 4G LTE standard were made.

This paper is organized as follows. Section 2 presents the implementation of the 4G LTE network. Section 3 presents the methodology for evaluating the implemented 4G LTE network. The results and discussion are presented in Sect. 4. Finally, Sect. 5 drawn the conclusions.

2 Network Implementation

In this section, we describe the implementation of a 4G LTE network driven by SDR and open-source LTE software. Figure 1 shows the implemented LTE network, whose hardware and software are detailed below.

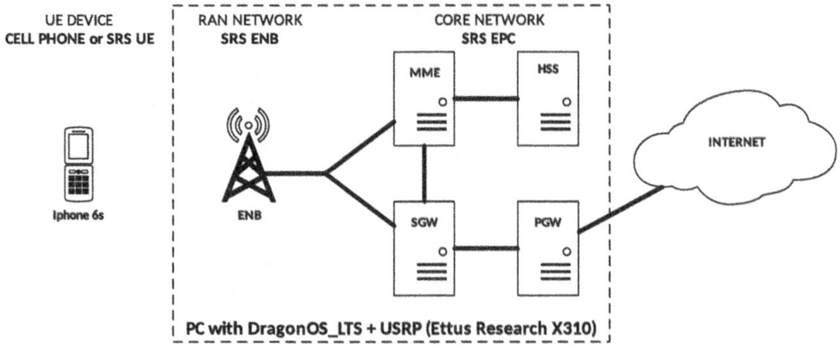

Fig. 1. 4G LTE network.

2.1 Hardware

In general, the hardware of a SDR platform consist of a general purpose processor (GPP), a digital signal processor (DSP) and a field-programmable gate array (FPGA) [14]. Depending on their capabilities and prices, there are several SDR commercial platforms. For the present work, we have used the Ettus Research USRP X310, due to its size, mobility and low energy consumption. Furthermore, this USRP works in conjunction with the daughterboard CBX-120, which is a full-duplex wideband transceiver.

The USRP X310 is used for the 4G srsLTE implementation, which communicates with a host (computer) via ethernet connection. Tables 1 and 2 summarizes the main features of the Ettus Research USRP X310 and Daughterboard CBX-120.

Regarding to the user equipment (UE), a smartphone iPhone 6S of the brand Apple was used, which operates on the 4G LTE bands. For the network implementation, the UE must use a universal subscriber identity module (USIM), in this implementation, the Anritsu W-CDMA/GSM Test USIM model P0035B was used [4].

Table 1. USRP X310 features.

Parameter	Value
Brand	Ettus
DC input	12 V
Output power	>10 dBm
FPGA model	Kintex 7 -410T
Interfaces	1G/10G ETH, PCIe x4
Host connectivity	Ethernet

Table 2. Daughterboard CBX-120 features.

Parameter	Value
Brand	Ettus
Frequency band	From 1.2 GHz to 6 GHz
Instantaneous bandwidth	120 MHz
Compatibility	Support applications with USRP X300/310

2.2 Software

There are several LTE open-source software alternatives [9]. In this work the srsLTE is used, due to its simple code, open-source library based on C and the plenty documentation as mentioned in the introduction. It should be noted that the eNB and the evolved-packet-core (EPC) are implemented by using srsLTE [21]. The UE can be also implemented using srsLTE. However, another SDR hardware is necessary; since the hardware cost is higher than a cell phone cost, we have used a cell phone.

The srsLTE is running in a single computer with DragonOS_LTS which is an out of the box of Lubuntu 18.04 [5]. Once the LTE network is deployed and operational, the corresponding software for EPC and eNB are then executed.

As a first step, the EPC and eNB applications are running onto two different terminals, using the commands:

- sudo srsepc/route_instalation
- sudo srsenb/route_instalation

Then, in order to generate the network route between the serving-gateway (S-GW) and the internet, the script *srsepc_if_masq* is used.

Once the srsenb software that drives the eNB is running, it connects to the EPC network, initiating the transmission between the mobile base station (BS) and the UE. After that, the UE is registered on the network and to check connectivity, a ping can be made between the UE and the IP address of the S-GW in EPC. At this point, a 4G LTE network is running, in our work we have called this network as LTE ESPE network.

For the evaluation LTE ESPE network, Iperf3 software is used, this application is installed in both the UE and the EPC. For the EPC is installed through the command *sudo apt install iperf3*, and for the UE using the Magic Iperf application.

Finally, for the server listening, the command *iperf3 -s* is executed and for the client to start the traffic injection, the command *iperf3 -c* is used; the Magic Iperf application in the iPhone 6S is used for the traffic injection in the upload and download, as shown in Fig. 3.

2.3 4G LTE Standard Parameters

The current srsLTE version implements a 4g LTE networks based on the third-generation-partnership-project (3gpp) release 14. Among the standard characteristic, we implement a single input single output (SISO) 4G LTE network using frequency division duplexing (FDD), on band 7 (i.e. a carrier frequency of 2600 MHz), with carrier bandwidth of 10M Hz, adaptive quadrature amplitude modulation (QAM) with 16 and 64 symbols for the downlink and quadrature phase shift keying (QPSK) for the uplink. Table 3 shows the theoretical throughput in the downlink and uplink [11].

Table 3. Theoretical information of the implemented 4G LTE Network.

Link	Modulation	Throughput
Uplink	QPSK	12.5 Mbits/s
Downlink	16-QAM	25 Mbits/s
Downlink	64-QAM	37.5 Mbits/s

3 Methodology

In this section, we present three scenarios for evaluating the network performance. The evaluated parameters are the throughput and latency by considering traffic injection. Furthermore, the measurements are performed using iPerf3 and the Fast web tool.

3.1 Scenario 1: UE and EPC

In this scenario, the link between the UE and EPC is analyzed, the uplink and downlink are measured. The Fig. 2 shows this scenario block diagram.

In this scenario, the Iperf3 server is hosted on the computer that contains the EPC, therefore, the throughput obtained in the different traffic injections is the throughput of the link between the UE and the EPC. An example of the traffic injection for the downlink and uplink are shown in Fig. 3a and Fig. 3b, respectively.

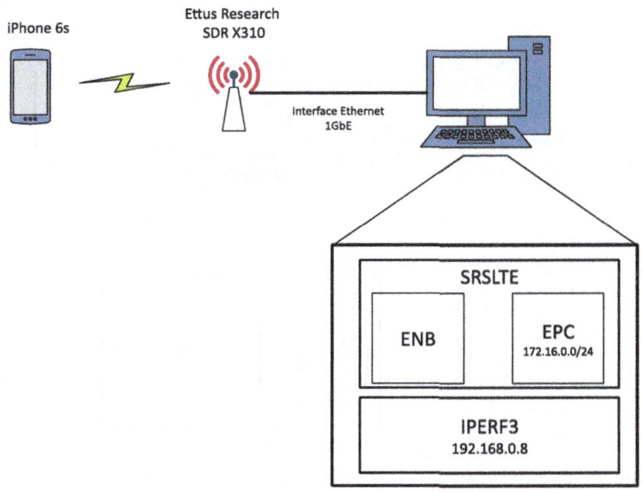

Fig. 2. Scenario 1: link between UE and EPC

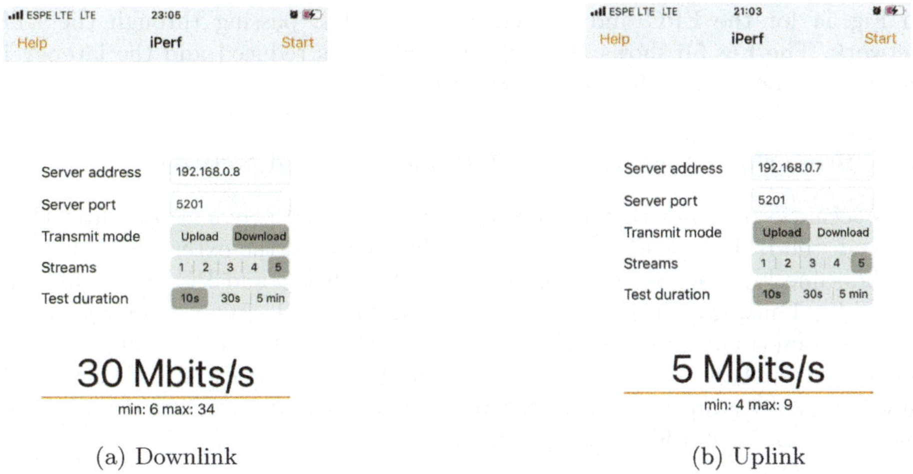

Fig. 3. Throughput between UE and EPC

3.2 Scenario 2: Link Between UE and EPC to the Internet

The second scenario is described in Fig. 4. This scenario proposes evaluating the internet upload and download throughput in the UE and the EPC. In this case, the network latency is also obtained.

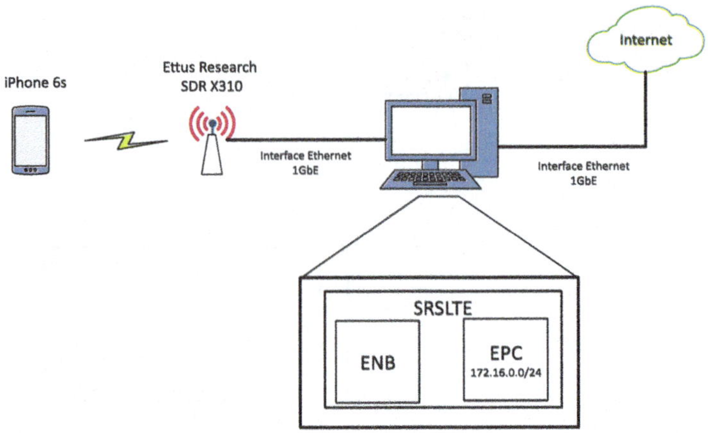

Fig. 4. Scenario 2: link between UE and EPC to the Internet

An example of the internet upload and download measurements are presented in Fig. 5a for the EPC and in Fig. 5b for the UE passing through the LTE network. The Fig. 5b shows that UE throughput is reduced and the latency is increased, since, the traffic goes through the EPC.

3.3 Scenario 3: Link Between UE and External Network

This scenario proposes that the UE reaches a external personal computer (PC) through the LTE network. Furthermore, the EPC and the external PC belongs to the same local access network (LAN), but both UE and PC are in different networks. Thus, the UE is reaching an external network. Figure 6 shows a block diagram describing this scenario. In this case the Iperf3 server is implemented on the external PC that is connected to the EPC through the Ethernet 1GbE interface. An example of the traffic injection for the downlink and uplink are shown in Fig. 7a and Fig. 7b, respectively.

4 Results and Discussion

In this section, we present the results of the obtained measures and compare with the theoretical values of the standard.

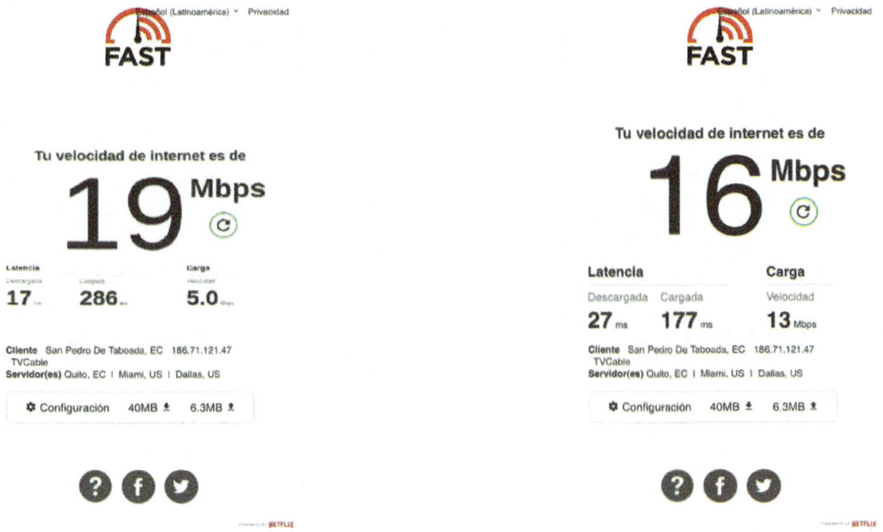

(a) EPC. (b) UE.

Fig. 5. Throughput and latency measurements from Internet.

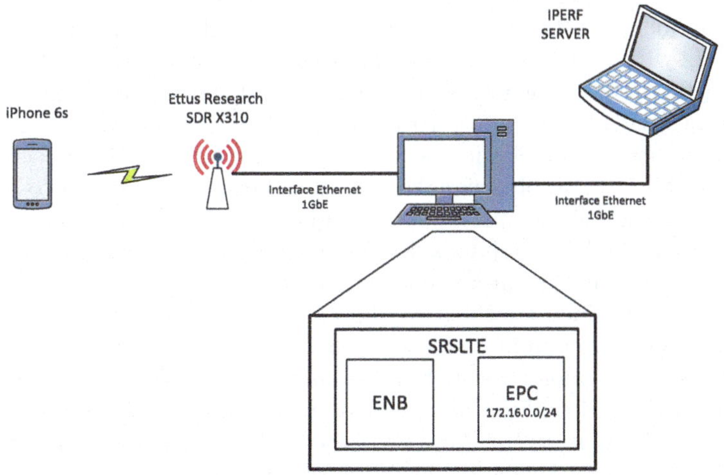

Fig. 6. Scenario 3: link between UE and external PC

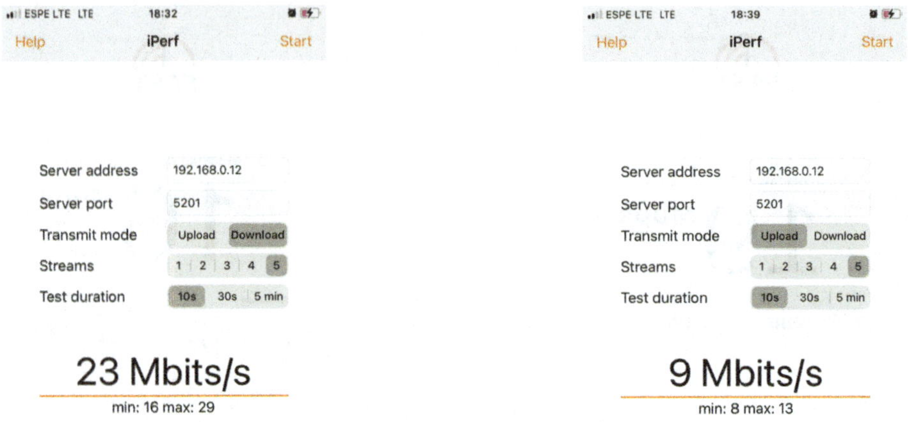

(a) Downlink (b) Uplink

Fig. 7. Throughput measurements between UE and External Network

Figure 8 shows the results of the downlink and uplink traffic injection in the first scenario, that is, the link between the UE and EPC, and vice versa. In the downlink the average throughput is 29 Mbits/s and in the uplink is 4.95 Mbits/s.

These results correspond to the use of a carrier frequency with bandwidth of 10 MHz, as was mentioned in Sect. 2.3. Furthermore, Fig. 9 shows the band 7 central frequency in the downlink, that is 2.685 GHz, the minimum frequency is 2.680 GHz with −116.34 dBm and the maximum frequency is 2.689 GHz with −115.43 dBm. Regarding to the resolution bandwidth (RBW) and the video bandwidth (VBW), they are of 10 MHz. RBW is the bandwidth of the intermediate frequency filter of the spectrum analyzer. If the RBW decreases, the selectivity (ability to resolve signals) increases because the signal-to-noise ratio increases, but the sweep time and the update rate of the trace degrades. On the other hand, the VBW allows smoothing the signal on the spectrum analyzer screen to discriminate low amplitude signals.

The results obtained for the second scenario are presented in Fig. 10, where the differences in throughput and latency are observed when measuring the Internet upload and download throughput from the UE and the EPC. Figure 10a, shows the Internet download and upload throughput from the UE, where the average throughput is 12.8 Mbits/s for the download and 6.1 Mbits/s for the upload. The average latency on downlink of 23.9 ms is also presented. On the other hand, in Fig. 10b, the Internet download and upload throughput from the EPC of the ESPE LTE network is presented, where the average downlink is 14.9 Mbits/s and the average uplink is 4.5 Mbits/s. The network latency is also shown Fig. 10. In this case, the average downlink latency between the UE and the Internet is 17.3 ms.

Figure 11 show the results of the third scenario. In this case, the UE inject traffic to an external PC. The average throughput for the downlink is 20 Mbits/s

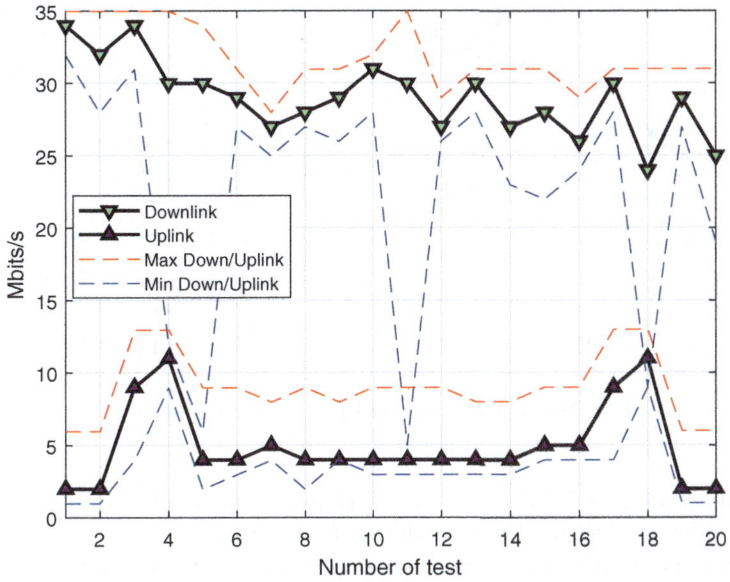

Fig. 8. Downlink and uplink throughput between UE and EPC.

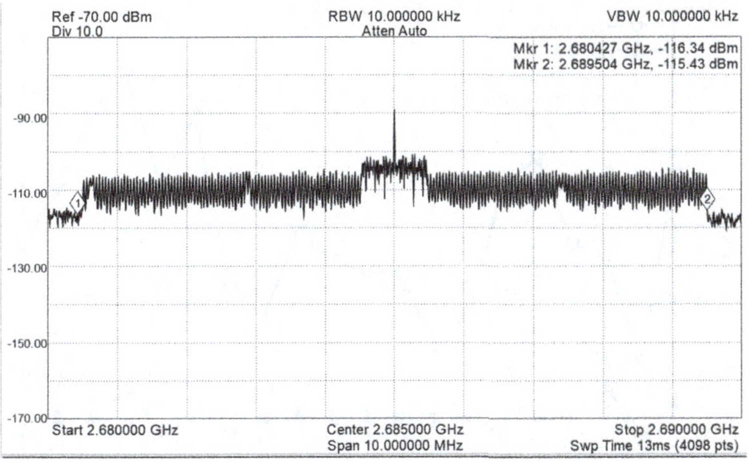

Fig. 9. ESPE LTE downlink spectrum in band 7.

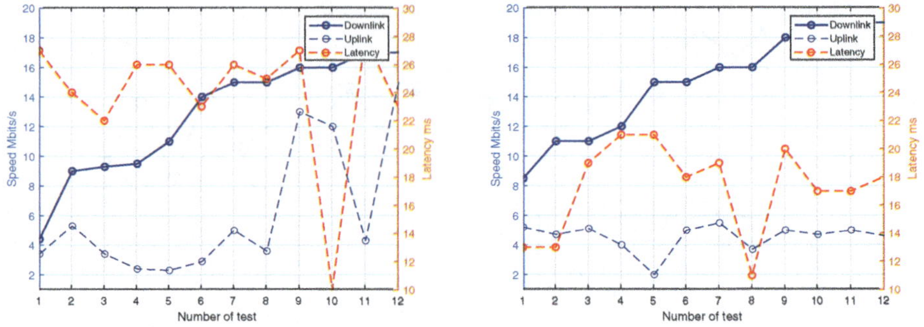

(a) UE (b) EPC

Fig. 10. Throughput and latency from Internet.

and for the uplink is 7 Mbits/s. Notice that the downlink throughput is lower than the downlink throughput of the first scenario, which is between UE and EPC, while the uplink throughput is higher compared to the first scenario. Once this scenario connects two networks, presents a greater variation between traffic injection measures than the first scenario.

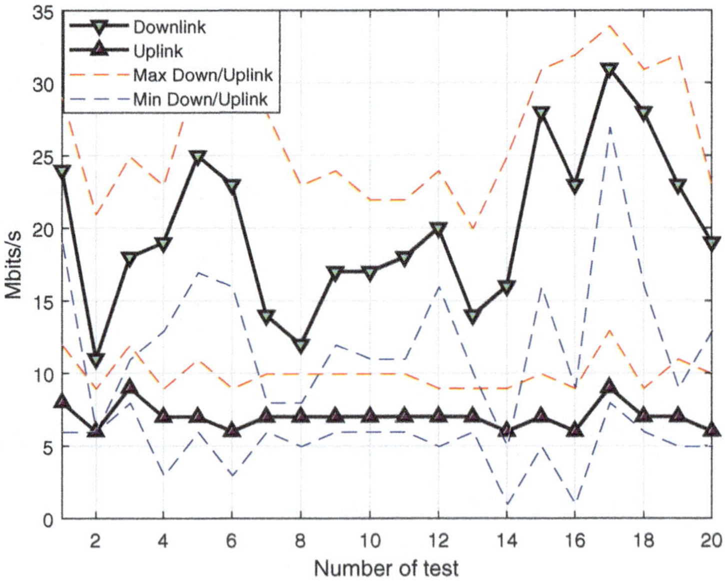

Fig. 11. Downlink and uplink rate between UE and External PC

5 Conclusions

A 4G LTE network was implemented by using SDR and open-source LTE software. Thus, the ESPE LTE network was deployed and can be used as a testbed and a platform for teaching the LTE standard.

The results in scenario 1 show that the normalized throughput of the implemented LTE ESPE network is 0.83 for the downlink and 0.39 for the uplink. The difference on the normalized throughput in the downlink and the uplink is due to the sensitivity of the SDR X310, this can be improved by using directional antennas and amplification.

Throughout the test scenarios, a stability of the network was obtained. However, the obtained throughput is lower than the theoretical parameters of 4G LTE networks, this can be improved by implementing the EPC on a computer with higher technical specifications than that used in our implementation.

According to the average latency results, the increased latency when measuring the download throughput from the UE is due to the entire structure added to generate the ESPE LTE network.

The use of SDR technology has become an alternative for the development of cellular communications, its use through the implementation of srsLTE provides viability, performance and reliability in different test scenarios.

Acknowledgment. The authors gratefully acknowledge the contribution of Universidad de las Fuerzas Armadas ESPE for the economical support in the development of this project under Research Grant 2020-PIS-012

References

1. Altamirano, C.D., Barragán, S., Grijalva, F., Paredes, M.: A portable GSM base station solution for military communications in hostile environments. In: Rocha, Á., Paredes-Calderón, M., Guarda, T. (eds.) MICRADS 2020. Smart Innovation, Systems and Technologies, vol. 181, pp. 319–329. Springer, Singapore (2020). https://doi.org/10.1007/978-981-15-4875-8_28
2. Avallone, S., Guadagno, S., Emma, D., Pescapè, A., Ventre, G.: D-ITG distributed internet traffic generator. In: First International Conference on the Quantitative Evaluation of Systems, QEST 2004. Proceedings, pp. 316–317. IEEE (2004)
3. Bang, S., Ahn, C., Jin, Y., Choi, S., Glossner, J., Ahn, S.: Implementation of LTE system on an SDR platform using CUDA and UHD. Analog Integr. Circuits Signal Process. **78**(3), 599–610 (2014)
4. Corporation, A.: Anritsu test UICC smartphones leaflet (2021)
5. DragonOSLTS: Dragonos_lts (2018). https://sourceforge.net/projects/dragonos-lts/
6. Dugan, J., Elliott, S., Mah, B.A., Poskanzer, J., Prabhu, K.: iPerf3, tool for active measurements of the maximum achievable bandwidth on IP networks (2012). https://github.com/esnet/iperf
7. Elkins, E.: Powered by Netflix: speed test services and video-on-demand's global development projects. Media Cult. Soc. **40**(6), 838–855 (2018)

8. Gelonch-Bosch, A., Marojevic, V., Gomez, I.: Teaching telecommunication standards: bridging the gap between theory and practice. IEEE Commun. Mag. **55**(5), 145–153 (2017). https://doi.org/10.1109/mcom.2017.1601231
9. Gomez-Miguelez, I., Garcia-Saavedra, A., Sutton, P.D., Serrano, P., Cano, C., Leith, D.J.: srsLTE: an open-source platform for LTE evolution and experimentation. In: Proceedings of the Tenth ACM International Workshop on Wireless Network Testbeds, Experimental Evaluation, and Characterization, pp. 25–32 (2016)
10. Gutiérrez, D., Giménez, F., Zerbini, C., Riva, G.: Measurement of 4G LTE cells with SDR technology. IEEE Latin Am. Trans. **18**(02), 206–213 (2020)
11. Laselva, D., Lopez-Perez, D., Rinne, M., Henttonen, T.: 3G PP LTE-WLAN aggregation technologies: functionalities and performance comparison. IEEE Commun. Mag. **56**(3), 195–203 (2018). https://doi.org/10.1109/MCOM.2018.1700449
12. Mitola, J.: Software radios: survey, critical evaluation and future directions. IEEE Aerosp. Electron. Syst. Mag. **8**(4), 25–36 (1993)
13. Mortimer, M.: iPerf3 documentation (2018)
14. Nikaein, N., Marina, M.K., Manickam, S., Dawson, A., Knopp, R., Bonnet, C.: OpenAirinterface: a flexible platform for 5G research. ACM SIGCOMM Comput. Commun. Rev. **44**(5), 33–38 (2014)
15. del Peral-Rosado, J.A., et al.: Software-defined radio LTE positioning receiver towards future hybrid localization systems. In: 31st AIAA International Communications Satellite Systems Conference, p. 5610 (2013)
16. del Peral-Rosado, J.A., et al.: Comparative results analysis on positioning with real LTE signals and low-cost hardware platforms. In: 2014 7th ESA Workshop on Satellite Navigation Technologies and European Workshop on GNSS Signals and Signal Processing (NAVITEC), pp. 1–8. IEEE (2014)
17. Raida, V., Svoboda, P., Kruschke, M., Rupp, M.: Constant rate ultra short probing (CRUSP): Measurements in live LTE networks. In: ICC 2019-2019 IEEE International Conference on Communications (ICC), pp. 1–6. IEEE (2019)
18. Raida, V., Svoboda, P., Rupp, M.: Constant rate ultra short probing (CRUSP) measurements in LTE networks. In: 2018 IEEE 88th Vehicular Technology Conference (VTC-Fall), pp. 1–5. IEEE (2018)
19. Rao, R.M., Ha, S., Marojevic, V., Reed, J.H.: LTE PHY layer vulnerability analysis and testing using open-source SDR tools. In: MILCOM 2017-2017 IEEE Military Communications Conference (MILCOM), pp. 744–749. IEEE (2017)
20. Shamaei, K., Khalife, J., Kassas, Z.M.: Exploiting LTE signals for navigation: theory to implementation. IEEE Trans. Wirel. Commun. **17**(4), 2173–2189 (2018)
21. srsLTE: srslte (2021). https://github.com/srsLTE/srsLTE
22. Stewart, B., Crockett, L., Barlee, K., Atkinson, D.: Model based design for 4G and 5G wireless communications software defined radio using MATLAB. In: 2016 Design, Automation & Test in Europe Conference & Exhibition (2016)

Modeling Wireless Propagation Channel: A Traditional Versus Machine Learning Approach

Fernando Lara[1,3(✉)], Román Lara-Cueva[1,2], Miguel Castillo[1,3], Jorge F. Arellano[1,3], and Luis Topón[1,3]

[1] Centro de Investigación en Redes Ad Hoc y WiCOM-Energy Research Group, Universidad de las Fuerzas Armadas ESPE, 171103 Sanqolquí, Ecuador
{mflara3,ralara,macastillo8,jfarellano,lxtopon}@espe.edu.ec

[2] Departmento de Eléctrica, Electrónica y Telecommunicaciones, Universidad de las Fuerzas Armadas ESPE, 171103 Sanqolquí, Ecuador

[3] Maestría de Investigación en Electrónica mención Telecomunicaciones, Universidad de las Fuerzas Armadas ESPE, 171103 Sanqolquí, Ecuador

Abstract. Wireless networks are one of the most important technologies worldwide, as modeling the different telecommunication channels is an active task of several groups of researchers. Obtaining a theoretical or semi-empirical model of a channel allows improving the planning of wireless networks. The proposed methodology allows channels to be modeled semi-empirically using machine learning tools. This article presents the use of Machine Learning techniques as a tool for describing wireless channels in the 2.4 GHz and 5.8 GHz bands with Zigbee and WiMAX technologies, respectively. This allows to develop a more accurate propagation model, in addition with the generation of heat maps and more reliable schedules. A synthetic data generation technique is proposed, in order to train the Support Vector Machines algorithms of Regression Learning, with this, two semi-empirical models are generated with a lower Root Mean Square Error of 1.4% for Zigbee, and 13,84% for WiMAX, with respect to traditional models.

Keywords: Heat maps · Regression learning · Synthetic data · Support vector regression · WiMAX · and Zigbee

1 Introduction

Wireless communications in recent years have expanded, evolved and acquired great relevance for society. The technological era that we are currently experiencing has allowed the incorporation of new services and solutions adopted in the field of the Internet, home automation, autonomous vehicles and others; the wireless networks are being the center of this digital transformation, since, they are the best way to share information in a fast, secure and reliable way without the presence of cables. This type of communication is based on a transmitter

M. Botto-Tobar et al. (Eds.): CIT 2021, LNNS 405, pp. 281–294, 2022.
https://doi.org/10.1007/978-3-030-96043-8_22

and receiver that form a wireless link, in order to transmit and receive information through electromagnetic signals that contain energy, which degrades due to reflection, refraction, dispersion and other phenomena, being of great value to understand the behavior of the electromagnetic signal while it is propagating [8].

For the design, planning and dimensioning of wireless technologies, the characterization of the propagation phenomenon has become crucial, in order to predict with high precision the behavior of a certain signal in a certain scenario. The characterization in question is carried out from mathematical models that closely resemble the real behavior of the signal. The trend is to determine the most optimal propagation models [10]. Generally, propagation models can be classified as empirical or statistical, theoretical or deterministic, and semi-empirical, which are a combination of the first models. While empirical models are based on measurements, theoretical models are based on the fundamental principles of electromagnetic wave propagation phenomena [3].

There are several previously proposed semi-empirical modeling works, for example, in [7], made an adjustment of the semi-empirical prediction method COST-231 Okumura-Hata, based on experiments at 2.3 GHz using WiMAX transmissions carried out in the west of India, better results were obtained, compared to the original COST-231 Okumura-Hata model; in another work [13], a semi-empirical propagation model is described based on reception power measurements obtained within the Universidad de las Fuerzas Armadas - ESPE in the 5.2 to 5.8 GHz band with WiMAX technology; and in [6] a semi-empirical propagation model based on Received Signal Strength Indicator (RSSI) measurements is presented for a point-to-point link in the 2.4 GHz band, using Zigbee technology, which presents a discontinuity of decreasing exponential behavior, caused by the height of the antennas of the devices, which is better explained in [6].

In this context, the use of Machine Learning arises as a powerful tool to improve channel models from data, therefore, by combining knowledge of wireless networks with Machine Learning techniques, efficient propagation models are reached for certain geographic areas. Some of the works related to this technique are: [16] where the precision of a model based on Random Forest, Adaboost, Neural Networks and K-Neareast-Neighbors Machine Learning with $42,157,324$ RSSI samples collected during one year with wireless sensor networks in the 2.4 GHz band, having a low noise level with Random Forest; moreover, the authors [2] designed a propagation model for the last-mile connectivity with a Low Altitude Platform technology (LAP) through previous modeling; In addition, the authors [1] proposes the use of Machine Learning as a tool for the calibration of propagation models.

Under our best understanding, the search for optimal propagation modeling continues, therefore, this article aims to present a semi-empirical modeling with Machine Learning techniques, this model is applied to point-to-point links and point-to-multipoint links with the generation of synthetic data based on the stochastic behavior of the channel, for this purpose, both semi-empirical propagation models based on Regression Learning (Coarse and/or Median Gaussian Support Vector Machines (SVM) [14]) are generated and evaluated, SVM technique is used to improve modeling [4]. Regression learning model allows to obtain the coverage areas for point-to-multipoint networks (WiMAX technology), furthermore, to determine the estimated power levels in the receiver for point-to-point networks (Zigbee technology). In both cases, a trained model based on Machine Learning is proposed, which estimates the losses produced by the medium with the a priori information available. In order to make it possible to train SVM models, a technique for generating synthetic data using stochastic knowledge of the channel is suggested. Finally, the heat maps of coverage areas are implemented.

Is necessary to mention that regression learning was chosen, specifically SVM with Gaussian or Radial Basis Function (RBF) kernel, due to the non-linearity that allows the channel modeling (curve) to better fit to the data, reducing the error between these two [4]. This linearity is controlled by the Kernel scale variation [14].

The rest of the article is organized as follows. In Sect. 2 the methodology used for data generation and algorithm training is presented. In Sect. 3 the case of a model obtained from a point-to-point link is presented. In Sect. 4 the application of the proposal to a point-to-multipoint link is presented. At the end the conclusions are established.

2 Methodology

Our proposal consists in to develop a semi-empirical model of wireless channel based on techniques of Regression Learning (Coarse and/or Median Gaussian SVM), using existing samples; the methodology have two stages, the first of generation of synthetic data that allows train the algorithms and the second where the model is generated, evaluated and compared with traditional models.

Two scenarios are established, to validate the approach: point-to-point scenario and point-to-multipoint, in both cases, the losses produced in the medium at a certain distance are determined, and in point-to-multipoint scenario, the generation of heat maps is presented to determine coverage areas.

2.1 Synthetic Data

The proposed model is based on Machine Learning techniques, therefore, a greater number of data is required to train the model and obtain an adequate description of the propagation channel, whereby, a data generation technique based on the stochastic behavior of the channel is proposed.

The channel can be described from the particularization of the log-normal loss model:

$$PL_D(d) = PL(d_0) + 10\alpha \log\left(\frac{d}{d_0}\right)$$

$$PL_E = X_\sigma$$

$$PL(d) = PL_D(d) + PL_E$$

We start from the premise that at each distance d is a deterministic estimate given by PL_D and a stochastic component PL_E, to generate the necessary data for training. The PL_E component is described by a Gaussian random variable [4], the probability density function of the random variable [9] is:

$$f_X(x) = \frac{1}{\sqrt{2\pi}\sigma_X} \exp\left(\frac{(x-\mu_X)^2}{2\sigma_X^2}\right)$$

Likewise, the Gaussian random variable X is described by its mean μ and variance σ, $X \sim N(\mu, \sigma^2)$. When a constant a is added to a Gaussian random variable, a new mean is obtained for the random variable [12]:

$$E[X] = \mu = \int_{-\infty}^{\infty} x f_x(x) dx$$

$$E[X + a] = E[X] + a = \mu + a$$

From this equation, we obtain $X + a \sim N(\mu + a, \sigma^2)$, therefore, PL_D to be a deterministic component, we have a constant in each position away, with this the channel model can be generalized as:

$$PL(d) = X_M \sim N(PL_D(d) + \mu, \sigma^2)$$

In the case of propagation channels, the random variable X_σ has a mean equal to zero [4,17], so the expression can be reduced to:

$$PL(d) = X_M \sim N(PL_D(d), \sigma^2)$$

In order to produce synthetic data for training, a Gaussian random variable of mean $PL_D(d)$ was generated at each distance position, see Fig. 1, generating N samples at each position of the random variable X_M.

2.2 Model Generation

Coarse and/or Median Gaussian SVM algorithm of Regression Learning is used depending the case, to generate the model that characterizes the telecommunications channel. The algorithm is trained with the synthetic data generated. Subsequently, a model is built to determine the expected loss in the environment for the two proposed scenarios with their respective particularities.

3 Point to Point Link

The database described in [6] is used, to verify the model, which are from a Zigbee link with 245 samples in a range from 2 m to 200 m, from 2 m a 30 m, when samples are taken every 2 m, and from 35 m to 200 m every 5 m is taken, with 5 samples at each point. This databases has measurements in two

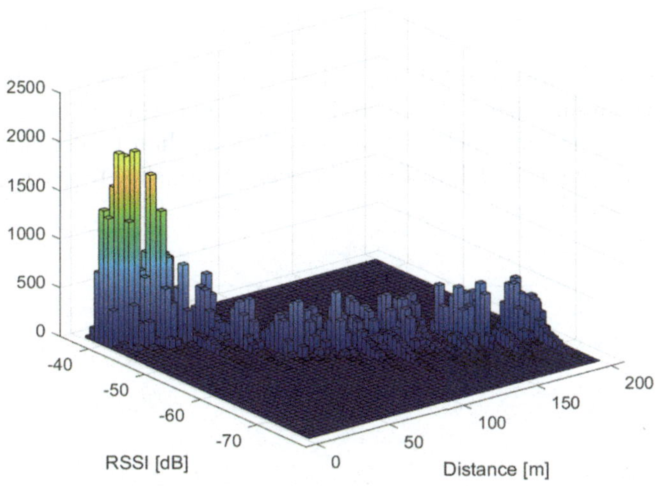

Fig. 1. Synthetic data: Gaussian random variable in each position d, with variance σ^2.

environments: rural (grass) and sub-urban (concrete). The propagation model proposed by the authors is:

$$L_{rural} = \begin{cases} -36.425 - 5.06\log(x), & 2 \leq x \leq 60 \\ 6.1147 - 31.19\log(x), & 60 < x \leq 200 \end{cases}$$

$$L_{subur} = \begin{cases} -37.97 - 4.42\log(x), & 2 \leq x \leq 60 \\ 15.55 - 35.69\log(x), & 60 < x \leq 200 \end{cases}$$

In this process, the RSSI parameter is known and the Log-Normal model [4] with adjustments in the coefficients is used. The Log-Normal model describes the propagation channel as:

$$PL(d) = PL(d_0) + 10\alpha\log\left(\frac{d}{d_0}\right) + X_\sigma$$

where, d represents the distance, $PL(d_0)$ represents the losses of the channel up to the reference distance d_0 and X_σ is the Gaussian random variable that represents the random channel.

3.1 Algorithm Training

The method to train the algorithms require, 1000 synthetic data per distance in the two environments (rural and sub-urban). Two features are used to train Machine Learning model.

The features used are: a boolean vector A to describe whether the environment is rural or sub-urban, and a distances vector D, besides, a output vector of the RSSI P, this get from the synthetic data, with a number of elements k.

$$F_{EX} = \left[A_k^T, D_k^T, P_k^T\right]$$

The Median Gaussian SVM algorithm is used to generate the automatic regression by training with F_{EX}. Once the algorithm is trained, we have $P_{RSSI} = \varphi(A, D)$, where φ is the Machine Learning with Median Gaussian SVM model [18]. From the trained model, in Fig. 2 two curves are shown to the propagation model from synthetic data.

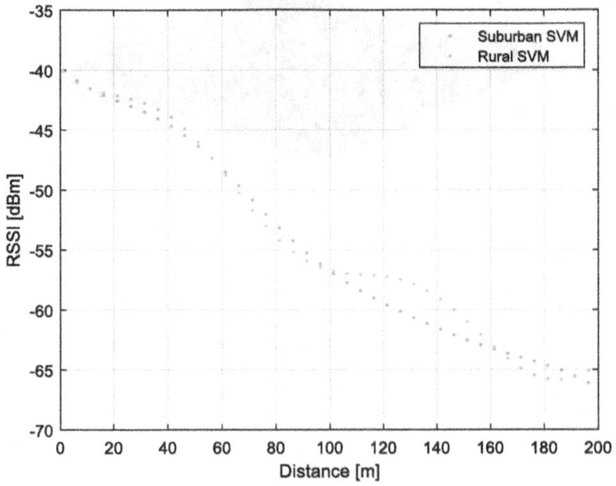

Fig. 2. Curves generated with machine learning SVM model for Zigbee propagation channel.

3.2 Model Evaluation

In the Fig. 3 the comparison between the models, Log-Normal model and SVM model is presented; as well, in Fig. 4, the obtained models and values taken from the link with Zigbee technology are shown, in effect, the comparison using the Root Mean Square Error (RMSE) [5] is made:

$$RMSE = \sqrt{\frac{1}{n}\sum_{i=1}^{n}\left(x[i] - \widetilde{x}[i]\right)^2}$$

where, $x[i]$ is measured value and $\widetilde{x}[i]$ is estimated value. The result in Table 1 is presented.

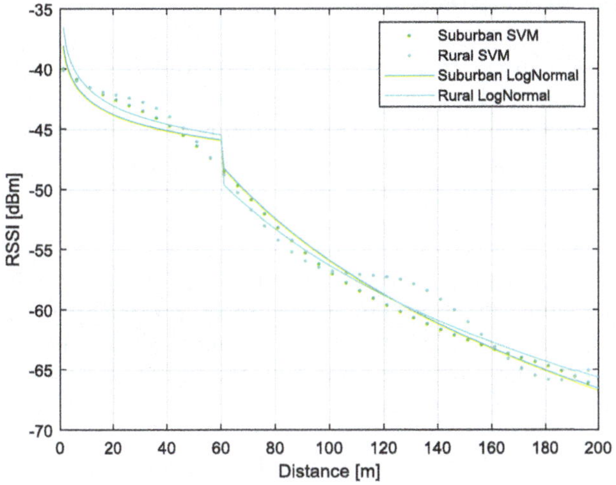

Fig. 3. Comparison between: Log-Normal model and median Gaussian SVM model

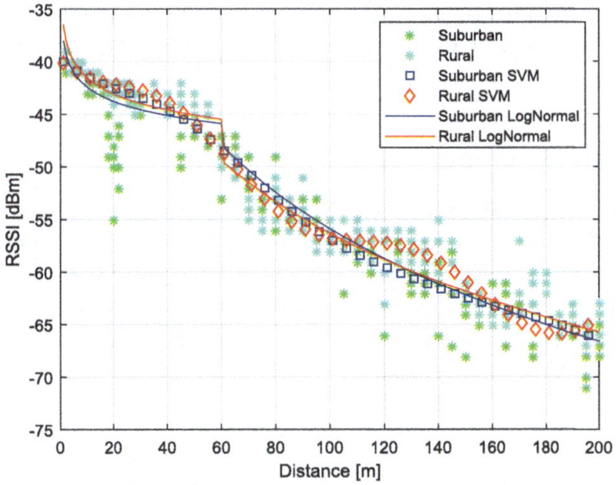

Fig. 4. Semi empirical models generated with data obtained and SVM

Table 1. Mean Square Error SVM model and Log-Normal model

RMSE	SVM model	Log-Normal model
Rural	2.40	2.38
Sub-urban	1.81	1.89
Average	2.11	2.14

4 Point to Multipoint Link

The database [13] is used, the samples were taken in 184 points and each point have 9 measurements, giving a total of 1656 measurements. The place where these measurements were made in ESPE campus, this allows the data set to be divided into environments. The link balance for these data is given by:

$$P_R = P_T + G_T + G_R - L_{\omega_L} - L_{\omega_R} - L_M$$

where: P_R are the measured powers at the receiver, $P_T = 22$ dBm is the power transmitter, the gains of the antennas G_R and G_T are 7 dBi, the losses of the cables L_{ω_R} and connectors L_{ω_L} are 1.75 dB, and the channel losses are represented by L_M [13]. The model presented by the authors is called ESPE-Model, described according to the expression:

$$L_M = 96.45 + 8.85 \times \log(f) + 11.39 \times \log(d)$$

The ESPE-Model is a particularization of the Log-Normal model adjusting the parameters to better describe the propagation channel. Our proposal consists of dividing the data into four environments: Urban (University Buildings), Sub-Urban (Streets-Parking), Vegetation (Green spaces) and High Vegetation (Forest), and use Median Gaussian and Coarse Gaussian SVM to generate a model that describes the channel.

4.1 Algorithm Training

A synthetic data generation process is carried out, with the data divided into environments, this similar to that described in Sect. 3. Here 1000 data is generated for each distance position of the which measurements have been taken.

The algorithm used to obtain the model of the wireless channel has two inputs and one output, the inputs are the distance D and the zone Z, and the output is the losses in the channel L:

$$F_{EX} = \left[Z_k^T, D_k^T, L_k^T\right]$$

The Median Gaussian and Coarse Gaussian SVM algorithm are used to generate the automatic regression and training the algorithm with F_{EX}, we have $L_M = \varphi(Z, D)$, where φ is the SVM model generated from Median Gaussian and Coarse Gaussian SVM. Finally, Coarse Gaussian SVM in the first meters and from 60 m onwards Median Gaussian SVM are implemented, this can be see in Fig. 5.

4.2 Model Evaluation

The RMSE value of the trained models is estimated with respect to the measured data, the RMSE expression for different areas is calculated from:

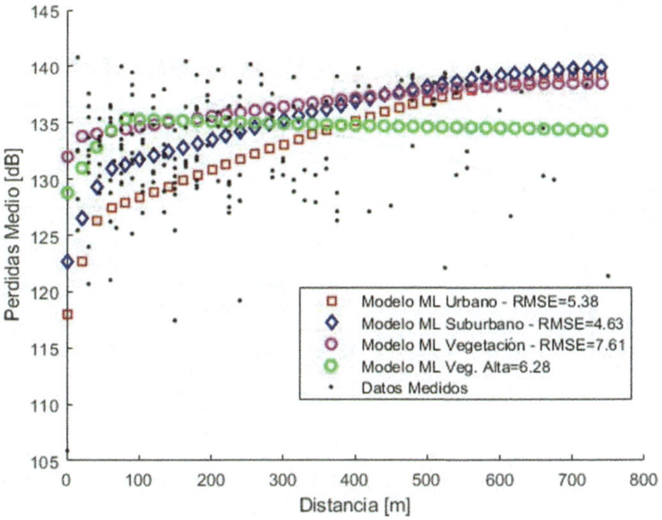

Fig. 5. Propagation model with SVM

$$RMSE_M = \frac{N_U}{N_T} \times RMSE_U + \frac{N_S}{N_T} \times RMSE_S + \dots$$

$$\frac{N_V}{N_T} \times RMSE_V + \frac{N_{Va}}{N_T} \times RMSE_{Va}$$

This allows comparing the RMSE result between the ESPE Model and the proposed SVM model, as shown in Table 2.

Table 2. RMSE SVE evaluating and comparing with RMSE ESPE model

Zone	RMSE zone	Total RMSE
Urban	5.38	5.40
Suburban	4.63	
Vegetation	7.61	
High vegetation	6.28	
Espe model	–	7.29

The SVM model is compared with the main large-scale propagation models (Free Space [15], Cost 231 Okumura-Hata [11] and the SUI Model [3]), to validate the proposed, the RMSE statistics is used to reference.

In Fig. 6 the comparison of the proposed model with those mentioned above is presented, verifying that the proposed model has a lower RMSE compared to traditional models.

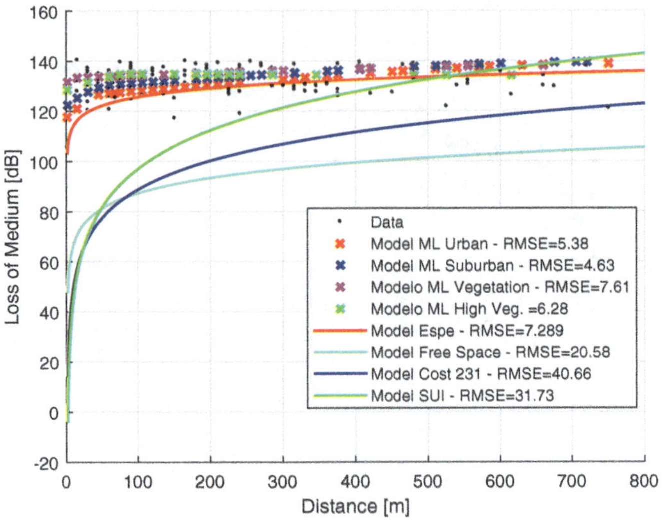

Fig. 6. SVM model comparison with traditional models

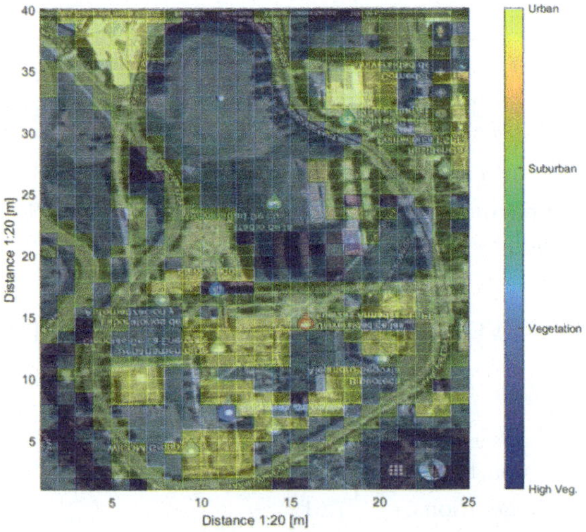

Fig. 7. Predominant areas in 20 × 20 m cells in ESPE facilities

4.3 Coverage Area

The coverage area can be established from the SVM Model and knowing the predominant area within 20 × 20 m cells. In the first instance, which area of the 20 × 20 m cells belong within of the ESPE Campus facilities is established, this is observed in Fig. 7.

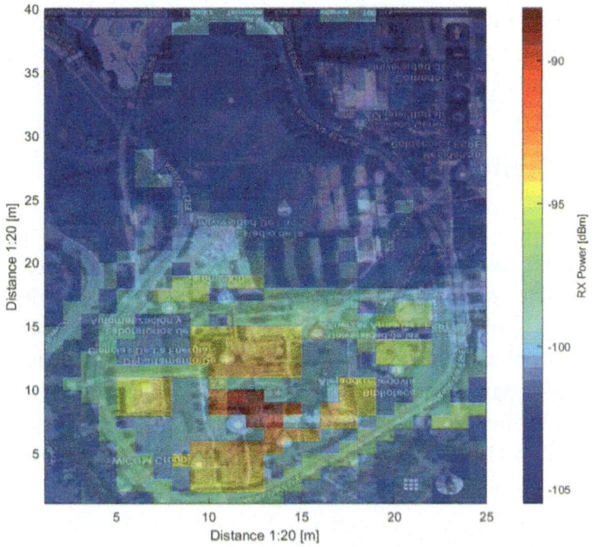

Fig. 8. Heat map with WiMAX station in original location

Once established, it is tried to place the transmitter in the position in which the measurements were made according to [13]. From this, with the trained algorithm the expected power in each cell within the University can be established, as is observed in Fig. 8, in this case the location of the WiMAX antenna corresponds to the current location in the university facilities.

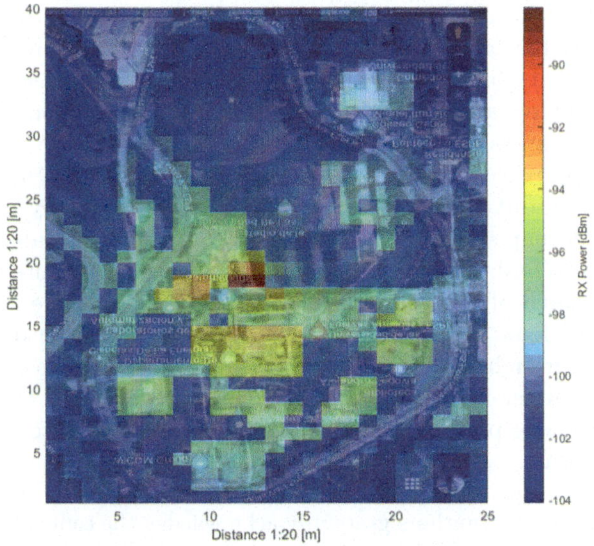

Fig. 9. Heat map with new WiMAX station location

The WiMAX station, with this simulator, can be located in any position and thus determine the new coverage area, as seen in Fig. 9, where the base station was located in a central position within the university campus, see Fig. 7.

5 Discussion and Conclusions

In [6], a propagation model with Zigbee technology at 2.4 GHz for outdoors with a logarithmic adjustment of up to 200 m is obtained, finding a discontinuity at 60 m. In the present work, through Machine Learning techniques, in Fig. 3, a propagation model is obtained without the discontinuity presented in the 60 m, also the significant change that the SVM model undergoes in said position is verified, achieving a general modeling for rural and suburban environments. The modeling used was generated based on synthetic data from the Gaussian random variable generated in the receiver. The Machine Learning methodology applied to modeling point-to-point wireless networks managed to reduce the RMSE by 1.4%, that is, to 2.11.

On the other hand, in [13] a semi-empirical propagation model called ESPE-Model is proposed, validating its effectiveness for future predictions of signal propagation in the proposed environment, substantially improving the RMSE compared to other propagation models analyzed. However, in this proposal when dividing by urban, sub-urban environments, vegetation and tall vegetation, as described in Fig. 5, the generation of synthetic data and application of Machine Learning techniques reduces the RMSE by 13.84% with respect to ESPE-Model, that is, to 5.4.

It has been proven that Machine Learning techniques contribute to the telecommunications area in the search for optimal models for specific areas. The Median Gaussian and Course Gaussian SVM techniques, depending on the technology, do better data analysis for propagation modeling in relation to other Machine Learning techniques, such as linear regression models, decision trees, among others, giving a basis for the proposal of propagation models and the study of coverage areas from Machine Learning techniques.

By knowing the accurate location from where the measurements were taken and after carrying out the modeling of the network with any of the featuring techniques, heat maps can be generated that allow to visualize the coverage areas, as well as to find the best location of the transmitting antenna, algorithms for automatic search of the optimal antenna position could be generated in the future; but if we look at it from another point of view, the determination of coverage areas through heat maps locates in a non-traditional way, the place where the transmitter is located, which is the foundation of Detection Fiding, namely, this principle presented in this article could be used to train algorithms for Detection Fiding.

Acknowledgment. The authors gratefully acknowledge the contribution of Universidad de las Fuerzas Armadas ESPE for the economical support in the development of this project under Research Grants 2008-PIT-WiMAX, 2010-PIT-009, and 2015-PIC-004.

References

1. Adeogun, R.: Calibration of stochastic radio propagation models using machine learning. IEEE Antennas Wirel. Propag. Lett. **18**(12), 2538–2542 (2019)
2. Almalki, F.A., Angelides, M.C.: A machine learning approach to evolving an optimal propagation model for last mile connectivity using low altitude platforms. Comput. Commun. **142**, 9–33 (2019)
3. Alonso Quintana, R., Bordón López, R., Montejo Sánchez, S.: Estudio comparativo de los modelos de propagación de canal inalámbrico. Ingeniería Electrónica, Automática y Comunicaciones **34**(1), 12–26 (2013)
4. Alsayyari, A., Kostanic, I., Otero, C.E.: An empirical path loss model for wireless sensor network deployment in a concrete surface environment. In: 2015 IEEE 16th Annual Wireless and Microwave Technology Conference (WAMICON), pp. 1–6. IEEE (2015)
5. Chai, T., Draxler, R.R.: Root mean square error (RMSE) or mean absolute error (MAE)?-Arguments against avoiding RMSE in the literature. Geosci. Model Dev. **7**(3), 1247–1250 (2014)
6. Cueva, R.A.L., Alarcón, K.G.G., Paredes, R.A.L.: Modelamiento del canal de propagación con zigbee para escenarios exteriores/propagation channel modeling with zigbee for outdoor scenarios. RECI Revista Iberoamericana de las Ciencias Computacionales e Informática **5**(9), 1–21 (2016)
7. Dalela, C., Prasad, M., Dalela, P.: Tuning of COST-231 Hata Model for Radio Wave Propagation Predictions. Academy & Industry Research Collaboration Center (2012)
8. Dignani, J.P.: Análisis del protocolo zigbee. Ph.D. thesis, Universidad Nacional de la Plata (2012)
9. Fernández, M.M.: Señales aleatorias y ruido. ETS de Ingenieros de Telecomunicación (1996)
10. García, J., Rodríguez, O., Castillo, J.: Desempeño de modelos de propagación en comunicación móvil para la zona de caldas parte 1: Modelos para áreas urbanas. In: X Congreso Internacional de Telecomunicaciones, Valdivia, Chile (2002)
11. Javornik, T., Kandus, G., Hrovat, A., Ozimek, I.: Comparison of WiMAX coverage at 450 MHz and 3.5 GHz. In: 2006 International Conference on Software in Telecommunications and Computer Networks, pp. 71–75. IEEE (2006)
12. Kay, S.: Intuitive Probability and Random Processes Using MATLAB. Springer, Boston (2006). https://doi.org/10.1007/b104645
13. Lara-Cueva, R.A., Aguilar, D., Leon, P., Almeida, D., Benitez, D.: A new propagation model in WiMAX-based wireless fixed networks. In: 2016 IEEE Colombian Conference on Communications and Computing (COLCOM), pp. 1–5. IEEE (2016)
14. MathWorks: Choose regression model options. https://la.mathworks.com/help/stats/choose-regression-model-options.html
15. Mompó, J., Corral, P., Segrelles, J., Pierucci, L.: Herramienta gráfica para la predicción de coberturas IEEE 802.11 a/b/g en interiores, exteriores y espacio libre. Università di Firenze, Florencia (Italia) (2005). consultada en. http://w3.iec.csic.es/URSI/articulos_gandia_2005/articulos/ED2/556.pdf. Consultado 7 de febrero de 2005
16. Oroza, C.A., Zhang, Z., Watteyne, T., Glaser, S.D.: A machine-learning-based connectivity model for complex terrain large-scale low-power wireless deployments. IEEE Trans. Cognit. Commun. Netw. **3**(4), 576–584 (2017)

17. Rappaport, T.S.: Wireless communications-principles and practice, (the book end). Microwave J. **45**(12), 128–129 (2002)
18. Shevade, S.K., Keerthi, S.S., Bhattacharyya, C., Murthy, K.R.K.: Improvements to the SMO algorithm for SVM regression. IEEE Trans. Neural Netw. **11**(5), 1188–1193 (2000)

Design of an Academic CSIRT – A Proposal Based on Strategic Planning Principles

Daniel Nuñez-Agurto, Mario Ron, Enrique V. Carrera, Freddy Tapia, Henry Cruz, Luis Recalde, and Walter Fuertes(✉)

Department of Computer Science, Universidad de las Fuerzas Armadas – ESPE, Sangolquí 171103, Ecuador
{adnunez1,mbron,evcarrera,fmtapia,hocruz,llrecalde,wmfuertes}@espe.edu.ec

Abstract. This work aims to design and implement an organizational model that supports the macro processes of an academic institution in response to computer incidents to raise the level of information security. This study intends to include the research, development, and innovation (R&D&I) in an academic CSIRT. To achieve this, we developed a systematic literature review using the guidance of Barbara Kitchenham through a broad search for information, which allows us to identify the characteristics of the different types of organizations, services, infrastructure, and procedures to design and implement Academic CSIRTs. We then apply the science of design with its specific guidelines for evaluation and iteration within research projects and carry out improvement and knowledge construction. Specifically, we employed the European Network and Information Security Agency (ENISA) guides and the Forum of Incident Response and Security Teams (FIRST) directions, as these organizations are dedicated to preventing and addressing network security and information security problems. We also applied the Strategic Planning Process principles to obtain the Organizational Model Proposal, the operation and services proposal of the Academic CSIRT, and the research areas.

Keywords: Academic CSIRT · CERT · Strategic planning process

1 Introduction

Cyber-security and cyber-defense are vital aspects of information security and the sovereignty of a nation [1]. In recent years, countries have been the victims of different types of cyber-attacks; this causes permanent concern in the population, business, industry, and government [2]. Every time, various threats are emerging with increasing frequency and complexity. These threats take advantage of existing vulnerabilities. Thus, the risk of a cyber-attack is growing day by day, both against people and nations [3]. Various technological and procedural solutions have been developed to reduce the impact. However, this problem persists and has instead been increasing.

M. Botto-Tobar et al. (Eds.): CIT 2021, LNNS 405, pp. 295–308, 2022.
https://doi.org/10.1007/978-3-030-96043-8_23

Among the best-known solutions are the Security Operation Center (SOC), the Computer Security Incident Response Team (CSIRT) [4], and the Computer Emergency Response Team (CERT). Each of them can be complementary to the other and they are duly defined. Both CERT and CSIRT are used interchangeably to describe incident response teams. However, the scope and some objectives may be different. The term CERT is a trademark of the Carnegie Mellon University Institute of Software Engineering. Generally speaking, SOCs and CSIRTs focus on preventing and responding to cyber-security incidents. SOCs address the incident's technical aspects, while the CSIRT's primary responsibility is to mitigate legal and business consequences. CERTs, for their part, rely more on intelligence gathering and information sharing. CSIRTs and CERTs develop guidelines for practical cyber-security activities, emphasizing reducing business impact and legal liabilities. The CERTs and CSIRTs investigate and recommend R&D&I solutions, coordinating and supporting the generation and implementation of strategies to improve the services they offer [5].

However, despite these essential efforts, cyber-crime persists on a global scale against individuals and businesses. The most alarming are those cyber-crimes that try to attack nations' sovereignty, which manage to collapse the states' critical mission infrastructures, such as electricity, telecommunications, health, banking, oil, airports, etc. Therefore, there is a need to design common strategies at the national and international level, which minimize emerging risks and threats from cyberspace [6]. In this sense, universities and research centers are called upon to develop information security solutions and the fight against crime, cyber-espionage, cyber-warfare, and cyber-terrorism. The North Atlantic Treaty Organization (NATO) recommends creating highly specialized teams for research, innovation, and technological development to guarantee the security of people, systems, and information networks of a nation [6].

This study aims to design and implement an organizational model, roles, and responsibilities that support the macro processes in response to computer incidents to raise information security in a public university and contribute to the improvement of cybersecurity in the country; all of this, through research, development, and innovation (R&D&I). To achieve this, we developed a systematic literature review using the guidance of Barbara Kitchenham [7] through a broad search for information, which allows us to identify the characteristics of the different organizations, services, infrastructure, and procedures to design and implement Academic CSIRTs. Also, we seek to know what factors influence this problem and what requirements will be needed to achieve the proposal. We then apply the Science of Design, an information technology research methodology that offers specific guidelines for evaluation and iteration within research projects and carries out improvement and knowledge construction. Specifically, we employed the European Network and Information Security Agency (ENISA) guides [8] and the Forum of Incident Response and Security Teams (FIRST) directions [9], since these are organizations dedicated to preventing and addressing network security and information security problems.

The rest of this paper is organized as follows. Section 2 describes the materials and methods combined with the problem statement and the systematic literature review. Section 3 describes the design of a proposal for an Academic CSIRT. Section 4 ends with the conclusions and describes our future work.

2 Materials and Methods

2.1 Problem Statement

During our research, we found various factors that affect the creation of Academic CSIRTs. For instance, insufficient resources to perform the implementation, the lack of equipment, software, and training would force significant investment of academic institutions. The absence of financial analysis, which justifies the Academic CSIRTs, generates doubts when deciding the university authorities' inadequate support. The insufficient education or training in computer security and the digital exclusion that our country currently has. The deficiency of orientation to the academic research nature of a CSIRT. The business orientation, from a purely commercial point of view that has been granted. The decrease in its fundamental functions from a technical and non-investigative perspective reduces itself to technological solutions-also, the lack of an organizational model that defines the functional and operational responsibilities [10].

Other problems concerning the creation of an Academic CSIRT/CERT are sufficient reasons to undertake this project, such as—there are no processes associated with managing IT incidents at the institution. Lack of orientation to the academic research nature of a CSIRT/CERT. Lack of knowledge of the scope and purpose of the CSIRT/CERT limits its possibilities. Also, there are no essential, preventive, and reactive services to manage information security alerts and incidents at the institution. There are no laboratories, equipment, software, and IT infrastructure to manage information security incidents at the institution. Also, no procedures, guides, or security policies are associated with managing the institution's information security incidents. In summary, there is absence of cyber-security specialists, defined processes, collaboration, resources, awareness, organization, and interest [11].

2.2 Literature Review

Faced with the problems indicated in Subsect. 2.1, we employ the Systematic Literature Review guide proposed by Kitchenham and Charters [7]. Precisely, we determine the relevant terms or keywords related to the problem. Some relative terms were computer security, CSIRT, computer forensics, assets, attacks, incidents, security standards, emergencies, CERT, early warning to cyber-attacks, incident management, alert management, cyber-security, proactive services, preventive services, data protection, access control, ITIL v4, ISO-27035, NIST, decision support, vulnerability analysis, business intelligence, vulnerability analysis. The most frequent terms were: CSIRT, CERT, incident and alert management, cyber-security, preventive and reactive services.

We only considered indexed databases such as IEEE Xplore, Springerlink, Elsevier, and Scopus for search string tuning, with exclusively those articles published since 2013. The search string included the following keywords: "CSIRT", "CERT", "informatics incidents", "preventive services", "basic services", "reactive services", "Academic CSIRT/CERT", "Research and Development on CSIRT/CERT", AND > 2013, IEEE, Springer, Elsevier, Scopus.

As an initial result, we found 531 articles, the same ones that, after applying the exclusion and inclusion criteria related to research and development purposes, focused on academic sectors, were limited to 27. In summary, very few studies have been developed on the design and dimensioning of an "Academic CSIRT/CERT". However, we found several works related to applying different methods, techniques, trends, and tools to improve commercial CSIRT/CERT services, but not necessarily academic. It is worth noting the efforts in countries such as Germany, Argentina, Australia, and the USA, where some Academic CSIRTs have been implemented [10].

We found other studies close to the CSIRT academic environment. For instance, in [6] with significant challenges in the structuring and institutionalizing of the communication. In [11], cyber-security education in the Ecuadorian environment. In [12], Incident Response Teams in IT Operations Centers. In [13] long-term experiences in balancing security and usability in research activities. Besides, Andrade and Fuertes in [14], developed the design and dimensioning of a CSIRT. In [15] and [16], the authors developed a comprehensive model for providing reactive and proactive services in a BI-based Academic CSIRT. In [17], information security management on the university campus uses cognitive security.

In more recent works, researchers attempt to promote solutions concerning different aspects. For instance, research and innovation [18], legal and police [19], historical to maintaining computer security [20], challenges to face the interoperable cyber-security of information [21], barriers that have been present in the implementation of CSIRTs [22].

Finally, at a local level, the ECU-CERT of the Ecuadorian government, the CSIRT of Ecuadorian Consortium for the Development of Research and Academia (CEDIA), and the CSIRT of the National Polytechnic School, which function with traditional services, are still in the implementation and improvement to reach their maturity levels. There are also other entities such as CSIRT UTPL, CERT ESPOL, CSIRT FFAA, Blue Hat CERT, CERT Radical, CSIRT GMS, CNT-Group, CSIRT Telconet, CSIRT Telefónica Ecuador, CSIRT Energía, CSIRT CELEC EP that already have experience in the CSIRT organization, processes, and services. They would be essential references to propose a new generation organization or services that can be more efficient.

In conclusion, as we can discern, the project of creating an Academic CSIRT/CERT is an institutional necessity. In the literature review, there is limited information on Academic CSIRTs. Our project would significantly impact this area since it is a contemporary, delicate problem with responsibility of the government and Armed Forces. Its development will provide new knowledge, innovation, and technological development, through multidisciplinary applied research, in solving real, current problems, with a guarantee of technical and research quality, which will implement solutions that will increase local and national cyber-security.

3 Proposal to Create an Academic CSIRT

3.1 Methodology

The methodology followed to prepare this proposal contemplates the steps of the Strategic Planning Process [23] and, in each one of them, the corresponding analysis or design and formulation activities. Reflection, systemic thinking, and participatory analysis have been the essential components applied during the process. The input information used to update the Situational Diagnosis comes from different sources, and in its compilation, members of the project have participated to create the Academic CSIRT.

It should be noted that during this study, the work team developed a SWOT analysis, which stands for Strengths, Weaknesses, Opportunities, and Threats. SWOT analysis is a technique used for the strategic planning of this project. Within our study, strategic planning was an exercise in the formulation and establishment of objectives and action plans to achieve the Academic CSIRT.

Philosophical Principles: As part of its strategic direction, the Academic CSIRT must respond to several philosophical principles:

- It is worth mentioning: THE CSIRT must promote research, professional training, and technological development;
- The development of its activities and processes is performed to satisfy its target community's requirements with quality and excellence;
- Excellence is sought through quality in all academic, research, extension, and management events carried out with moral, ethical, and civic values;
- The legal provisions and human rights must be respected, with a deep citizen conscience that contributes to searching for the truth and the formation of men of honor, free and disciplined;
- The Organization's first responsibility is the interest and training of university students and academics to contribute to national development in cyber-security and cyber-defense;
- It must be a dialectical, formative, investigative, humanistic, comprehensive, scientific, technical, and cultural organization, immersed in the dynamics of social, scientific, and technological changes;
- It must be an Institution that generates knowledge, where flexibility and planned change are the habits and the basis for producing creative alternatives in the context of cyber-security and cyber-defense.

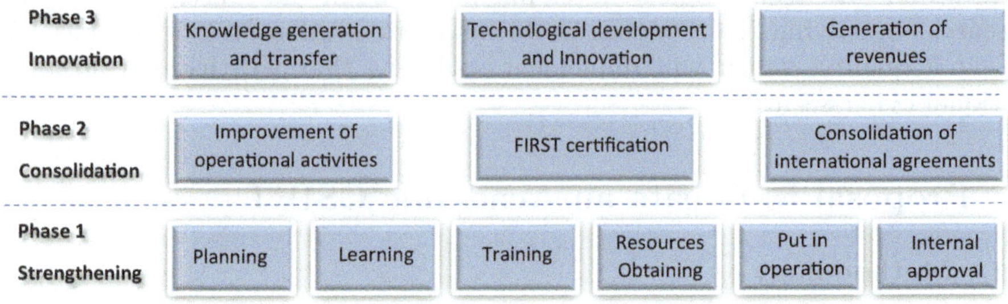

Fig. 1. Strategic phases to achieve the proposed vision.

Mission: To be an academic Computer Security Incident Response Team that proposes R&D&I alternatives, training, and specialized cyber-security training. To demonstrate excellence, scientific ethics, human quality, and social awareness of cyber-protection at the Ecuadorian academy and society service, based on advanced technology, financial autonomy, and highly qualified technical personnel.

Vision: To achieve leadership at the national level and international recognition as a response team to computer incidents of an academic and research nature in cyber-security and cyber-defense before 2025.

General Strategy: To implement an integrated management system focused on quality, which allows achieving a flexible, dynamic, and proactive organization through continuous improvement of processes, teamwork, development of human talent, and permanent technological innovation.

3.2 Strategic Phases

The strategic effort will be applied in three consecutive phases in such a way that ensures that the proposed vision is achieved (see Fig. 1).

- Phase 1 - Strengthening: It includes initial development, establishment, learning, training, organization of resources, and implementation of the Academic CSIRT.
- Phase 2 - Consolidation: In this phase, the operational activities of the CSIRT are perfected, and the initial errors are corrected to search of the international certification of the Forum of Incident Response and Security Teams (FIRST) [9]. We aspire to achieve association and consolidation of national and international agreements.
- Phase 3 - Innovation. Technological products are developed, and the more specific work of research and innovation is entered, with activities that allow it to achieve leadership at the national level.

3.3 Proposal for Operations and Services

A CSIRT is a team of IT security experts that responds to information security threats or incidents. Its primary purpose is to provide response services to computer security incidents.

An important aspect to consider during the establishment of any CSIRT is the services to be provided [24]. These services must be aligned with the resources available in the CSIRT and can be grouped into three main categories [25] as presented in Table 1. Based on the guidelines established by ENISA [8] and FIRST [9], the operations and services proposed for the Academic CSIRT are described below (see Fig. 2).

Table 1. Services provided by a CSIRT.

Reactive services	Proactive services	Quality management services
Alerts and warnings	Announcements	Risk analysis
Incident handling	Security audits	BC and DR planning
Vulnerability handling	Tools development	Security awareness
Artifact handling	Intrusion Detection	Training
	Technology watch	

The *reactive services* consist of post-incident reports in the community or other events related to threats or attacks, such as compromised hosts, malware, vulnerabilities, or other similar incidents [26]. Our proposal is particularly interested in alert and warning management activities, response to malicious artifacts, and incident and vulnerability handling.

On the other hand, *proactive services* aim to detect and prevent attacks before a tangible impact on the production systems. In this category, the information generated by the Academic CSIRT is disseminated to the institution and partners in order to protect their assets and avoid being targets of an attack. Specifically, the Academic CSIRT will provide security audits (including penetration testing, monitoring, and alert management), and R&D&I, threat intelligence, tool development, and technology watch.

Finally, the community can demand *quality management services* (*i.e.*, aggregated value services) to review and improve their organizations' security posture. This category of services is not time-dependent. In the case of the Academic CSIRT, it will provide risk analysis, business continuity planning (BCP), training, awareness, and cyber-security entrepreneurship, including security consultancies.

In addition to all the services mentioned above, this work's main objective is concentrated on creating a broader type of Academic CSIRT [25]. Thus, we have defined a new service that could be part of the third category: quality management services. The new service is defined as Research, Development, and Innovation (*i.e.*, R&D&I) for cyber-security and cyber-defense [27]. This service requires that the Academic CSIRT provides facilities for creating research and elaboration environments for security practices within the CSIRT. Specifically, this service will be focused on the following actions:

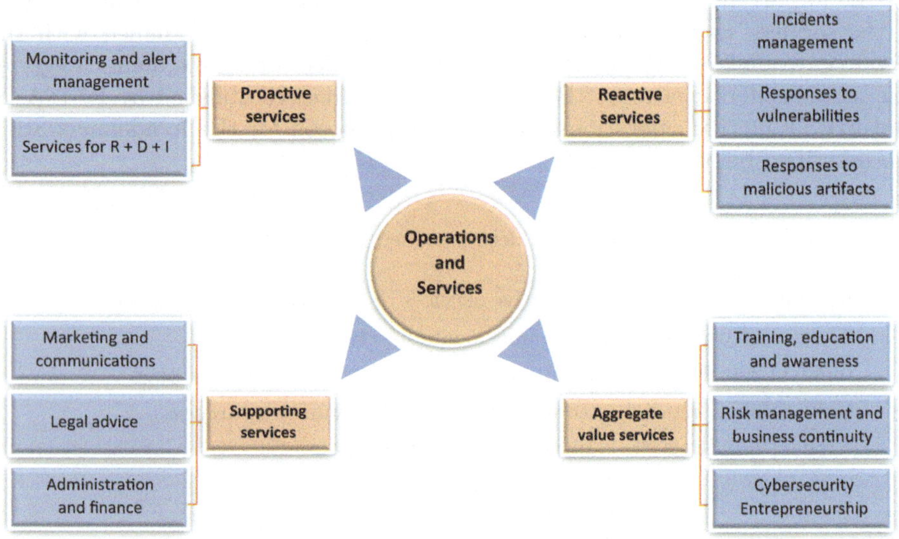

Fig. 2. Proposal for operations and services of the academic CSIRT.

- Incident analysis: it will be possible to identify the cause and scope of each incident, collect evidence and propose solutions through a technical and statistical analysis of the data. Also, recommendations and analytical reports of the incidents could be presented, based on a forensic analysis of the events;
- Vulnerability analysis: its objective is to identify and verify the behavior of each vulnerability, propose solutions and test them through a technical and statistical analysis of the proposals. Moreover, according to the tests and analysis carried out, recommendations and analytical reports of vulnerabilities could be generated;
- Threat analysis: the purpose of this activity is to identify possible threats to the community and propose solutions or mitigation plans [8]. Specifically, one particular threat that needs to be studied is human behavior and social engineering; this will allow us to present recommendations and analytical reports of various threats and the alternatives to face them;
- Tool development: *e.g.*, the development of algorithms to identify incidents and vulnerabilities and recreate security incidents; this will also allow us to create prototypes based on the proposed algorithms to improve the institution and partners' computer security;

- Support for R&D&I: Take care of the generation, organization, and maintenance of the CSIRT operation logs. These logs will be used to support the analysis of incidents and vulnerabilities and the development of new computer security tools. As mentioned above, it is essential to access all the information generated by the CSIRT for research purposes to improve computer security research. In addition to our philosophical principles and vision 2025 (see Subsect. 3.1) established in the Strategic phases (see Fig. 1), Stage 3, we will achieve our strengthening through the generation of knowledge and transfer and technological development innovation.

3.4 Proposal of the Research Areas

Given that the Academic CSIRT will promote Research, Development, and Innovation (R&D&I) activities, Fig. 3 shows the proposed research areas directly related to the Services Framework Version 2.1.0 of FIRST [9], published in November 2019.

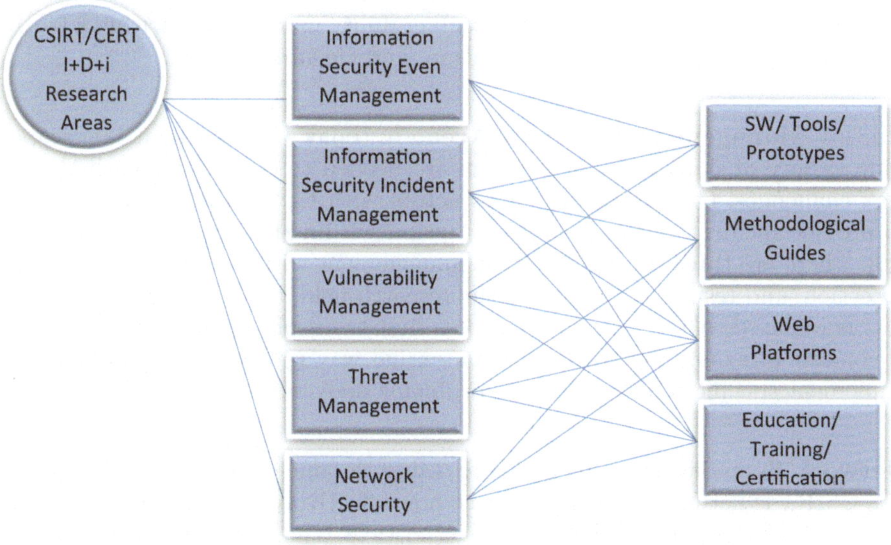

Fig. 3. Proposal of the research areas of the Academic CSIRT.

Information Security Event Management Research Area: According to FIRST [9], it aims to identify information security incidents based on the correlation and analysis of security events from a wide variety of event and contextual data sources. Its main processes include monitoring and detection, and event analysis.

Information Security Incident Management Research Area: As established by FIRST [9], this area consists of services that can collect and evaluate information security incident reports, analyze relevant data, and perform detailed technical analysis of the incident itself and any artifacts used. Among the main events, we find information security incident report acceptance, information security incidents analysis, artifact and forensic evidence analysis, mitigation and recovery, information security incident coordination, and crisis management support.

Vulnerability Management Service Research Area: According to FIRST [9], this area includes services related to the discovery, analysis, and handling of new or reported security vulnerabilities in information systems. It also includes services related to the detection of and response to known vulnerabilities to prevent them from being exploited. Its primary services include vulnerability discovery/research, vulnerability report intake, vulnerability analysis, vulnerability coordination, vulnerability disclosure, and vulnerability response. Also, in [28] are mentioned the five areas of vulnerability management: discovery, reporting, analysis, coordination, disclosure, and response. That means that several research fields must be considered for vulnerability management within a CSIRT, from the beginning, detection, mitigation, and closure.

Threat Management Research Area: In today's cyber threat landscape, a wide variety of skills and coordination are needed to combat increasingly complex challenges. CSIRTs are essential actors with the technical skills necessary to provide incident response and prevention within this changing environment [29]. There are three stages for threat management within a CSIRT: prevention, detection, and response. Finally, in the research field, the *threat intelligence* process is required. It is a screening process that takes place among heaps of data. It consists of examining them in a contextual way to detect real problems and implement specific solutions for the problems detected.

Network Security Research Area: Likewise, one of the essential objectives of a CSIRT is network security. That is, how effectively we guarantee the regular operation of the network has become a critical issue. To avoid network damage and loss of information resources, network administrators must accurately understand the entire network's current state and future security trends. They must also detect attacks and harmful behaviors on time, give an emergency response, make reasonable emergency strategies for the security configuration of the network, and the allocation of resources to achieve an early prediction [30]. To strengthen network security, actors increasingly accumulate their knowledge of vulnerabilities and other threat-related information that could help CSIRTs to prevent and respond to incidents [29].

3.5 Organizational Model Proposal

The organizational model will be internal, centralized in the headquarters, and distributed to the rest of the existing campuses in the country. Due to IT services, the academic institution manages from its headquarters to its different locations and extensions. Therefore, the Academic CSIRT and its resources will be centrally located, gathering information from a wide variety of sources for appropriate treatment. The team will maintain responsibility for handling all the reports, analyzing incidents and vulnerabilities that arise, and the authority to publish warnings, best practices, response steps, and security recovery.

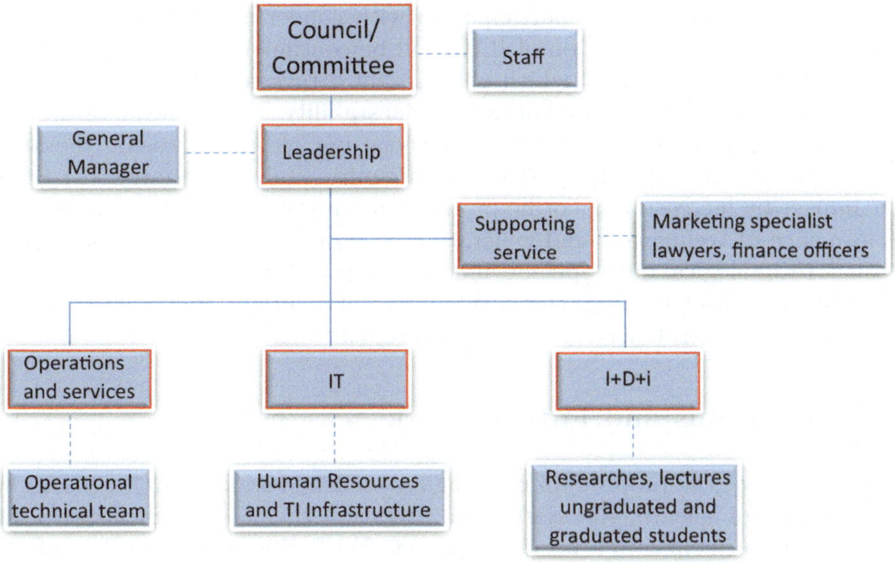

Fig. 4. Organizational model proposal for the academic CSIRT.

The proposed CSIRT will work closely with the University's IT platform and systems experts as needed. As the institution is geographically dispersed, it must reasonably respond to incidents in collaboration with the responsible technical staff to efficiently provide incident and vulnerability response coordination services.

Following the organizational model of the institution, our proposal will have the organization by processes. Figure 4 outlines a proposed structural organization chart representing the organizational structure by processes with its responsible administrative units and their dependencies. Within the organizational network of the institution, the proposed Academic CSIRT will be hierarchically dependent on the Department of Computer Science (DCCO) with a technical dependence on the Vice-Rectorate for Research through the hierarchical line of the DCCO. The initial services recommended for a CSIRT are incident, vulnerability, threat management, system monitoring, alert publication, and R&D&I processes.

In addition, within this organizational structure, we established some attributions, responsibilities, and processes such as (1) Strategic government services and administration control; (2) Administration and management, organization, and control of the provision of proactive, reactive, and security quality management services; (3) Management of CSIRT operations and services; (4) R&D&I management about information security incidents; (5) Administrative management; (6) Technology management.

Finally, regarding the impact, sustainability, and projection of the Academic CSIRT, we should note that the project's execution will open many possibilities for direct academic research participation with gender equality. The processes associated with the application of ICTs will be continuously improved. All this will optimize security levels, which will prevent attacks on systems and infrastructure, guaranteeing its economic and social stability. The following result indicators are expected to be obtained: (1) Social impact by increasing the levels of information security, the academic community, and the country. (2) Scientific impact by increasing scientific production, improvement of the professional performance of technicians and generation of knowledge. (3) Economic impact through technological development, innovation, possibility of providing services. (4) Political impact through the development of solutions that help safeguard mission-critical infrastructures and, therefore, national sovereignty.

4 Conclusions and Future Work

This project has the proposal of creating an academic and research CSIRT. We used the design science, the systematic literature review, and Strategic Planning Process principles for its development. With this, the execution of this project was planned in three phases: strengthening, consolidation and innovation. Likewise, it was possible to identify the primary operations and services of an Academic CSIRT, in which proactive, preventive, support, and value-aggregated services stand out. Additionally, the CSIRT's research areas and possible deliverables for technology transfer were determined. Finally, the organizational model that would allow to put into operation the Academic CSIRT in the educational institution was designed.

As future work, we plan to fulfill the consolidation phase, that is, to improve the operational activities, the consolidation of international agreements, and to reach the FIRST certification.

Acknowledgments. The authors would like to thank the Universidad de las Fuerzas Armadas – ESPE of Sangolquí, Ecuador, for the resources granted for the development of the research project entitled: "Design and Implementation of the IT infrastructure and service management system for the ESPE Academic CERT", coded as PIC-ESPE-2020-ESPE-CERT.

References

1. Smeets, M.: NATO members' organizational path towards conducting offensive cyber operations: a framework for analysis. In: 2019 11th International Conference on Cyber Conflict (CyCon), vol. 900, pp. 1–15. IEEE (2019)
2. Efthymiopoulos, M.P.: A cyber-security framework for development, defense and innovation at NATO. J. Innov. Entrep. **8**(1), 1–26 (2019). https://doi.org/10.1186/s13731-019-0105-z
3. Yannakogeorgos, P., Lowther, A.: Conflict and Cooperation in Cyberspace: The Challenge to National Security. CRC Press, Boca Raton (2013)
4. Ahmad, A., Maynard, S., Desouza, K., Kotsias, J., Whitty, M., Baskerville, R.L.: Cybersecurity incident response in organizations: an exploratory case study and process model of situation awareness. Comput. Secur. **101**, 102–122 (2021)
5. Cascavilla, G., Tamburri, D., Van Den Heuvel, W.: Cybercrime threat intelligence: a systematic multi-vocal literature review. Comput. Secur., 102258 (2021)
6. Hellwig, O., Quirchmayr, G., Huber, E., Goluch, G., Vock, F., Pospisil, B.: Major challenges in structuring and institutionalizing CERT-communication. In: 2016 11th International Conference on Availability, Reliability and Security (ARES), pp. 661–667. IEEE (2016)
7. Kitchenham, B., Brereton, O.P., Budgen, D., Turner, M., Bailey, J., Linkman, S.: Systematic literature reviews in software engineering-a systematic literature review. Inf. Softw. Technol. **51**(1), 7–15 (2009)
8. ENISA: Exploring the opportunities and limitations of current Threat Intelligence Platforms (2017). https://www.enisa.europa.eu/publications/exploring-the-opportunities-and-limitations-of-current-threat-intelligence-platforms. Accessed 31 Mar (2021)
9. Kossakowski, K.P.: Computer Security Incident Response Team (CSIRT) Services Framework (2019)
10. Steinke, J., et al.: Improving cybersecurity incident response team effectiveness using teams-based research. IEEE Secur. Priv. **13**(4), 20–29 (2015)
11. Catota, F.E., Morgan, M.G., Sicker, D.C.: Cybersecurity education in a developing nation: the Ecuadorian environment. J. Cybersecur. **5**(1), tyz001 (2019)
12. Brown, J.M., Greenspan, S., Biddle, R.: Incident response teams in IT operations centers: the T-TOCs model of team functionality. Cognit. Technol. Work **18**(4), 695–716 (2016). https://doi.org/10.1007/s10111-016-0374-2
13. Murakam, T., et al.: Long-term experiences in keeping balance between safety and usability in research activities in KEK. In: EPJ Web of Conferences, vol. 214, p. 08001 (2019)
14. Andrade, R., Fuertes, W.: Diseño y dimensionamiento de un equipo de un equipo de respuesta ante incidentes de seguridad informática (CSIRT). Caso de estudio: ESPE (2013)
15. Fuertes, W., Reyes, F., Valladares, P., Tapia, F., Toulkeridis, T., Pérez, E.: An integral model to provide reactive and proactive services in an academic CSIRT based on business intelligence. Systems **5**(4), 52 (2017)
16. Reyes, F., Fuertes, W., Tapia, F., Toulkeridis, T., Aules, H., Pérez, E.: A BI solution to identify vulnerabilities and detect real-time cyber-attacks for an academic CSIRT. In: Arai, K., Kapoor, S., Bhatia, R. (eds.) SAI 2018. Advances in Intelligent Systems and Computing, vol. 857, pp. 1135–1153. Springer, Cham (2019). https://doi.org/10.1007/978-3-030-01177-2_82

17. Andrade, R.O., et al.: Information security management in university campus using cognitive security. Int. J. Comput. Sci. Secur. (IJCSS) **13**(4), 124 (2019)
18. Felkner, A., et al.: Research and innovation aspects. In: Felkner, A., et al. (eds.) Cybersecurity Research Analysis Report for Europe and Japan. Studies in Big Data, vol. 75, pp. 37–97. Springer, Cham (2021). https://doi.org/10.1007/978-3-030-62312-8_3
19. Felkner, A., et al.: Legal and policy aspects. In: Felkner, A., et al. (eds.) Cybersecurity Research Analysis Report for Europe and Japan. Studies in Big Data, vol. 75, pp. 3–35. Springer, Cham (2021). https://doi.org/10.1007/978-3-030-62312-8_2
20. Slayton, R., Clarke, B.: Trusting infrastructure: the emergence of computer security incident response, 1989–2005. Technol. Cult. **61**(1), 173–206 (2020)
21. Rantos, K., Spyros, A., Papanikolaou, A., Kritsas, A., Ilioudis, C., Katos, V.: Interoperability challenges in the cybersecurity information sharing ecosystem. Computers **9**(1), 18 (2020)
22. Espín, F.V.: Guidelines and their challenges in implementing CSIRT in Ecuador. In: Botto-Tobar, M., Gómez, O.S., Rosero Miranda, R., Díaz Cadena, A. (eds.) ICAETT 2020. AISC, vol. 1302, pp. 239–251. Springer, Cham (2020). https://doi.org/10.1007/978-3-030-63665-4_19
23. Bowen, G., Sethi, A.: Internal marketing cybersecurity-conscious culture. In: Handbook of Research on Innovations in Technology and Marketing for the Connected Consumer, pp. 135–154. IGI Global (2020)
24. Ortiz-Garces, I., Sánchez-Viteri, S.: Proposal for an implementation guide for a computer security incident response team on a university campus. Computers **10**(8), 102 (2021)
25. Rigoni, A., Veneziano, D.: An introduction to CERT types, services and organization models. Next Gener. CERTs **54**, 5 (2019)
26. Settanni, G., et al.: A collaborative cyber incident management system for European interconnected critical infrastructures. J. Inf. Secur. Appl. **34**, 166–182 (2017)
27. Tanczer, L.M., Brass, I., Carr, M.: CSIRT s and global cybersecurity: how technical experts support science diplomacy. Glob. Policy **9** 60–66 (2018). https://www.overleaf.com/project/60678a8d5472c797de363eca
28. Spring, J.M., Galyardt, A., Householder, A.D., VanHoudnos, N.: On managing vulnerabilities in AI/ML systems. In: New Security Paradigms Workshop 2020, pp. 111–126 (2020). https://doi.org/10.1145/3442167.3442177
29. Bradshaw, S., Governance, G.C.O.I.: Combatting Cyber Threats: CSIRTS and Fostering International Cooperation on Cyber Security, Cyber Security in a Volatile World, pp. 105–120. Centre for International Governance Innovation (2017). https://www.jstor.org/stable/resrep05239.13
30. Kong, D., Dong, H., Li, H., Wang, Z., Li, J.: Research on situation analysis technology of network security incidents. In: Proceedings of the 2020 International Conference on Cyberspace Innovation of Advanced Technologies, pp. 213–218 (2020). https://doi.org/10.1145/3444370.3444574

Educational Programming as a Strategy for the Development of Logical-Mathematical Thinking

Itaty Albán Bedoya[1(✉)] and Mauro Ocaña-Garzón[1,2]

[1] Universidad de Las Fuerzas Armadas ESPE, Sangolqui, Ecuador
{ijalban,mhocana}@espe.edu.ec
[2] Universidad Técnica de Manabí, Portoviejo, Ecuador

Abstract. The purpose of educational programming is to teach children to write code so a computer can perform the task autonomously. Multiple skills participate in this process and lead the student to develop different types of thinking one of those being logical-mathematical. This study focuses on describing how such thinking develops, through the use of educational coding activities. To this end, an online educational programming course was created for 313 children aged 6 to 12 that involved students and coding in a fun way. All of this was done to measure how this influences the development of mathematical skills. The results show that there is a direct relationship between the aforementioned thinking and the direct engagement in coding practices. Problem-solving skills showed notorious improvements. The usage of a block-based programming platform was needed to achieve these results. In conclusion, educational programming is a potential new tool for 21st century education practices, mainly due to the improvement found by applying the same logical thinking needed for coding in daily life.

Keywords: Coding · Education · Logic · Mathematics · Children · Online course

1 Introduction

Logical-mathematical thinking, usually accompanied by linguistic thinking, are considered to be a core element in the classroom since they are the basis of IQ tests which measure the intelligence of a person. Logical-mathematical thinking is shown in the ability to use numbers correctly, recognize patterns and reason properly [1]. Everyone develops it in his mind through the relationships that exist between objects. The learning process to develop logical-mathematical thinking is gradual, depending on the existing stimuli, according to Piaget, as cited by [2]. It involves reasoning, imagining, discovering, intuiting, testing, generalizing, estimating, and checking results for understanding reality [3].

In recent years, the use of educational programming has increased. Which involves the participation of children in the creation of detailed step-by-step instructions, in the form of code, that a machine must interpret and follow [4]. When children code, they involve an algorithm, which is a list of ordered acts that allows achieving a task. Learning

M. Botto-Tobar et al. (Eds.): CIT 2021, LNNS 405, pp. 309–323, 2022.
https://doi.org/10.1007/978-3-030-96043-8_24

to program is seen as a way in which the child understands the world and awakens their curiosity to explore, create, question and think about their actions [5, 6]. Therefore, coding can change not only the way a learner thinks but also the way they see themselves as part of society [7].

The teaching of educational programming currently contains various problems: 1) there are almost no proper studies, with preschool-age children in mind, about coding and no study evaluates the effects of coding on cognitive ability since all are directed to explain the level of learning in terms of programming skill [8]. 2) Another difficulty is that the field of early computer science education has only just begun to emerge and, therefore, the coding stages are not yet clearly defined or thoroughly investigated, as proposed by [7]. 3) When using an application of any kind, digital native generations place children as consumers rather than creators of digital media [9, 10]. In an analogy, children are being taught to read more but not to write. This occurs for cultural reasons, there is still no predisposition to participate in modern teaching technologies [11].

Seeking to contribute to the solution of the problems found, the current study aims to describe how the use of educational coding activities exert influence on the development of logical mathematical thinking. Through the creation of an asynchronous online course called "Programming for children" which aimed at children from 6 to 12 years old. For this, the following research questions will be investigated:

1. What types of thinking are developed through Educational Programming
2. What skills pertaining to the development of the logic mathematical field are displayed with the Educational Programming?
3. How does educational programming impact in the development of logical - mathematical thinking?
4. What activities related to Educational Programming favor the domain of logical mathematical thinking?

The benefit of having a child in contact with programming is not only in the use of technology for their enjoyment. It also involves the mental capacity to analyze problems and conceive executable algorithmic procedures for their solution [12, 13], finding a causal link between coding activities and planning and inhibition skills, improving logical thinking and fostering the creativity of children's response [6, 13, 14]. Also, this includes the development of reasoning and listening skills, observation and working memory, using non-verbal organization [15, 16]. It is important to note that all students will not become programmers but they will certainly have a strong asset [17].

Students enjoy the learning activities associated with coding because they provide opportunities to explore mathematical and probabilistic concepts and participate in social interactions, as well as an increase in the use of ordering skills [8, 18]. That is why it is recommended to integrate coding as part of solving concurrency problems in the student's curriculum. Connecting with the topics of the education curriculum, and at the same time teaching values and inclusion [19].

It is important that the educator and their students show interest in the area of technology and feel motivated to integrate these activities in their classroom [5]. From this perspective, it is suggested to introduce computational thinking through programming in the short term [12]. That is why it is essential to start from meaningful contexts where children currently live by helping them to become familiar with programming through practical activities with concrete representations. This will allow for the coding practice to be meaningful [20].

As for the way in which coding should be taught to children, the ages of the students influence their cognitive development during the teaching of programming [21]. There must be a curriculum designed according to the development stages or learning trajectories and the developmental model of how children operate within a domain [7, 21]. Therefore, the concern is not whether children are exposed to coding but how to expose them appropriately, in a playful way, for the benefit of the students' development [20]. To this end, different approaches are used to teach programming to young children, including disconnected activities, block-based programming platforms[1] and physical programming[2] [22, 23].

Taking into account that the main objective is not to teach a specific programming language but to show that the logic is the same for all languages [17], visual or block programming is preferable for younger age groups [24] as these are attractive to children [25, 26]. New technological tools and programming environments offer new opportunities [27] where practice sessions are an important factor due to the age of the learners [28]. Taking this into account as the main component, the time used in each activity is key for their personal involvement: intellectually and emotionally (Papert cited by [27]).

In the Ecuadorian and Latin American contexts, educational programming is widely unknown and unused, especially in school curriculums, despite that, for [29], the Ministry of Education must take actions to guarantee that learners are prepared for the demands of the 21st century. Currently, the approaches that have occurred between early childhood education and programming in Ecuador, have come from the hand of the academy, the same as through initiatives such as "The Hour of Code" (Code.org 2013), which involved the participation of children from the community. The most prominent is the one carried out by ESPOL from 2015 to 2018. Outside of this, there are Extracurricular Studies Centers, focused on the teaching of coding which, due to its high cost, are inaccessible to certain sectors of society. Finally, the ideal in this discipline would be to follow in the footsteps of some countries that promote it such as Estonia, England, the United States, Israel, among others. All of them have included coding, either within the curriculum or as support for other topics [30].

---

1 Block programming: Programming language composed of visual blocks which control structures, as well as data and values. Each type of block has a specific shape, this allows children to learn about logic and problem solving without worrying about syntax [36].

2 Physical programming: Tangible languages give children the opportunity to directly manipulate objects that form a code, they also allow children to interact with objects and transform the logic of the world into the logic of the program, converting the symbolic and concrete procedures [37].

2 Method and Materials

2.1 Research Design

This study has a quantitative non-experimental design since none of the variables were deliberately manipulated. In addition, participants in the course were observed in their natural environment in order for their learning to be analyzed [31].

The type of sampling was non-probabilistic in order to answer the research questions and fulfill the objective by creating an online programming course for children between 6 and 12 years old. The population of this study was made of 313 children (184 boys and 129 girls) from different provinces of Ecuador in the following age ranges (Table 1):

Table 1. Number of children according to their age

Age	6 years	7 years	8 years	9 years	10 years	11 years	12 years
No. of children	71	44	92	44	28	15	19

The course was open to children older than five years since previous studies indicate that older preschool children demonstrated a higher average level of mastery in all programming concepts than younger preschool children (less than 5 years old years) [32]. This can be due to a variety of factors including increased working memory, attention span, and the ability to plan.

The course was hosted in an institutional Moodle platform (https://academy.espe.edu.ec/course/view.php?id=138). This virtual environment allowed for the recording of all the interactions that children had with the different activities proposed in the course. These activities were designed for the development of programming skills and their connection with logical mathematical thinking. Moodle is expected to help students learn, expand their knowledge and complete tasks online [33].

The current investigation put into place the following procedures:

a) A literature review was carried out to establish the appropriate activities for the creation of the Online Course named "Programming for Children". In this way, it was determined that the activities would be based on the tools found in Code.org. This platform is the most effective option for children from preschool to primary school, since it is no different from the tangible environments that are used in educational robotics, for solving coding problems in a virtual environment [12].

From this platform, activities called Hour of Code are derived, which became a de facto standard for the first encounter with programming [23]. It is important to note that these activities have a focus on visual block programming. These are easier and faster to learn for young children, and they are also used to learn programming for the first time, since students can adapt it to other programming languages more easily [11]. All of this is accompanied by linking the educational context of the learner with words or directional arrows, sequential words and the combination of them. Based on strategies where programming logic is used to focus on pre-existing problems in the students' disciplines [17, 20].

b) An asynchronous course with 96 activities was structured within the Moodle platform and divided into four modules: "Introduction to Programming", "Programming logic", "Programming Fundamentals", "Computer's Science". These covered five topics each, related to computing, computational thinking, and coding. At the end of the course there was a formative evaluation, which takes into account all the activities seen previously.

The proposed activities vary between games such as word search, crossword puzzles, find the pairs, order images, and others as well as motivational videos for each topic. Regarding the explanations of concepts, mobile cards and interactive presentations were used. For the practical part, we used games that implied the use of visual programming in blocks with their respective tutorial such as: "Lightbot", "Kodable", "Corre Marco". At the end of each topic there was a questionnaire to find out how much students learned about each topic (Figs. 1, 2 and 3).

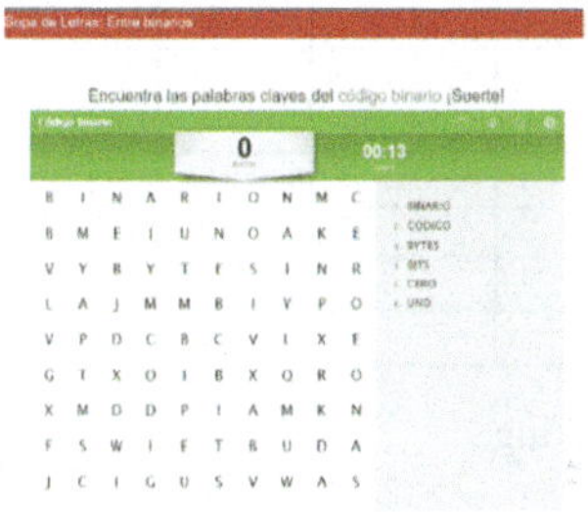

Fig. 1. Motivational activity "alphabet soup"

Fig. 2. Explanation activity "mobile cards"

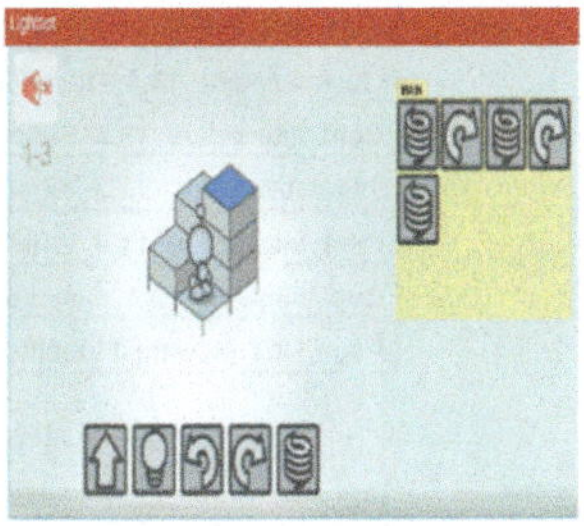

Fig. 3. Practice activity "block programming"

c) With the information gathered 20 days after the opening of the course, preliminary quantitative data were obtained for the multivariate analysis of PCA (Principal Component Analysis). The treatment of the information obtained was processed through Excel and Minitab for the analysis and interpretation of the variables studied.

2.2 Results

The asynchronous course "Programming for children" allowed us to have a first approximation on how Programming and Logical-Mathematical Thinking were related. Thus, 20 days after its opening, the following results were obtained:

I. What thinking are developed through Educational Programming?

To answer this question, the PCA multivariate analysis (Principal Component Analysis) was used. PCA allowed us to group activities according to common characteristics, obtaining as a product the clustering of 5 components, which are shown below in descending order of significance:

i. Numerical thinking (Fig. 4)

Rotated Component Matrix					
	Component				
	1	2	3	4	5
Page: 14.3. What is different?	,930				
Page: 15.6. Word Search: Between binaries	,923				
Interactive Cont.: 15.4. What is it?	,911				
Page: 16.3. I take care of myself!	,903				
Questionnaire: 16.7. Question XVI	,895				
Page: 16.5. Crossword: The Rules	,887				
Quiz: What We Learned	,879				
Interactive Cont.: 14.1. Among many mistakes	,875				
Questionnaire: 15.7. Question XV	,867				
Questionnaire: 14.6. Question XIV	,854				
Cont. Interactive: 13.4. Did you know...	,852				
Questionnaire: 13.8. Question XIII	,848				
Cont. Interactive: 13.6. Searching for Pixels	,812				
Interactive Cont.: 13.2. What is it?	,808				
Cont. Interactive: 11.4. What is it?	,697				
Questionnaire: 12.4. Question XII	,679				
Page 12.1: Solving it together...	,637				

Fig. 4. Component 1: numerical thinking

Component 1, there are the activities that develop numerical thinking, they propose the use of exercises where numbers are essential to be able to create new knowledge, in addition to experiencing emotions on issues that have to do with empathy. Putting into practice skills in the use of numbers to reason quantitatively through counting and series. These items are related because mathematics is applied to new situations.

ii. Logical thinking (Fig. 5)

Component 2, is made of the activities that develop logical thinking. They propose the use of exercises where student must create their own ordered sequences with a defined objective. All of this must start from a previous explanation, putting into practice cause-effect reasoning and concluding from premises. These activities also show the deployment of analysis, including classification, through temporal sequencing and understanding of patterns. These items are related because they create reasoning challenges for children.

iii. Conceptual thinking (Fig. 6)

Rotated Component Matrix					
	Component				
	1	2	3	4	5
Interactive Cont.: 7.4. What is it?		,845			
Interactive Cont.: 8.4. Why solve problems?		,829			
Interactive Cont.: 6.4. Let's order the algorithm		,809			
Questionnaire: 8.5. Question VIII		,790			
Page: 8.1. Solving Problems		,781			
Questionnaire: 6.6. Question VI		,746			
Interactive Cont..: 6.3. What is it?		,737			
Page: 9.2. CodeSpark Tutorial		,735			
Interactive Cont.: 9.4. What is it?		,717			
Questionnaire: 7.7. Question VII		,711			
Questionnaire: 9.5. Question IX		,646			
Pag. 10.3: Run Marco Tutorial		,621			
Questionnaire: 5.6. Question V		,598			
Questionnaire: 10.6. Questioner X		,573			

Fig. 5. Component 2: logical thinking

Rotated Component Matrix					
	Component				
	1	2	3	4	5
Page: 3.4. Crossword: Programming			,864		
Questionnaire: 4.5. Question IV			,856		
Page: 4.3. Knowing programming languages			,838		
Interactive Cont..: 4.4. Fun facts...			,814		
Interactive Cont..: 3.1. Guess guesser			,806		
Page: 5.2. Workshop Where are they? - Part I			,772		
Questionnaire: 3.5. Question III			,763		
Interactive Cont.: 4.2. Most Popular Programming			,753		
Page: 2.4. How to Install a Program			,744		
Page: 5.4. Counting steps			,726		
Interactive Cont.: 2.3. Learning about the program			,724		
Questionnaire: 2.5. Question II			,671		

Fig. 6. Component 3: conceptual thinking

Component 3 is made of activities that develop conceptual thinking. They propose the use and knowledge of new words, related to programming. They facilitate the understanding of the exercises assigned in the different topics, putting the observation skill into practice as well as creativity and enrichment of the vocabulary in learners. These activities too used the discrimination of words, the verification and storage of knowledge. These items are related because they are explanatory and generally used mobile cards.

iv. Reactive thinking (Fig. 7)

Component 4 is composed by activities that develop reactive thinking. They propose to remember themes that the children probably already identified, in addition to creating

Rotated Component Matrix					
	Component				
	1	2	3	4	5
Page. 1.3: Word Search: Parts of the Computer				,720	
Interactive Cont..: 1.4. Computer: Uses				,691	
Quiz: Let's Learn I				,603	
Interactive Cont.: 1.1. Guess guesser				,601	
Questionnaire: 1.5. Question I				,509	

Fig. 7. Component 4: reactive thinking

curiosity about the themes that will be seen later. This set of activities puts working memory into practice through the use of experiences and representation. These items are related because they deal with issues that are found in the educational context easily.

v. Algorithmic thinking (Fig. 8)

Rotated Component Matrix					
	Component				
	1	2	3	4	5
Interactive Cont.: 10.5. What is it?					,402

Fig. 8. Component 5: algorithmic thinking

In component 5, there is an activity that develops algorithmic thinking. It proposes the use of exercises that allow the child to create ordered sequences based on examples, to reach a fixed objective, in which conditionals intervene, as well as loops. This activity puts into practice algorithmic procedures, planning, and inhibition as well as problem-solving, through imitating, inferring, and ordering. This item is characterized by being completely practical.

2. What skills pertaining to the development of the logic mathematical thinking are displayed with the Educational Programming?

To answer this question, we used the statistical analysis of the scores obtained from the comprehension questionnaires applied at the closing of each unit. Obtaining the following results:

Figures 9 and 10 display the average grade ($\overline{X}$ 8,997 $\approx$ 9) obtained in each questionnaire whose maximum grade is 10. This means that students have mastered the required learning according to the scale of ratings from the Ministry of Education (2016) [34, 35]. The lowest grades in the questionnaires were displayed in the topics which did not used a visual tool for learning. The learners with the highest scores in their questionnaires, are those that resort to memory, that is, those children who have previous knowledge in these topics.

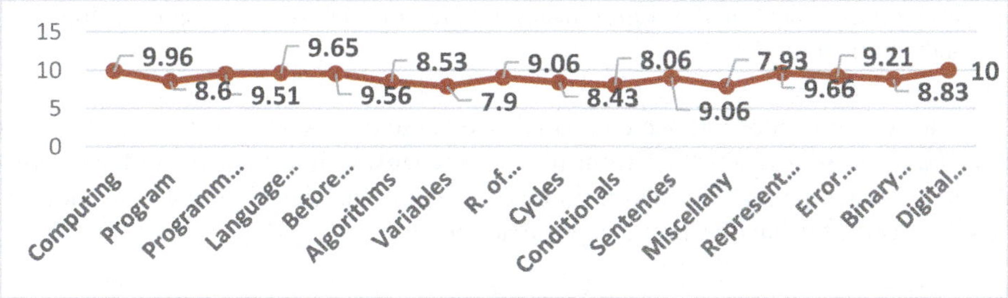

Fig. 9. Analysis of the averages of the qualifications obtained in the questionnaires

Variable	Arithmetic Average	Minimum	Maximum	Median
Average Scores on Questionnaires	8,99/10	7,90/10	10/10	9,06/10

Fig. 10. Analysis of the averages of the qualifications obtained in the questionnaires

Based on Results displayed on Figs. 9 and 10, show the topics with the best scores in the comprehension questionnaires. These topics have a direct relationship with the development of logical-mathematical thinking which are ordered according to the curriculum of the Ministry of Education [34] as shown in Fig. 11.

Topics of Programming	Related Skills
Before programming & Lightbot	M.1.4.4. Distinguish the location of objects in the environment according the notions up and down, front / back and above / below plus right and left
Cycles	M.1.4.10. Describe and build simple patterns by grouping quantities of up to ten items
Binary numbers	M.1.4.14. Identify quantities and associate them with the numerals 1 to 10 and 0.
	M.1.4.15. Write the natural numbers, from 0 to 10, in meaningful contexts
Algorithms	M.1.4.16. Use ordinal numbers, from first to fifth, in the location of elements of the environment.
Sentences	M.1.4.29. Compare and relate activities with the notions of time: yesterday, today, tomorrow, afternoon, night, before, now, after and days of the week in everyday situations
Conditionals	M.1.4.33. Identify probable and non-probable events in everyday situations.
Resolution of problems	M.1.4.34. Collect and represent information from the environment in pictograms, solving simple problems.

Fig. 11. Logical mathematical skills related to programming [34]

3. How does educational programming impact on the development of logical - mathematical thinking?

To answer this question, we compared two parameters shown in Fig. 12. First, we used the mean scores obtained from the comprehension questionnaires of the topics mentioned in the previous question (in blue). Second, the average number of views of the topics addressed in question 2 is presented in red.

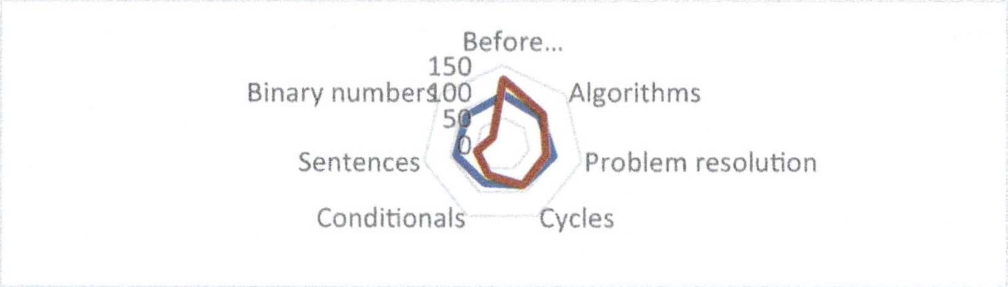

Fig. 12. Relationship between the grade averages of the questionnaires and the visualizations of the topics related to logical- mathematical thinking

Variable	Total	Minimum	Maximum	Median
Averages number of views of the topics related to logical-mathematical thinking	522	24	125	81

Fig. 13. Averages scores of the questionnaires and number of views of the topics related to logical-mathematical thinking

Variable	Arithmetic Average	Minimum	Maximum	Median
Averages scores of the questionnaires of the topics related to logical-mathematical thinking	8,86/10	8,06/10	9,56/10	8,83/10

Fig. 14. Averages scores of the questionnaires and number of views of the topics related to logical-mathematical thinking

Figure 12 shows the relationship that exists between the number of views of the activities proposed in the educational programming topics focused on the development of logical-mathematical thinking with the qualifications of the questionnaires of the same topics. As seen in Fig. 12, the high scores go hand in hand with the high number of views of coding activities. In other words, it may be presumed that the more educational programming activities are carried out, the greater the development of logical-mathematical thinking skills.

On the other hand, Fig. 13 shows the number of views on the programming topics related to logical mathematical thinking. These views totaled 522 up to the data collection cut-off point. Of this total, most views (125) concentrated on the topics which used practical activities. Meanwhile, in Fig. 14, we can see the average of the marks in the final questionnaires of the topics related to logical mathematical thinking. With an average of 8.86, which means that, so far, the students have reached the required learning. The topic with the highest qualification is the one which has a higher number of visualizations per activities, confirming what is reflected in Fig. 12.

4. What activities related to Educational Programming favor the development of logical mathematical thinking?

We first analyzed the average number of views in each unit and obtained the following results:

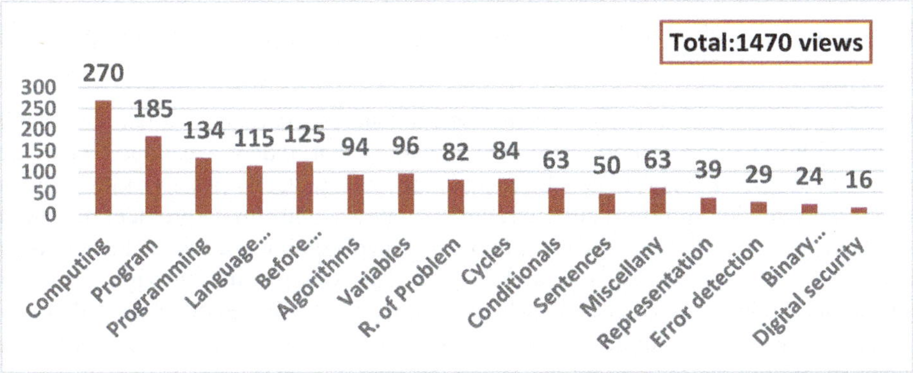

Fig. 15. Average number of total views.

Figures 15, shows the average number of times that children viewed each activity with a sum of 1470 for the entire course. This was considered good since the data were collected approximately one month after the start of the course. It should be taken into account that the **topics** with the highest number of visualization are those that have practical activities. Activities which used block programming were the ones with the highest numbers of views.

Finally, Fig. 16 shows the topics previously related to the skills of logical-mathematical thinking based on the number of views. This figure shows that practical activities were the most popular ones. These activities referred to games that involved block programming. The second most favored are the explanatory activities. These activities usually seek to give a simple and summarized definition which is linked to the context of the child. The third most favored were motivational activities which involved simpler items that encouraged student to develop the topics.

2.3 Discussion

Based on the results presented so far, it seems evident that different kinds of thinking (e.g. numerical, logical, and algorithmic) are developed through educational coding activities designed for children. Coinciding with [8] who affirm that programming in students

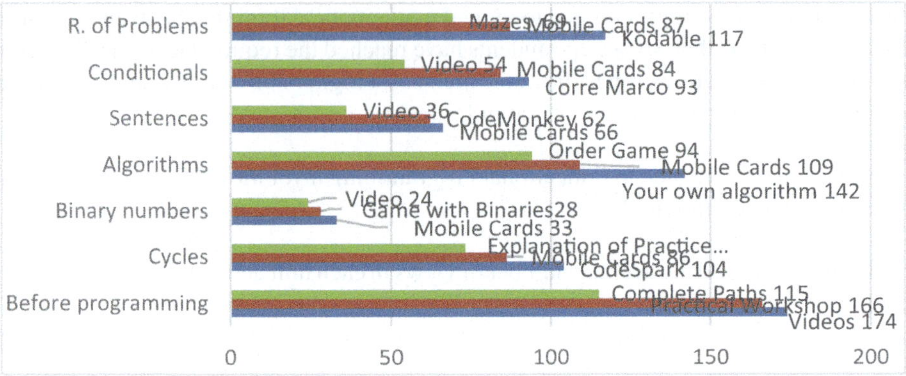

Fig. 16. Activities with more views in topics related to logical mathematical thinking

provides opportunities to explore mathematical concepts, in addition to an increase in the use of skills to order. Additionally, our study found that students developed skills such as: spatial location of oneself and objects in the environment, contrast activities based on time, representation of information through pictograms and simple problem resolution which, in turn, are closely related to the improvement of computer skills, abstract thinking and problem-solving strategies of a child [13]. This was possible through the use of educational programming activities which are linked to logical-mathematical thinking. Regarding this type of thinking, it was found that the more visualizations, the more engagement with the topics, and the higher the scores in questionnaires about logical-mathematical thinking. The relationship between number of views and high achievement has also been reported by previous studies, e.g. [38].

So far, our students have mastered the required learning in the area of coding through the proposed topics. This has been achieved through the use of visual tasks that draw the child's attention, agreeing with [18, 25] on the importance of graphic interfaces for the development of algorithmic thinking. Along the same line, the activities that had the greatest impact on the development of logical - mathematical thinking were the practical ones over those tasks which focused on theory. This is in accordance with what affirmed [28] on practice as an important factor in the programming learning process. In addition to practice, there are other elements that may help to motivate student participation. Even in more informal environments it is necessary to foster certain human factors such as curiosity, interest and even the teacher guidance [39].

3 Conclusions

Educational Programming is gaining strength throughout the world as a tool for education in the 21st century. It engages the child with the platform that used teaching-learning processes based on coding. Yet, what is really important is not to teach so that they code as engineers, but as a way for them to apply programming logic skills in their daily life. In this study, we found that educational coding was generally linked to numerical, logical, conceptual, algorithmic and reactive thinking. These types of thinking are considered necessary for the successful development of the students' academic performance.

Our findings support what is stated by similar studies, educational programming has a direct relationship with the development of logical-mathematical thinking. In our case, this relationship was most evident in the amount of coding activities completed, the larger the amount the higher the average score in evaluations related to logical-mathematical thinking and its related skills: spatial location, temporal-spatial relationships and problem solving. Graphical interfaces and block - based programming were fundamental tool for the development of these skills.

Some limitations appeared in the research. One of them was the lack of monitoring for activities that used external URLs. Future studies may follow up on this course to obtain more precise data regarding those external activities or, instead, use the course to investigate how it influences other types of thinking (e.g. linguistic thinking). Also, when the possibility of a face-to-face course becomes available, we can measure its results and compare them with those obtained online. Thanks to this this research, more information was found on how coding affects children's cognition, thus narrowing the gap that exists in the topic.

References

1. Gardner, H.: Inteligencias múltiples. In: La inteligencia reformulada: las inteligencias múltiples en el siglo XXI, Barcelona, España, Paidos, pp. 33–34 (2001)
2. Linares, R.: Master en Paidopsiquiatría Módulo I Desarrollo Cognitivo: Las Teorías. Master en Paidopsiquiatría. Bienio 07–08, **I,** 10–11 (2008)
3. Medina, M.: Estrategias Metodológicas Para El Desarrollo Del Pensamiento Lógico Matemático. UNESUM-Ciencias. Rev. Científica Multidisciplinaria, vol. 1, no. 3, pp. 73–80 (2017). ISSN 2602-8166. https://doi.org/10.47230/unesum-ciencias.v1.n3.2017.28
4. Lee, J.: Coding in early childhood. Contemp. Issues Early Child. (2019). https://doi.org/10.1177/1463949119846541
5. Pinto, M., Osório, A.: Learn to program in preschool: analysis with the participation scale. Pixel-Bit, Revista Medios y Educación (55), 133–156 (2019). https://doi.org/10.12795/pixelbit.2019.i55.08
6. Alkadhi, B., Alsaif, S., Alangri, A., Alkallas, F., Aljadou, H., Altamimi, N.: Labenah: an arabic block-based interactive programming environment for children. the journey of learning and playing. In: Stephanidis, C. (ed.) HCII 2019. CCIS, vol. 1034, pp. 179–187. Springer, Cham (2019). https://doi.org/10.1007/978-3-030-23525-3_23
7. Bers, M.U.: Coding as another language: a pedagogical approach for teaching computer science in early childhood. J. Comput. Educ. **6**(4), 499–528 (2019). https://doi.org/10.1007/s40692-019-00147-3
8. Çiftci, S., Bildiren, A.: The effect of coding courses on the cognitive abilities and problem-solving skills of preschool children. Comput. Sci. Educ. **30**(1), 3–21 (2020). https://doi.org/10.1080/08993408.2019.1696169
9. Figueiredo, M., Cifredo-Chacón, M.-Á., Gonçalves, V.: Learning programming and electronics with augmented reality. In: Antona, M., Stephanidis, C. (eds.) UAHCI 2016. LNCS, vol. 9739, pp. 57–64. Springer, Cham (2016). https://doi.org/10.1007/978-3-319-40238-3_6
10. Price, C.B., Price-Mohr, R.M.: Stories children write while coding: a cross-disciplinary approach for the primary classroom. Camb. J. Educ. **48**(6), 735–747 (2018). https://doi.org/10.1080/0305764X.2017.1418834
11. Ayman, R., Sharaf, N., Ahmed, G., Abdennadher, S.: MiniColon; teaching kids computational thinking using an interactive serious game. In: Göbel, S., et al. (eds.) JCSG 2018. LNCS, vol. 11243, pp. 79–90. Springer, Cham (2018). https://doi.org/10.1007/978-3-030-02762-9_9

12. Arfé, B., Vardanega, T., Ronconi, L.: The effects of coding on children's planning and inhibition skills. Comput. Educ. **148** (2020). https://doi.org/10.1016/j.compedu.2020.103807
13. Panskyi, T., Rowinska, Z., Biedron, S.: Out-of-school assistance in the teaching of visual creative programming in the game-based environment – case study: Poland. Think. Creat. **34** (2019). https://doi.org/10.1016/j.tsc.2019.100593
14. Fatourou, E., Zygouris, N.C., Loukopoulos, T., Stamoulis, G.I.: Teaching concurrent programming concepts using scratch in primary school: methodology and evaluation. J. Eng. Pedagogic **8**(4), 89–105 (2018). https://doi.org/10.3991/ijep.v8i4.8216
15. Kahn, K., Winters, N.: Child-friendly programming interfaces to AI cloud services. In: Lavoué, É., Drachsler, H., Verbert, K., Broisin, J., Pérez-Sanagustín, M. (eds.) EC-TEL 2017. LNCS, vol. 10474, pp. 566–570. Springer, Cham (2017). https://doi.org/10.1007/978-3-319-66610-5_64
16. Strawhacker, A., Bers, M.U.: What they learn when they learn coding: investigating cognitive domains and computer programming knowledge in young children. Educ. Technol. Res. Dev. **67**(3), 541–575 (2018). https://doi.org/10.1007/s11423-018-9622-x
17. Miranda Júnior, A., Botega Tavares, D., Caires Carvalho, J.: Programming teaching tools feature assessment associated with Brazilian curriculum base obtained through BPL platforms analysis. In: Zaphiris, P., Ioannou, A. (eds.) HCII 2019. LNCS, vol. 11590, pp. 380–389. Springer, Cham (2019). https://doi.org/10.1007/978-3-030-21814-0_29
18. Akpinar, Y., Aslan, Ü.: Supporting children's learning of probability through video game programming. J. Educ. Comput. **53**(2) 228–259 (2015). https://doi.org/10.1177/0735633115598492
19. Bers, M., González-González, C., Armas-Torres, M.: Coding as a playground: Promoting positive learning experiences in childhood classrooms. Comput. Educ. **138,** 130–145 (2019). https://doi.org/10.1016/j.compedu.2019.04.013
20. Lee, J., Junoh, J.: Implementing unplugged coding activities in early childhood classrooms. Early Child. Educ. J. **47**(6), 709–716 (2019). https://doi.org/10.1007/s10643-019-00967-z
21. Chou, P.: Using ScratchJr to foster young children's computational thinking competence: a case study in a third-grade computer class. J. Educ. Comput. Res. **58**(3), 570–595 (2020). https://doi.org/10.1177/0735633119872908
22. Kalelioglu, F., Sentance, S.: Teaching with physical computing in school: the case of the micro:bit. Educ. Inf. Technol. **25**(4), 2577–2603 (2019). https://doi.org/10.1007/s10639-019-10080-8
23. Effenberger, T.: Towards adaptive hour of code. In: Isotani, S., Millán, E., Ogan, A., Hastings, P., McLaren, B., Luckin, R. (eds.) AIED 2019. LNCS (LNAI), vol. 11626, pp. 339–343. Springer, Cham (2019). https://doi.org/10.1007/978-3-030-23207-8_62
24. Blancas, M., Valero, C., Mura, A., Vouloutsi, V., Verschure, P.F.M.J.: "CREA": an inquiry-based methodology to teach robotics to children. In: Merdan, M., Lepuschitz, W., Koppensteiner, G., Balogh, R., Obdržálek, D. (eds.) RiE 2019. AISC, vol. 1023, pp. 45–51. Springer, Cham (2020). https://doi.org/10.1007/978-3-030-26945-6_4
25. Deng, X., Wang, D., Jin, Q.: TLogic: a tangible programming tool to help children solve problems. In: El Rhalibi, A., Pan, Z., Jin, H., Ding, D., Navarro-Newball, A.A., Wang, Y. (eds.) Edutainment 2018. LNCS, vol. 11462, pp. 255–262. Springer, Cham (2019). https://doi.org/10.1007/978-3-030-23712-7_35
26. Pila, S., Aladé, F., Sheehan, K., Lauricella, A., Wartella, E.: Learning to code via tablet applications: an evaluation of Daisy the Dinosaur and Kodable as learning tools for young children. Comput. Educ. **128**, 52–62 (2019). https://doi.org/10.1016/j.compedu.2018.09.006
27. Papavlasopoulou, S., Giannakos, M., Jaccheri, L.: Exploring children's learning experience in constructionism-based coding activities through design-based research. Comput. Hum. Behav. **99**(7491), 415–427 (2019). https://doi.org/10.1016/j.chb.2019.01.008

28. Plaza, P., et al.: Scratch as educational tool to introduce robotics. In: Auer, M.E., Guralnick, D., Simonics, I. (eds.) ICL 2017. AISC, vol. 715, pp. 3–14. Springer, Cham (2018). https://doi.org/10.1007/978-3-319-73210-7_1
29. Navarrete, G., Mendieta, R.: Las tic y la educación ecuatoriana en tiempos de internet: breve análisis. Espirales **2**(15), 123–136 (2018). http://www.revistaespirales.com/index.php/es/article/download/220/165
30. López, L.: ¿Por qué trabajar la programación de computadoras en la escuela? Preguntas, sugerencias y herramientas. Para el Aula **8**, 9–11 (2013). https://www.usfq.edu.ec/sites/default/files/2020-07/0004_para_el_aula_06.pdf
31. Sampieri, R., Mendoza, C.: Metodología de la investigación. Las rutas cuantitativa, cualitativa y mixta, VIII. Ciudad de México, McGraw-Hill (2018)
32. Elkin, M., Sullivan, A., Bers, M.: Programming with the KIBO robotics kit in preschool classrooms. Comput. Sch. **33**(3), 169–186 (2016). https://doi.org/10.1080/07380569.2016.1216251
33. Clare, C., Bradley, N., August, D., James, K.: Calculating the success of moodle at the University of Belize. In: Proceedings of the Third Annual Research for National Development Conference, vol. 3, pp. 1–17 (2019)
34. MINEDUC: Instructivo Para La Aplicación De La Evaluación Estudiantil, pp. 1–44 (2016). https://educacion.gob.ec/wp-content/uploads/downloads/2016/07/Instructivo-para-la-aplicacion-de-la-evaluacion-estudiantil.pdf
35. MINEDUC: Educación General Básica, pp. 66–69 (2016) https://educacion.gob.ec/wp-content/uploads/downloads/2016/03/PREPATORIO.pdf
36. Milne, L.: Blocks4All: making block programming languages accessible for blind children (117), 26–29 (2017). http://delivery.acm.org/https://doi.org/10.1145/3060000/3051525/p26omilne.pdf
37. Papavlasopoulou, S., Giannakos, M.N., Jaccheri, L.: Reviewing the affordances of tangible programming languages: implications for design and practice. In: IEEE Global: Engineer and Education Conference (EDUCON), pp. 1811–1816 (2017). https://doi.org/10.1109/EDUCON.2017.7943096
38. Ocaña, M., Khosravi, H., Bakharia, A.: Profiling language learners in the big data era. presented at the ASCILITE, Singapore, Conference Paper (2019). https://espace.library.uq.edu.au/view/UQ:eb11b3a
39. Ocaña, M., Mejía, R., Larrea, C., Analuisa, C., Freire, C.: Informal learning in social networks during the COVID-19 pandemic: an empirical analysis. In: Botto-Tobar, M., Cruz, H., Díaz Cadena, A. (eds.) CIT 2020. AISC, vol. 1327, pp. 400–414. Springer, Cham (2021). https://doi.org/10.1007/978-3-030-68083-1_30

Computational Modeling and Characterization of Composites

Neurotoxic Effects on Banana Workers Exposed to Agrochemicals: Ecuador Case Study

Gabriela Zambrano-Ganchozo[1(✉)], Andrea Rodriguez-Ramos[1], Kenny Escobar-Segovia[1,2], Luis Duque-Cordova[1], and Daniela Guzmán-Cadena[2]

[1] Universidad Espíritu Santo, Samborondón, Ecuador
{gzambranog,alrodriguezr,luisduque}@uees.edu.ec, kescobar@espol.edu.ec
[2] Escuela Superior Politécnica del Litoral, Guayaquil, Ecuador
ddguzaman@espol.edu.ec

Abstract. Organophosphorus insecticides (OPs) are pest and disease control agents used worldwide to improve agricultural processes. Erythrrhytic acetylcholinesterase is the biomarker of choice for the biological monitoring of the working population exposed to organophosphorous pesticides, because the combination of agrochemicals with this enzyme acetylcholinesterase in the nerve endings of the brain and nervous system prevent the transmission of nerve impulses, causing poisoning. The objective of this study was to determine the neurotoxic effects associated with chronic exposure to pesticides by applying the Neurological-Psychological Questionnaire (PNF) and laboratory analysis (erythritemia acetylcholinesterase levels). A cross-sectional descriptive study was conducted on workers from six banana plantations who are occupationally exposed to organophosphorous pesticides, represented by 103 individuals out of 138 who constituted the universe, as inclusion criteria were considered agricultural workers exposed for a period longer than six months, with apparent good health condition, without psychosocial habits (smoking, alcohol intake). The prevalence of persistent symptoms was 6 per 10 workers in men and 5 per 10 workers in women, in turn, using linear regression models, it was possible to predict the values of neurological (N) symptoms and psychoneurovegative instability (NP) compared to the independent variable of erythrctarian acetylcholinesterase (AchE) levels. This research showed a marked decrease in the values of erythrctarian acetylcholinesterase (AchE) showing signs and symptoms that characterize pesticide overexposure and likely chronic poisoning in banana workers.

Keywords: Acetylcholinesterase · Organophosphorus · Neurological-psychological questionnaire

1 Introduction

Organophosphorous (OP) compounds represent a large amount of synthesized chemicals, since the 19th century, for various purposes, such as ectoparasiticides, chemical weapons, drugs, petroleum additives, solvents, electrical insulators, but mainly as agrochemicals for the control of vectors of sanitary importance. In underdeveloped countries,

M. Botto-Tobar et al. (Eds.): CIT 2021, LNNS 405, pp. 327–337, 2022.
https://doi.org/10.1007/978-3-030-96043-8_25

about 25% of the world's existing pesticides are used; in addition, they are responsible for 50% of cases of poisoning and 99% of deaths from exposure to these compounds [1, 2].

In Ecuador, it is estimated that 1 in 7 workers are intoxicated using pesticides, however, accurate information on the extent of the problem is limited. Information on the neurotoxic potential of most of these substances is scarce (<0.4% of all chemicals), long-term toxicity effects are unknown and rural areas exist under-registration of existing cases [3–5].

The Centre for Toxicological Information and Advice (CIATOX) of ecuador's Ministry of Public Health in 2018 reported 2,113 cases of toxic substance poisoning of which 425 cases were caused by pesticides. The age group between the age of 20 and 49 was most affected by these events, however; under-registration and the absence of epidemiological surveillance systems transform this data into unreliable [6].

According to the Food and Agriculture Organization of the United Nations (FAO), 99% of poisoning cases occur in countries with poor education, regulation, health and control systems, their main causes being those related to education, communication, risk regulation and lack of participation in decision-making, as well as problems in pesticide storage and final packaging disposal [7].

Nationally, the province of Los Ríos is the main producer of bananas, generating a daily export of $4 million, however, different reports by Human Rights Watch (2002) and IFA-Corporation for the Development of Labour Production and Environment in 2006 and 2009, states that there is a precarious work regime characterized by long working hours and measures to protect and prevent poor or non-existent occupational hazards. Staff who perform chemical mixtures for weed control and phytosanitary diseases are considering as a high-risk group of occupational and environmental exposure, presenting variable levels of erythritary acetylcholinesterase, due to exposure time, working seniority, dosage, combination of chemicals used and the extent of area where it is applied [8, 9].

In the agricultural sector, pesticide poisonings are caused mainly by carbamates and organophosphates [6]. These compounds produce genotoxic effects characterized by the presence of symptoms such as decreased concentration, fatigue, memory disturbances, and headache [10]. On the other hand, they modify the characteristics of the environmental components (water, air, soil) when there are direct and indirect discharges in the environment, becoming a risk factor for terrestrial and aquatic ecosystems [11].

The objective of this study was to determine the neurotoxic effects associated with chronic exposure to pesticides through the application of the Neurological-Psychological Questionnaire (PNF) and laboratory analysis (erythrocyte acetylcholinesterase levels) in banana farms in the Province of Los Ríos, Ecuador.

2 Methodological Design

A descriptive cross-sectional study was conducted to identify the prevalence of persistent symptomatology associated with exposure to pesticides and their relationship to erythrtite acetylcholinesterase levels. The research was carried out in the Province of Los Ríos, located in the center of Ecuador. Sampling was carried out on the population of workers occupationally exposed to pesticides in six local banana trees, represented by 103 individuals out of 138 who constituted the universe, as criteria of inclusion agricultural workers were considered exposed for a period longer than six months, with apparent good health condition, without psychosocial habits (smoking, alcohol intake).

To carry out the study, ethical standards and principles were applied for medical research in humans of the Helsinki Declaration, including the informed letter of consent. To determine symptomatology related to occupational exposure, the Neurological-Psychological Questionnaire (PNF), developed in the 1970s by the Central Institute of Occupational Medicine in Berlin, was used and adapted in 1983 by the Institute of Labour Medicine of Cuba, which presents 38 questions that allow to record the neurotoxic effects of harmful substances manifested through symptoms and displacing states, through five scales or dimensions: psychoneurovegative instability, neurological symptoms, asthenia, irritability and deficit concentration and memory.

Laboratory determinations of the levels of erythritial acetylcholinesterase measured in U/L by the semi-automated MICROLAB method were made in May to November 2019, the reference values established by the laboratory were 5,400–13,200 U/L, according to the Standardized Method of Buririlcoline.

All data were analyzed through the SPSS Version 25 statistical program, which enabled a descriptive analysis by constructing frequency distribution tables and calculating summary measures (average and standard deviation [SV]). To establish the correlation between AchE levels and PNF questionnaire scores, Spearman's correlation coefficient was used, considering that the data did not meet a normal distribution. Through linear regression analysis, predictor variables were determined for those dimensions of the NFP questionnaire that had an association with the AchE. The levels of erythrtite acetylcholinesterase and as an independent perception of symptomatology and exposure level were considered as a dependent variable, taking as indicators working age, age and sex.

3 Results

3.1 Demographic Characteristics and Perceived Symptomatology

Of the 103 workers, 31% (32) are women and 69% (71) men, the average age was 40 ± to 10 years (range: 21 to 58), with an average working age of 33 years. The prevalence of persistent symptomatology was 6 per 10 male and 5 workers in women. No statistically significant differences were found between the risk of persistent symptomatology according to age and working seniority. However, the highest proportions of symptoms were found in male workers aged 30 to 55 years, with working age ranges of 6 months to 5 years (Table 1).

As Table 2 shows the results obtained in women: 41.2% frequently had irritability symptoms, neurological symptoms once had 23.5%, 20.6% very often refer to instability; continuing asthenia at 14.7% at some time. Somewhat different results were seen in male staff: 44.9% sometimes showed memory deficit and concentration, instability very often showed 27.5%; continuing sometimes refer to neurological symptoms (24.6%) asthenia (20.3%).

Table 1. Demographic characteristics and perceived symptomatology

Variable	Symptomatology									
	N		PN		E		A		K	
	N	%	N	%	N	%	N	%	N	%
Sex										
Male	22	21.4	22	21.4	17	16.5	5	4.9	23	22.3
Woman	43	41.7	33	32.0	53	51.5	20	19.4	56	54.4
Age (age group)										
<30	24	23.3	19	18.4	31	30.1	11	10.7	32	31.1
30–55	37	35.9	32	31.1	35	34.0	12	11.7	42	40.8
>55	4	3.9	4	3.9	4	3.9	2	1.9	5	4.9
Working time										
6 months - 5 years	28	27.2	26	25.2	31	30.1	12	11.7	33	32.0
6–15 years	26	25.2	20	19.4	26	25.2	9	8.7	29	28.2
>15 years	11	10.7	9	8.7	13	12.6	4	3.9	17	16.5
Instruction level										
Primary	34	33.0	30	29.1	37	35.9	15	14.6	43	41.7
Secondary	21	20.4	25	24.3	33	32.0	10	9.71	36	35.0

Note: Neurological symptoms (N), Psychoneurovegative instability (PN), Asthenia (A), Irritability (E), Concentration and memory deficits (K)

77% of the population (women n = 26, men n = 53) had alterations in the cognitive sphere, including symptoms such as: easily distracting, difficulty remembering simple things (names, people), lack of memory, difficulty concentrating, causing alterations in the performance of neurological functions attributable to possible intoxication, while more than 50% had symptoms in the areas of neurovegetative irritability and instability, these results differ from those presented by Corella & Gamarra; and, Palacios et al., who noted that the absence of clinical, behavioral and neurotic alterations in exposed workers relates to the tolerance developed by the body over time, and normal acetylcholinesterase values can be found. In the study conducted by Luna, et al., there were prothermal clinical manifestations [12–14] related to continued exposure to pesticides. Hurtado & Gutierrez, claim that symptoms of organophosphate poisoning appear when AChE activity has decreased by more [15] than 50%.

The highest prevalence of perceived symptomatology was presented in the age group of 30 to 55 years, with an age of 6 months to 5 years, considering as the main causal the inexperience and unknownness of the doses required for the formulation and application of pesticides in agricultural plantations. A prospective epidemiological-toxicological study aimed at understanding the exposure of the Honduran population to pesticides and their effect on occupational health, states that younger workers have a higher incidence

Table 2. PNF neurotoxic effects

			Never or rarely (0)	Sometimes (1)	Frequently (2)	Very common (3)
Psychoneurovegative instability (PN)	Woman	N	15	8	4	7
		%	44.1%	23.5%	11.8%	20.6%
	Man	N	23	15	12	19
		%	33.3%	21.7%	17.4%	27.5%
	Total	N	38	23	16	26
		%	36.9%	22.3%	15.5%	25.2%
Neurological symptoms (N)	Woman	N	21	8	4	1
		%	61.8%	23.5%	11.8%	2.9%
	Man	N	31	17	12	9
		%	44.9%	24.6%	17.4%	13.0%
	Total	N	52	25	16	10
		%	50.5%	24.3%	15.5%	9.7%
Asthenia (A)	Woman	N	26	5	2	1
		%	76.5%	14.7%	5.9%	2.9%
	Man	N	43	14	6	6
		%	62.3%	20.3%	8.7%	8.7%
	Total	N	69	19	8	7
		%	67.0%	18.4%	7.8%	6.8%
Irritability (E)	Woman	N	11	9	14	0
		%	32.4%	26.5%	41.2%	0.0%
	Man	N	22	18	23	6
		%	31.9%	26.1%	33.3%	8.7%
	Total	N	33	27	37	6
		%	32.0%	26.2%	35.9%	5.8%
Concentration and memory deficit (K)	Woman	N	8	18	3	5
		%	23.5%	52.9%	8.8%	14.7%
	Man	N	16	31	5	17
		%	23.2%	44.9%	7.2%	24.6%
	Total	N	24	49	8	22
		%	23.3%	47.6%	7.8%	21.4%

of symptoms, due to the disinterest of complying with precautionary measures and non-compliance with the use of protective equipment provided by the contracting institutions [13, 16].

In relation to the level of instruction, 54% of respondents had a full primary level and 46% were high schoolers, of these more than 50% had symptoms in neurological dimensions, psychoneurovegative instability, irritability, concentration and memory deficits, the values of which are similar to those reported by Marrero and Palacios, who state that in populations with low sociocultural level there is a deficit in protection measures due to non-compliance with the time of entry into the fields, neglect of health measures, intake of food in the work area or because the employer does not provide the necessary equipment. While Caraballo & White, refers to the need for the application of a test battery to detect early the possible neurobehavioral effects of pesticides on occupationally exposed workers [6, 17, 18].

The highest incidence of neurotoxic alterations was in men, because the study population is dominated by this genus. These results differ with the study conducted by Carmona, in the Aburrá Valley and the Near East of Antioquia, where the female group had the highest and lowest levels of acetylcholinesterase, considering as limiting factors the alteration of hormones due to pregnancy, menstruation, menopause or the use of hormone or anovulatory therapies, while Sierra and collaborators, claim that testosterone has some protective factor [16, 19].

3.2 Erythr Thirteenth Acetylcholinesterase Values

As mentioned in Table 3, normal erythrctarian acetylcholinesterase levels had an average of 10,668.31 u/l with a standard deviation of ±2371.05 u/l, in 32 women equivalent to 94.1% and high in 5.9% (2), in contrast in the male gender high values were determined at 5.8% (4), low 8.7% (6) and normal 85.5% (59). The prevalence of lower than 5,400 u/l and higher than 13,200 u/l was 1 per 20 workers.

Table 3. Erythrocyte acetylcholinesterase values.

			Under	Normal	High	Total
Sex	Woman	N	0	32	2	34
		%	0%	94.1%	5.9%	100.0%
	Man	N	6	59	4	69
		%	8.7%	85.5%	5.8%	100.0%
Total		N	6	91	6	103
		%	5.8%	88.3%	5.8%	100.0%

The determination of blood levels of erythrititarian acetylcholinesterase is used as a pre-emergence prevention and diagnostic method in workers exposed to pesticide use. The variability of 10% of the above Acetylcholine in occupationally exposed individuals does not cause pathological problems, however, the reduction from 25 to 30%, is an indicator of acute (low) exposure, while levels below 50% are related to chronic poisoning. In this study 88% of workers had normal acetylcholinesterase levels, however, these values could underestimate the percentage of reduction of the above because in Ecuador there are no baseline determinations of acetylcholinesterase levels. Similar results presented Lessenger, et al., who stated that workers with typical organophosphorous poisoning tables in the medical consultation had normal ranges of acetylcholinesterase, in addition in the grounds of residence or work no pre-occupational medical assessment is performed to know the comoarities that may alter the values of tests or basal and in this way achieve better individual follow-up [20–22].

5.8% (n = 6) of the population studied had low levels of erythrctarian acetylcholinesterase, as well as 5.8% (n = 6) had high values of this above, with a prevalence of 1 per 20 workers. Jimenez & Schosinsky, state that workers with low acetylcholinesterase levels should be removed from exposure and the test should be repeated weekly until levels remain constant, while elevated acetylcholinesterase values are indicators of chronic degenerative diseases such as Diabetes, Hyperlipemia, Nephrosis, Reticulosis, among others [23, 24].

Workers with symptomatology in neurological dimensions, psychoneurovegative instability, irritability, concentration, and memory deficits, respectively, had a normal acetylcholinesterase level, while 5.8% of workers with likely levels of intoxication (AchE > 5400 u/l) had symptoms in the neurological and cognitive spheres. In a study conducted by Palacios and collaborators reported the presence of persistent symptoms such as headache, irritability, insomnia, weakness and libido disturbances; without alteration of acetylcholinesterase or with a very discreet decrease [13].

3.3 Correlation Erythrocyte Acetylcholinesterase Values vs Perceived Symptomatology

As mentioned in Table 4 using Spearman's correlation test, the association between AchE values and symptomatology on the neurological scale ($r_s = -0.944^{**}$) and neurovegetative instability ($r_s = -0.857^{**}$) was evident.

In Fig. 1, the relationship between acetylcholinesterase levels and perceived symptomatology can be observed according to the dimensions set out in the PNF questionnaire. In turn, using linear regression models, to predict N and PN values, AchE (X) levels and predictor variables are identified.

There was a significant correlation between acetylcholinesterase levels with neurological dimensions and psychoneurovegative instability, i.e. the reduction of acetylcholinesterase tends to increase the onset of symptoms such as headaches, vahydes, dizziness, sleep disorders, weakness, tiredness, exhaustion, feeling cold or hot, dry mouth, dizziness, vomiting, loss of muscle strength, balance disturbances, insecurity when walking, tingling in the feet and hands, disorders in sex, heaviness in the joints and tremors in the arms and legs. The absence of statistical significance in the deficit

Table 4. Spearman's correlation.

Neurological symptoms (N)	Correlation coefficient	−0.944**
	Sig. (bilateral)	0
	N	103
Psychoneurovegative instability (PN)	Correlation coefficient	−0.857**
	Sig. (bilateral)	0
	N	103
Asthenia (A)	Correlation coefficient	−0.171
	Sig. (bilateral)	0.084
	N	103
Irritability (E)	Correlation coefficient	0.093
	Sig. (bilateral)	0.352
	N	103
Concentration and memory deficits (K)	Correlation coefficient	0.102
	Sig. (bilateral)	0.304
	N	103

Note: **The correlation is significant at level 0.01 (2 queues).
*The correlation is significant at level 0.05 (2 queues).

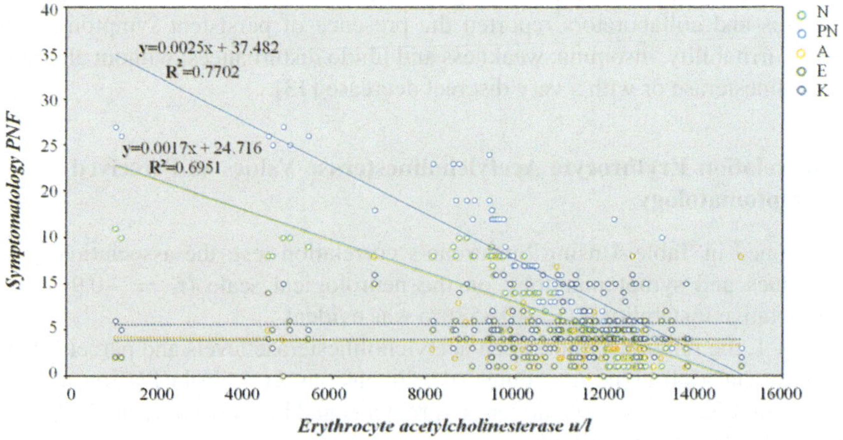

Fig. 1. Correlation of acetylcholinesterase levels and PNF questionnaire

dimensions of concentration, asthenia and irritability could be due to the absence of comparisons with different exposure magnitudes. However, these results differ with the study conducted by Toro and collaborators, who found that workers occupationally exposed to pesticides with acetylcholinesterase levels outside normal limits did not exhibit any symptomatology [13, 25].

However, organophosphate pesticide poisoning is not limited to the handling of the mixture in the workplace, there are other sources of contamination derived from the dispersed chemical due to rain or irrigation system, which reaches the water table and transports the pollutant through infiltration into the groundwater network until entering the trophic (food) chains causing the death of several species necessary for the balance of ecosystems [25, 26].

Some authors establish that the main causes of contamination by agrochemicals in the environmental components (soil, air and water) are related to ignorance about the required doses, the handling of empty containers, leaks in storage tanks and accidental spills, which threatens the stability of ecosystems and represents a danger to public health [25, 27]. In addition, the pathway of organophosphate pesticides in the environment is related to their physicochemical properties, the geomorphological conditions of the soils, and the hydrogeological and meteorological conditions of the areas [28].

4 Conclusions

This study established the relationship between chronic occupational exposure to organophosphorous agents and the emergence of neurotoxic effects on neurological dimensions and psychoneurovegative instability.

In the occupationally exposed population, there was no marked decrease in the activity values of erythrititarian acetylcholinesterase, however, most workers had signs and symptoms characterizing overexposure to pesticides and likely chronic poisoning.

The high percentages of symptomatology perceived show the need for stricter programs of intervention of hygienic measures, monitoring and comprehensive monitoring, including incorporation of medical examination pre-exposure occupations, newspapers and post in workers exposed to agrochemicals, considering as priority group women in gestation period or with liver problems, diabetics or addictions, in addition to intervene in the incorporation of a culture based on habits, to reduce the risks inherent in the use and exposure of chemical compounds.

The study presented limitations due to the lack of pre-entry examinations and the establishment of a control population, which allow the establishment and comparison of baseline acetylcholinesterase values to determine the variability presented by acetylcholinesterase in workers occupationally exposed to pesticide use, in addition to the suppression of contraceptive use as an independent variable, which could influence the results obtained.

The results obtained allow to create a significant base to initiate exhaustive studies on chronic alterations in the health of workers exposed to acetylcholinesterase inhibitor pesticides; however, it is recommended to consider open populations to establish levels of variability between exposed and unexposed. Likewise, estimate the dispersion of pesticides in the environment and their effects on natural resources such as air, water, soil, and biota.

References

1. Mangas, I., Vilanovab, E., Estévezc, J., França, T.: Neurotoxic effects associated with current uses of organophosphorus compounds. J. Braz. Chem. Soc. **27**(5), 809–825 (2016)

2. Bernardino, H., Mariaca, R., Nazar, A., Álvarez, J., Torres, A., Herrera, C.: Conocimientos, conductas y síntomas de intoxicación aguda por plaguicidas entre productores de tres sistemas de producción agrícolas en los Altos de Chiapas. México. Revista Internacional de Contaminación Ambiental **35**(1), 7–23 (2019)
3. Nivia, E.: Mujeres y plaguicidas. Una mirada a la situación actual, tendencias y riesgos de los plaguicidas. Rapalmira, Colombia (2000)
4. Naranjo, A.: La otra guerra: la situación de los plaguicidas en el Ecuador. Acción Ecológica, Ecuador (2017)
5. Tellerias, L., Paris, E.: Impacto de los tóxicos en el neurodesarrollo. Rev. Chil. Pediatr. **79**(1), 55–63 (2008)
6. Marrero, S., González, S., Guevara, H., Eblen, A.: Evaluación de la exposición a organofosforados y carbamatos en trabajadores de una comunidad agraria. Comunidad y Salud **1**(15), 30–41 (2017)
7. Boletín de Temas de Salud: Asociación de médicos municipales de la ciudad de Buenos Aires, Suplemento del Diario del Mundo Hospitalario 2005. http://www.trabajosocial.unlp.edu.ar/uploads/docs/accidentes_de_transito._temas_de_salud_nov_dic_2005.pdf
8. Martínez, C., Gómez, S.: Riesgo genotóxico por exposición a plaguicidas en trabajadores agrícolas. Int. Contam. Ambient **23**(4), 185–200 (2007)
9. Vitali, S.: Precariedad en las condiciones de trabajo y salud de los trabajadores del sector bananero. Salud de los Trabajadores **25**(1), 9–22 (2017)
10. Vásquez, C., León, S., González, R., Preciado, M.: Exposición laboral a plaguicidas y efectos en la salud de trabajadores florícolas de Ecuador. RevSalJal **1**(3) (2016)
11. Martínez, C., Romano, G., Cuadras, A., Ortega, L.: Plaguicidas, impacto en salud y medio ambiente en Sinaloa (México): implicaciones y retos en gobernanza ambiental. TraHs **4**(1), 103–122 (2019)
12. Corella, A., Gamarra, R.: Síntomas neurotóxicos y relación de los niveles de colinesterasa eritrocitaria debido al nivel de exposición en el uso de plaguicidas en personal de una florícola ubicada en el cantón Cayambe en el periodo comprendido entre octubre a noviembre del 2016, tesis de licenciatura, PUCE (2016)
13. Palacios, M., Paz, P., Hernández, S., Mendoza, L.: Sintomatología persistente en trabajadores industrialmente expuestos a plaguicidas organofosforados. Salud Pública de México **41**(1), 55–61 (1999)
14. Luna, J., Hanna, M., Amado, C.: Condición clínica y niveles de colinesterasa de trabajadores informales dedicados a la fumigación con plaguicidas. NOVA **17**(31), 67–77 (2019)
15. Hurtado, C., Gutiérrez, M.: Enfoque del paciente con intoxicación aguda por plaguicidas organofosforados. Rev. Fac. Med. **53**(4), 244–258 (2005)
16. Sierra, A., Padilla, G., Cambar, P.: Niveles de colinesterasa sérica y sintomatología observados en obreros que trabajan con insecticidas organofosforados. Revista Médica Hondur. **54**(1), 9–22 (1986)
17. Palacios, M., Paz, M.: Sintomatología persistente en trabajadores agrícolas expuestos a plaguicidas órgano-fosforados. Rev. Fac. Nac. Salud Pública **29**(2), 153–162 (2011)
18. Caraballo, S., Blanco, G.: Evaluación neuropsicológica de trabajadores expuestos a solventes orgánicos en una empresa de transporte público. Revista de la Facultad de Medicina **28**(1), 79–88 (2005)
19. Palacios, M., García, G., Paz, M.: Determinación de niveles basales de colinesterasa en jornaleros agrícolas. Rev. Fac. Med. UNAM **52**(2), 63–68 (2009)
20. Lessenger, J., Estock, M., Younglove, T.: An analysis of 190 cases of suspected pesticide illness. J. Am. Board Fam. Pract. **8**(4), 278–282 (1995)
21. Restrepo, B., Londoño, A., Sánchez, J.: Valores de colinesterasa plasmática y eritrocitaria con ácido 6–6'-ditiodinicotínico (DTNA) como indicador. Rev. Colomb. Quim. **46**(1), 13–19 (2016)

22. Jiménez, M., Schosinsky, K.: Valores de referencia de colinesterasa plasmática y eritrocítica en población costarricense. Comparación del desempeño clínico de ambas enzimas. Rev. Costarric Cienc. Méd. **21**(3) (2000)
23. Mejía, G., Ramelli, M.: Interpretación clínica del laboratorio, Séptima ed. Editorial Médica Panamericana, Colombia (2006)
24. Carmona, J.: Colinesterasas eritrocitaria y plasmática en trabajadores con enfermedades crónicas controladas y en usuarios de medicamentos. IATREIA **19**(1), 16–28 (2006)
25. Toro, B., Rojas, A., Díaz, J.: Niveles de colinesterasa sérica en caficultores del Departamento de Caldas. Rev. Salud Pública **19**(3) (2017)
26. Unidad de Limnología y Recursos Acuáticos: Organismos Acuáticos como Indicadores de Contaminación por Pesticidas Organofosforados (Chapare-Cochabamba), Primera ed. Publisher (2017)
27. Rodríguez, B., Martínez, L., Peregrina, A., Ortiz, C., Cárdenas, O.: Análisis de residuos de plaguicidas en el agua superficial de la cuenca del río Ayuquila-Armería, México. Terra Latinoamericana **37**(1), 151–161 (2019)
28. Del Puerto, A., Suárez, S., Palacio, D.: Efectos de los plaguicidas sobre el ambiente y la salud. Revista Cubana de Higiene y Epidemiología **52**(3) (2014)

Nonparametric Geostatistical Prediction of Daily $PM_{2.5}$ Concentrations Based on Satellite Measurements of Aerosol Optical Depth

Sergio Castillo-Páez(✉), Paul Medina Vazquez, and Vicente García-Mancero

Departamento de Ciencias Exactas, Universidad de las Fuerzas Armadas ESPE, Sangolqui, Ecuador
sacastillo@espe.edu.ec
https://dece.espe.edu.ec/

Abstract. The prediction of particulate matter ($PM_{2.5}$) has been a widely studied topic, and most of the techniques proposed are based on correlating the concentrations of this variable, obtained by sensors on the ground, with satellite measurements of *Aerosol optical depth* (AOD). In the present work, a non-parametric geostatistical method based on the local linear estimator is proposed in order to obtain flexible predictions in positions not observed by the monitoring network. This procedure was applied to real daily data, and the results obtained by this approximation were compared with those values derived from models based on linear regression and parametric kriging for a validation sample.

Keywords: Aerosol optical depth · $PM_{2.5}$ · Kriging · Local linear estimation

1 Introduction

It is known that airborne concentrations of fine particulate matter ($PM_{2.5}$), even at low levels, have adverse effects on the health of the population [8]. For this reason, different methods have been developed to predict this variable in a given region and instant of time. Several of these techniques make use of the existing correlation between satellite measurements of *Aerosol optical depth* (AOD) with the corresponding mass of $PM_{2.5}$ reported by in situ air quality monitoring networks [13]. A basic approach involves using simple linear regression models between observations of both variables, previously subjected to a space-time validation process [15].

Supported by Universidad de las Fuerzas Armadas ESPE.

M. Botto-Tobar et al. (Eds.): CIT 2021, LNNS 405, pp. 338–349, 2022.
https://doi.org/10.1007/978-3-030-96043-8_26

Based on the previous idea, according to [6], different techniques have been used such as Multiple Linear Regression, Mixed Effects Models, Chemical Transport Models, Geographically weighted Regression Models, among others. Alternatively, geostatistical techniques have also been developed based on kriging methods, which take into account the spatial information of the sample data to obtain these predictions [1]. However, these last procedures resort to parametric structures to model spatial dependence, which in certain cases can lead to problems of model misspecification and inflexible predictions with respect to the observed values.

In the present work it is proposed to use the information from satellite AOD data together with data derived from ground monitoring networks to model the spatial dependence of $PM_{2.5}$ concentrations. In order to obtain predictions of particulate matter in unobserved positions, a non-parametric approach based on core-type estimates is proposed. The following section details both the sample data collection procedures as well as the validation procedures used. The parametric and non-parametric approaches analyzed are also indicated. Section 3 presents the results of the application of these procedures to a sample of real daily data in England, and the prediction errors of each method are compared from a validation sample. Finally, Sect. 4 discusses some of the relevant results obtained in this study.

2 Analyzed Data and Methods Applied

2.1 Collection and Validation of Satellite and Terrestrial Data

Atmospheric aerosols are solid or liquid particles that are suspended in the air. On the other hand, *Aerosol Optical Depth* is a dimensionless measure that represents the total aerosol load in an atmospheric column. This variable is affected by a variety of factors, including the wavelength of the light λ, though for studies like this one, measurements with $\lambda = 550$ nm are usually considered [13].

In the present study, we used *AOD* data obtained by the Moderate-Resolution Imaging Spectroradiometer (MODIS) sensor on board of NASA's Terra and Aqua satellites. These satellites provide various atmospheric products (MYD or MOD for Aqua and Terra, respectively), each one with a different processing level. In our case we will use the product MOD04_L2, which has a spatial resolution in each pixel of 10×10 km, and which provides daily measurements of aerosols obtained at 5-min intervals. This implies that each product generally contains a grid of measurements of 135×203 pixels [16]. After selecting the region and date of analysis, the data for this product is downloaded from the NASA website. The downloaded files are in *Hierarchical Data Format* (HDF), which allows to obtain a large amount of information about the product. Specifically, we are interested in retrieving data such as: Satellite scan start time, Latitude, Longitude, and AOD measurements at 550 nm.

However, since the above measurements are dependent on several climatic factors, in situ measurements are needed to validate them. For this, aerosol measurements are used in terrestrial monitoring networks, such as the Aerosol

Robotic Network (AERONET). This global network of over 499 stations records AOD measurements for a variety of wavelengths (1020, 870, 675, 500, and 440 nm) [14].

[15] proposed the following procedure for satellite AOD validation: First, measurements from the AERONET network are downloaded for all sensors within the analysis region on the date considered. Then, based on the satellite scan start time, for each station only the measurements obtained in the 30 min before and after that hour are averaged (1 h time bandwidth). With these data, the AOD values for 550 nm are obtained by linear interpolation. Subsequently, the satellite pixels that geographically contain each of the ground stations are identified, and the satellite AOD observations of an observation grid composed of the 5 × 5 pixels centered on said station are averaged (50 km^2 spatial bandwidth). Finally, the linear correlation coefficient between both averages is calculated (space-time validation).

For the case of $PM_{2.5}$ measurements of this variable (usually in $\mu g/m^3$) are obtained from terrestrial environmental monitoring networks in the region and date of interest, and through the space-time validation process explained above, comparable AOD measurements are obtained at each terrestrial measurement point of the particulate material.

2.2 Parametric Modeling and Estimation

Assume we have the vector $\mathbf{Y}$ corresponding to n space-time validated observations of $PM_{2.5}$. The parametric model on which the relationship between these variables is analyzed is usually the following:

$$\mathbf{Y} = \boldsymbol{\mu} + \boldsymbol{\epsilon}, \tag{1}$$

where usually $\boldsymbol{\mu} = \mathbf{X}\boldsymbol{\beta}$ is the trend function, which depends on the vector of parameters $\boldsymbol{\beta}$ and where $\mathbf{X}$ is the design matrix that collects the validated AOD observations. The vector $\boldsymbol{\epsilon}$ corresponds to the independent error terms, normally distributed and with zero mean.

The basic approach is usually to estimate model (1) using ordinary least squares linear regression:

$$\hat{\boldsymbol{\beta}}_{ols} = (\mathbf{X}^t\mathbf{X})^{-1}\mathbf{X}^t\mathbf{Y}. \tag{2}$$

Based on this approach, different techniques have been proposed such as generalized linear models, mixed effects models, among others, which include in $\mathbf{X}$ the information on covariates related to climatological variables [6].

Nevertheless, there are also approximations that consider including the spatial positions of the observation points as relevant information. In this study, we will consider this last case and assume the following model:

$$\mathbf{Y}(\mathbf{s}) = \boldsymbol{\mu}(\mathbf{s}) + \varepsilon(\mathbf{s}), \tag{3}$$

where $\mathbf{s} \in \mathbb{R}^2$ corresponds to the spatial position of the observed point, usually measured in latitude and longitude. We will call $\boldsymbol{\mu}(\mathbf{s})$ as the large-scale variability, which we will consider deterministic, and which depends on the AOD measurements. Likewise, $\varepsilon(\mathbf{s})$ represents the small-scale variability, and corresponds to a realization of a stationary second-order process with null mean, in which it is established that

$$Cov[\varepsilon(\mathbf{s}), \varepsilon(\mathbf{s}+\mathbf{u})] = C(\mathbf{u}),$$

where the function $C(\cdot)$ is called a covariogram and determines the spatial dependence between the error terms as a function of their spatial separation $\mathbf{u}$. However, in practice, the (semi) variogram is commonly used function to estimate the spatial dependence. This latter one is given by [7]:

$$\gamma(\mathbf{u}) = C(0) - C(\mathbf{u}).$$

The estimation of model (3) implies, first, estimating the trend $\hat{\boldsymbol{\mu}} = \mathbf{X}\hat{\boldsymbol{\beta}}_{ols}$, and then estimating $\gamma(\mathbf{u})$ from the residuals $\hat{\varepsilon} = \mathbf{Y} - \hat{\boldsymbol{\mu}}$. On the other hand, the process of estimating spatial dependence is usually carried out in two steps: (1) constructing a pilot estimator of the variogram from the residuals. For this purpose, the empirical variogram proposed by [17] can be used. (2) Then, a valid parametric model of variogram $\hat{\gamma}_\theta(\cdot)$ is fitted to this pilot estimator. There are different types of models [7], for example, the spherical variogram is given by

$$\gamma_\theta(\mathbf{u}) = \begin{cases} 0; & if\ ||\mathbf{u}|| = 0 \\ c_0 + c_1 \left\{ \frac{3}{2}\frac{||\mathbf{u}||}{a} - \frac{1}{2}\left(\frac{||\mathbf{u}||}{a}\right)^3 \right\}; & if\ 0 < ||\mathbf{u}|| < a \\ c_0 + c_1; & if\ ||\mathbf{u}|| > a \end{cases} \tag{4}$$

where $\theta = (c_0, c_1, a)$ is the vector of parameters, where c_0 represents the nugget effect, c_1 the partial sill, and a is the range. These variogram parameters determine the degree of spatial dependence between the data [5]. The adjustment method for these parameters can also be performed by ordinary, weighted, or generalized least squares.

Once the model (3) has been estimated, the $PM_{2.5}$ values can be predicted at an unobserved position $\mathbf{s}_0$ from

$$\hat{\mathbf{Y}}(\mathbf{s}_0) = \hat{\boldsymbol{\mu}} + \hat{\varepsilon}_{ko}(\mathbf{s}_0)$$

where $\hat{\varepsilon}_{ko}(\mathbf{s}_0)$ represents the prediction by ordinary kriging obtained from $\hat{\gamma}_\theta(\cdot)$. This prediction is constructed by a linear combination of the observed values, such that

$$\hat{\varepsilon}_{ko}(\mathbf{s}_0) = \sum_{i=1}^{n} \lambda_i \hat{\varepsilon}(\mathbf{s}_i), \quad \sum_{i=1}^{n} \lambda_i = 1$$

where λ_i are called kriging weights and are obtained by minimizing the mean square prediction error. Finally, it can be verified that these weights depend on the values of the variogram $\hat{\gamma}_\theta(\cdot)$ calculated from the separations between the observed positions and with respect to the prediction position $\mathbf{s}_0$ [7].

There are several studies that use ordinary kriging for the prediction of particulate matter with AOD data [1,18], however the values obtained depend on the estimated parameters. For example, the estimation of β_{ols} depends on the degree of correlation between $PM_2.5$ and AOD, which in some cases can be very low, especially when atmospheric conditions do not allow us to have sufficiently reliable aerosol data, and there may even be a large amount of missing data. In this regard, for model (3) it can be assumed that all the information that is not collected by the large-scale variability will be reflected in the error variability. However, the corresponding kriging predictions depend on an adequate selection of the parametric variogram model $\gamma_\theta(\cdot)$ and the corresponding estimate $\hat{\theta}$, and therefore, their predictions are exposed to problems of model misspecification. At this point, it is proposed to use non-parametric approaches, which allow building flexible predictions, which better fit the observed data and do not depend on the estimation of parametric models.

2.3 Non-parametric Geostatistical Prediction

Firstly, from the residuals $\hat{\varepsilon}$, a nonparametric pilot variogram $\tilde{\gamma}(\cdot)$ is constructed, based on the local linear estimator [12], as follows

$$\tilde{\gamma}(\mathbf{u}) = \frac{1}{2\sum_{i,j} w_{ij}(\mathbf{u})} \sum_{i,j} w_{ij}(\mathbf{u}) \left(\varepsilon(\mathbf{s}_i) - \varepsilon(\mathbf{s}_j)\right)^2, \tag{5}$$

where $w_{ij}(\mathbf{u}) \geq 0, \forall i, j$.

Under the isotropy assumption, the weights are given by

$$w_{ij}(u) = K\left(\frac{\|\mathbf{s}_i - \mathbf{s}_j\| - u}{g}\right) \times$$

$$\sum_{k<l} K\left(\frac{\|\mathbf{s}_k - \mathbf{s}_l\| - u}{g}\right) \left(\|\mathbf{s}_k - \mathbf{s}_l\| - u\right) \left(\|\mathbf{s}_k - \mathbf{s}_l\| - \|\mathbf{s}_i - \mathbf{s}_j\|\right).$$

where K is a kernel function and g is a smoothing bandwidth. This bandwidth is usually obtained through cross-validation techniques, minimizing the quadratic or relative quadratic error of the estimates of the respective variogram.

Subsequently, based on these nonparametric pilot estimates, the valid variogram $\hat{\gamma}(\cdot)$ is obtained from a flexible Shapiro-Botha (S-B) model based on the spectral representation of the covariogram [19]. This variogram has been extensively studied, verifying its good behavior and statistical properties in comparison with the traditional parametric procedure [4,9]. It is worth mentioning that these non-parametric approaches have been used for the construction of risk maps [11], in geostatistical bootstrap methods [2], and have been extended to more complex models considering spatial heteroskedasticity [3]. On the other hand, the use of $\hat{\gamma}(\cdot)$ in ordinary kriging makes it possible to obtain more flexible estimates, and that better fit the sample observations. Finally, it is important to mention that the adjustment procedure for this type of non-parametric models is implemented in the *npsp* package of the R software [10].

3 Case of Application to Real Data

In order to analyze the behavior of the estimation procedures presented in the previous section, the $PM_{2.5}$ and AOD data pertaining to the England region, corresponding to February 26, 2019, were selected. This date corresponds to the fact that in the near days a significant number of forest fires were detected in the United Kingdom, so it was expected that the satellite detection of aerosols generates less missing data. Indeed, when obtaining the information of the product MOD04_L2 for this region and date, the AOD measurements cover almost the entire country, as shown in Fig. 1a.

According to the satellite scan time, it is established that these measurements were made between 11:45 am and 11:50 am (5 min duration). Considering this time interval and considering the time bandwidth in 60 min from the passage of the satellite, the measurements of the AERONET network corresponding to 8 sensors on the ground available within the region of analysis were obtained. After the space-time validation process, a coefficient of determination of 0.641 was found between both measurements, which indicates a relatively acceptable level of validation of the satellite measurements of AOD.

Furthermore, $PM_{2.5}$ data was obtained from the *UK Air information Resource of the Department of Environment Food and Rural Affairs (DEFRA)* website (https://uk-air.defra.gov.uk/data/). Data on latitude, longitude and concentrations of particulate matter were downloaded for 55 ground stations, in the same time bandwidth around the time of the satellite scan. The spatial distribution of the corresponding temporal averages of $PM_{2.5}$ (in $\mu g/m^3$) are shown in Fig. 1b.

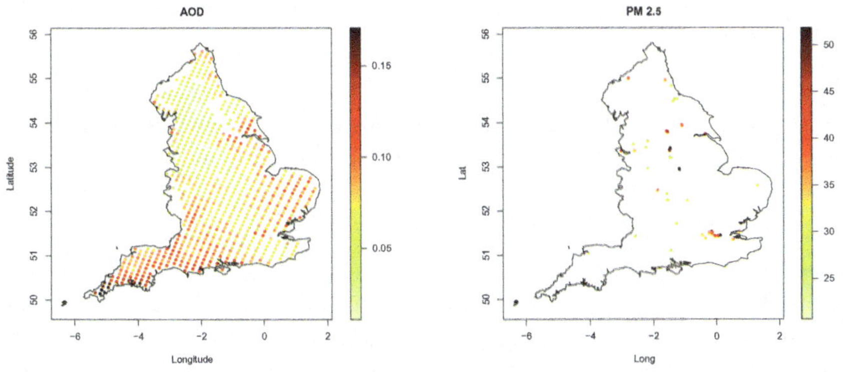

(a) (b)

Fig. 1. (a) Satellite measurements of AOD over England, and (b) averages of $PM_{2.5}$ from the 55 ground monitoring stations.

To make the comparison between the satellite data of AOD and the terrestrial measurements of $PM_{2.5}$, the space-time validation process was applied explained in the previous section, and which is illustrated in Fig. 2a. For each point on the ground (represented by a square in said figure), the closest satellite measurement was identified (green point), and around this location, a spatial validation grid of 5×5 pixels (e.g., $50\,\mathrm{km}^2$) was built, shown in blue. Finally, the average of the AOD measurements of this grid is obtained, which is assigned to the terrestrial measurement. This process is repeated at each $PM_{2.5}$ measurement point, as shown in Fig. 2b. To validate the behavior of the models proposed in the previous section, a random validation sample was selected from the original data set (approximately 10% of the total observations), to compare the prediction errors of each model at these points (represented in blue in this last Figure).

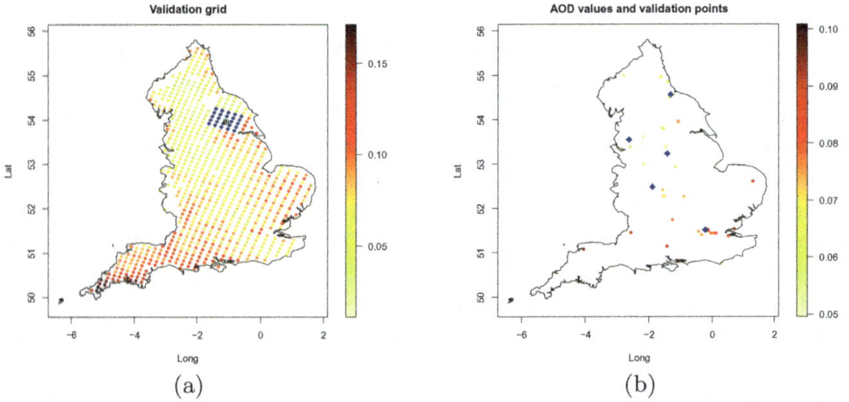

Fig. 2. (a) Example of validation grid of satellite AOD data with ground measurement of $PM_{2.5}$, and (b) averages of AOD assigned to ground measurement points, and validation points.

With these averaged and validated values of AOD and $PM_{2.5}$, we first proceeded to estimate model (1) by means of (2), obtaining a

$$\hat{\boldsymbol{\beta}}_{ols} = (37.517, -23.293)^t .$$

However, the t tests indicate that the second coefficient is not significant. (P-values $= 5.9 \times 10^{-6}$ and 0.816 respectively), and also the value obtained for $R^2 = 0.001$ indicates that, for these data, the relation between AOD and $PM_{2.5}$ is almost non-existent. As mentioned above, these results tend to be presented in practice since the measurements depend on many uncontrollable factors (e.g., climatological variables). Therefore, it is expected that these problems are extensible to those models that use the correlation between both variables as a basis for analysis.

Taking into consideration the previous outcomes, it was decided to estimate $\mu(\mathbf{s})$ of model (3) from the sample mean $\bar{Y} = 35.85\ \mu\mathrm{g/m}^3$. With this value, the corresponding residuals were obtained, and the parametric variogram $\hat{\gamma}_\theta(\cdot)$

was estimated as follows. First, the pilot variogram (proposed by Matheron) was calculated, and then, and then, based on these estimates, three possible models were selected: exponential, circular and spherical, the latter being chosen because it presented the lowest squared error for the adjustment. With this model we obtained the estimate of $\hat{c}_0 = 15.979$, $\hat{c}_1 = 71.138$ and $\hat{a} = 0.864$. Both the pilot variogram and the parametric model fit for the variogram were obtained using the *gstat* package, and their estimates are presented in 3a.

Subsequently, the non-parametric procedure was applied, for which the pilot variogram $\tilde{\gamma}(\cdot)$ was first calculated using (5). The selection of the parameter g was carried out by cross validation. From these estimates, the valid S-B variogram was constructed, which is presented in Fig. 3b. These nonparametric approximations were obtained using the routines of the *npsp* package. As can be seen, the nonparametric estimates present more flexible and smoothed values than the corresponding parametric versions, which is usual when using type-kernel estimates.

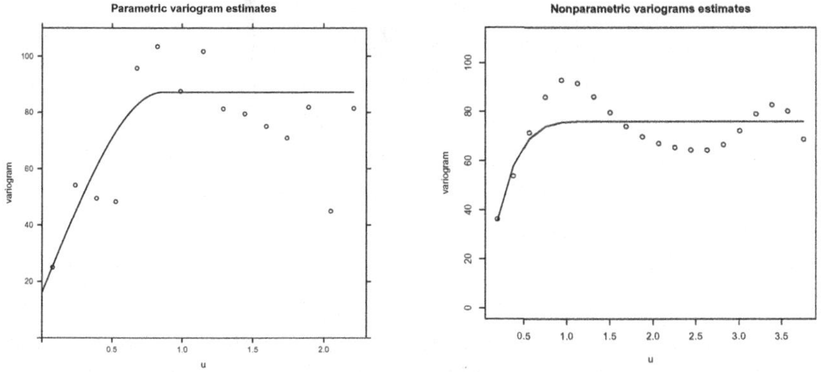

(a) (b)

Fig. 3. (a) Pilot variogram (points) and estimated parametric variogram $\hat{\gamma}_\theta(\cdot)$ (solid line), and (b) Local linear variogram $\tilde{\gamma}(\cdot)$ (points) and non-parametric variogram S-B $\hat{\gamma}(\cdot)$ (solid line).

From all previous estimates, it was feasible to construct $PM_{2.5}$ prediction maps for the entire observation region at the points where satellite AOD observations exist. For the case of model (1) a flat prediction is obtained, due to the almost null correlation between the variables (see Fig. 4a). On the other hand, the parametric and non-parametric predictions were obtained by ordinary kriging, using $\hat{\gamma}_\theta(\cdot)$ as $\hat{\gamma}(\cdot)$ in the corresponding cases (Figs. 4b and 4c respectively). The parametric predictions appear to present values that are closer to the mean value (given by the OLS estimation), and with a low tendency towards the high values of the scale. On the other hand, non-parametric approximations present more flexible predictions, especially when compared with the $PM_{2.5}$ sample measurements presented in Fig. 1b.

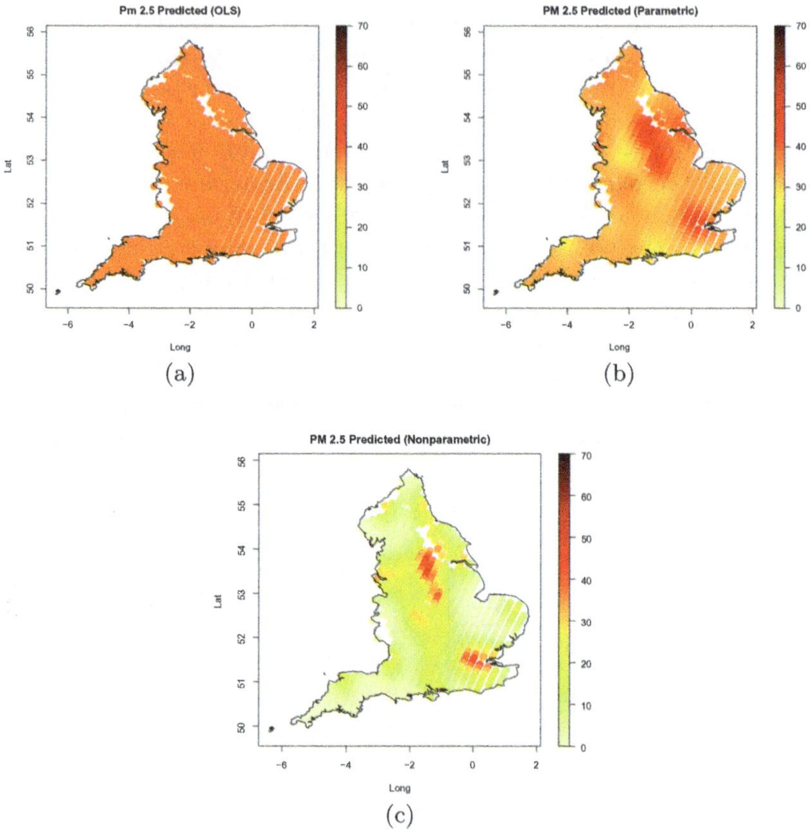

Fig. 4. $PM_{2.5}$ prediction map using: (a) the OLS-based estimation, (b) the parametric variogram $\hat{\gamma}_\theta(\cdot)$ and (c) the non-parametric variogram $\hat{\gamma}(\cdot)$.

Finally, to compare the performance of these procedures, their respective predictions were compared with the actual values of $PM_{2.5}$ for the validation sample (see Fig. 2b). In each case, the respective absolute and square errors were calculated, and the corresponding statistics were obtained. As an example, Fig. 5 shows the box plots of the absolute errors for each case. As can be seen, these errors present a better behavior for the non-parametric case (NP), compared to linear regression (OLS) where their dispersion is greater, and with the parametric case (PAR) where even outliers are observed. On the other hand, Table 1 compares the means and standard deviations of the squared errors obtained by each method. When analyzing these values, it is observed that the values obtained by the non-parametric estimation yield better results in both statistics, showing the best performance of this procedure in comparison with the parametric alternatives (Fig. 5).

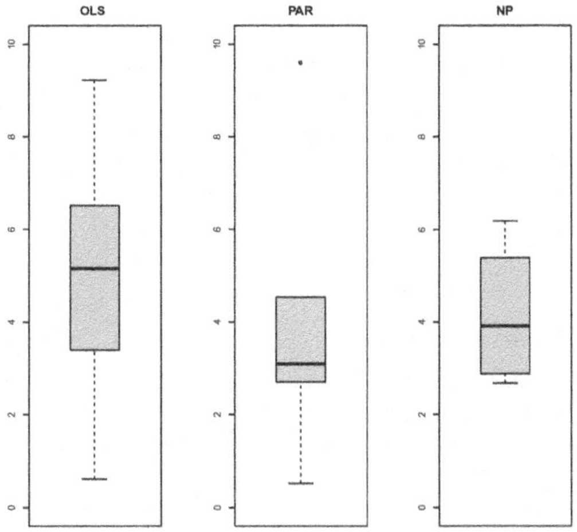

Fig. 5. Boxplot of absolute errors obtained with the validation sample for the linear (OLS), parametric (PAR) and non-parametric (NP) cases.

Table 1. Statistics of the squared validation errors

Method	Mean	Std. Dev.
OLS	33.193	33.017
PAR	25.993	37.686
NP	19.603	13.538

4 Discussion

The present study made it possible to compare various alternatives to predict particulate matter concentrations in a region of interest, combining information from aerosol satellite data with in situ measurements from $PM_{2.5}$ monitoring networks. However, as could be observed in the analyzed data set, those methods that try to establish estimates based mainly on the correlation between these variables will not necessarily have adequate predictive power. This explains why in some studies other covariates are included in the analysis, especially those related to climatological factors, in order to obtain more reliable predictions.

On the other hand, the estimates obtained in a parametric way depend on many elements that are subject to the criteria of the analyst. This includes from the model used to estimate the trend, and in the case of small-scale spatial variability, an adequate selection of the valid variogram and its subsequent adjustment process. However, even when it can be assumed that the parametric models fit adequately, their predictions they try to reproduce the selected functional form and do not necessarily fully adapt to the observed sample behavior.

The present study therefore constitutes an interesting non-parametric approximation to obtain more flexible estimates, without the need to depend on a predetermined model. It is even possible to extend these procedures by including information on other variables, to improve the reliability of the predictions obtained. Another important point to note is that these techniques can be used under the parametric approach, as their estimates can be used as a first exploratory analysis to help select and adjust more complex models.

References

1. Alexeeff, S.E., Schwartz, J., Kloog, I., Chudnovsky, A., Koutrakis, P., Coull, B.A.: Consequences of kriging and land use regression for PM2.5 predictions in epidemiologic analyses: insights into spatial variability using high-resolution satellite data. J. Expos. Sci. Environ. Epidemiol. **25**(2), 138–144 (2015)
2. Castillo-Páez, S., Fernández-Casal, R., García-Soidán, P.: A nonparametric bootstrap method for spatial data. Comput. Stat. Data Anal. **137**, 1–15 (2019)
3. Castillo-Páez, S., Fernández-Casal, R., García-Soidán, P.: Nonparametric bootstrap approach for unconditional risk mapping under heteroscedasticity. Spat. Stat. **40**, 100389 (2020)
4. Cherry, S.: An evaluation of a non-parametric method of estimating semi-variograms of isotropic spatial processes. J. Appl. Stat. **23**(4), 435–449 (1996)
5. Chilès, J., Delfiner, P.: Geostatistics: Modeling Spatial Uncertainty, 2nd edn. Wiley, New York (2012)
6. Chu, Y., et al.: A review on predicting ground PM2.5 concentration using satellite aerosol optical depth. Atmosphere **7**(10), 129 (2016)
7. Cressie, N.: Statistics for Spatial Data. Wiley, New York (1993)
8. Feng, S., Gao, D., Liao, F., Zhou, F., Wang, X.: The health effects of ambient PM2.5 and potential mechanisms. Ecotoxicol. Environ. Saf. **128**, 67–74 (2016). https://doi.org/10.1016/j.ecoenv.2016.01.030. https://www.sciencedirect.com/science/article/pii/S014765131630029X
9. Fernández-Casal, R., González-Manteiga, W., Febrero-Bande, M.: Space-time dependency modeling using general classes of flexible stationary variogram models. J. Geophys. Res. **108**, 8779 (2003)
10. Fernandez-Casal, R.: npsp: Nonparametric spatial (geo)statistics (2018). http://CRAN.R-project.org/package=npsp. R package version 0.7-2
11. Fernández-Casal, R., Castillo-Páez, S., Francisco-Fernández, M.: Nonparametric geostatistical risk mapping. Stoch. Environ. Res. Risk Assess. **32**(3), 675–684 (2017). https://doi.org/10.1007/s00477-017-1407-y
12. García-Soidán, P.H., González-Manteiga, W., Febrero-Bande, M.: Local linear regression estimation of the variogram. Stat. Probab. Lett. **64**, 169–179 (2003)
13. Hoff, R.M., Christopher, S.A.: Remote sensing of particulate pollution from space: have we reached the promised land? J. Air Waste Manag. Assoc. **59**(6), 645–675 (2009)
14. Holben, B.N., et al.: AERONET - a federated instrument network and data archive for aerosol characterization. Remote Sens. Environ. **66**, 1–16 (1998)
15. Ichoku, C., et al.: A spatio-temporal approach for global validation and analysis of MODIS aerosol products. Geophys. Res. Lett. **29**(12), MOD1-1 (2002)
16. Levy, R., Hsu, C., et al.: MODIS atmosphere L2 aerosol product. NASA MODIS Adaptive Processing System, Goddard Space Flight Center, USA (2015)
17. Matheron, G.: Traité de géostatistique appliquée. Éditions Technip (1962)

18. Rush, A.C., Dougherty, J.J., Engel-Cox, J.A.: Correlating seasonal averaged in situ monitoring of fine PM with satellite remote sensing data using geographic information system (GIS). In: Remote Sensing in Atmospheric Pollution Monitoring and Control, vol. 5547, pp. 91–102. International Society for Optics and Photonics (2004)
19. Shapiro, A., Botha, J.D.: Variogram fitting with a general class of conditionally nonnegative definite functions. Comput. Stat. Data Anal. **11**(1), 87–96 (1991)

Defense Engineering

Comparative Analysis of Defense Industry Models in South American Countries

Dolores Borsic Laborde Zlata[1](✉), Angie Fernández-Lorenzo[1], Darwin Manolo Paredes Calderón[2], and Stéfanny Nicole Montoya Loor[1]

[1] Department of Economic, Administrative and Commercial Sciences, Universidad de Las Fuerzas Armadas-ESPE, Sangolquí, Quito, Ecuador
zdborsic@espe.edu.ec

[2] Soldier Training School "ESFORSE", Military Applications Research Center, Universidad de Las Fuerzas Armadas-ESPE, Sangolquí, Ecuador

Abstract. The work carried out focused on the development of a comparative analysis between South American countries similar to Ecuador, in terms of their defense industry models, based on a set of aspects valued as fundamental in the related theory. The main techniques applied were analysis and synthesis and content analysis, based on consulting bibliographic sources. The experiences of nearby countries with similarities to Ecuador were studied, with a view to strengthening the national development of the defense industry. It was evidenced that the countries analyzed have a greater development of R + D + i in the defense sector than Ecuador, despite the fact that from the point of view of public policy, institutional system and regulations there are no notable differences, not so in the case of the assigned budget, in which differences are evident between Ecuador and the rest of the countries analyzed. The analysis carried out will allow us to continue deepening the theoretical and practical aspects that will drive the development of this strategic sector in the future.

Keywords: Industry · Defense · Technology · Models · South American countries

1 Introduction

The defense system of a country constitutes a strategic issue when it comes to safeguarding the sovereignty of nations, which, at the same time, is directly related to the development of both technology and national industry, for design and production. complex systems, such as: radars, navigation or the self-supply of key inputs, such as: solid propellants, which involves having freedom to make decisions; It should be noted that innovations used in everyday life, such as GPS or the internet, had their origin in developments for Defense [1]. The viability and sustainability of the Defense industry is a cross-cutting and significant task for various State portfolios, the weapons and material programs, in addition to being a guarantee of power, sovereignty and security for the citizens, constitute an opportunity for the development and training of the industry in a

M. Botto-Tobar et al. (Eds.): CIT 2021, LNNS 405, pp. 353–366, 2022.
https://doi.org/10.1007/978-3-030-96043-8_27

national context, initially, since the possibility of generating regional or global development cannot be limited. In any case, it should be noted that the defense industry market is a captive, complex, competitive and monopolized market.

Consequently, the synchronized work between various actors from the public and private spheres constitutes the engine that energizes the scaffolding of the industry. Institutional support is essential and ends up being a key element to encourage government and industrial cooperation and achieve the harmony required in the systems and rationalize defense spending. Note that size and partnerships are key factors in positioning the industry internationally. Likewise, research is the key to success to achieve concepts of sustainable and sustainable development.

So, being R + D + i (Research, Development and innovation) an activity that is executed under the perspective of a cyclical process, whose duly validated results, constitute important inputs for the industry and strengthens an academic-industry relationship [2, 3], the need to strengthen each edge of the industrial development process is decisive, since it feeds on the strength of the academy, and in turn, during its production and commercialization stages, detects variables, requirements and useful needs for strengthen investigation processes. These concepts gave rise to philosophies of the triple helix, which has traditionally been handled in countries that have boosted their industry to become true world powers [4].

Therefore, it is important to broaden the spectrum of cause-effect analysis of the true strategies that the various actors in a society must follow, allowing for the need to characterize societies, under genetic, cultural and behavioral approaches; This is how the proposal to energize the defense industry arises, under the approach of the five-fold helix [5].

1.1 Model Dimensions in the Defense Industry

Briones [6] mentions that the modernization of the Armed Forces arises from the new conception of defense as a function of the State for the benefit of society, where its continued evolution is essential; both in relation to its structures and the resources necessary for the proper development of its activity. When analyzing the Defense industry and the development of strategic capabilities, it is recognized that more and more investment in the Defense sector must be optimized, prioritizing the areas with the greatest capacity for innovation and productivity, observing strategic criteria such as the existence of a base based on research-development, the existence of both public and private sources of financing and the characteristics of the human personnel [7].

The defense industrial development model is defined as an orderly process of industrialization of goods and services destined for the defense of a country; what is evidenced through formal institutions, both in their structure and organization and in the public policies that energize them [4].

This theoretical analysis seeks to disaggregate the central dimensions that make up a defense industrial development model: public policy, budget, defense development and the level of management of the countries analyzed, given that the viability and sustainability of the defense industry It is a shared task between the administration and the defense, where the weapons and material programs constitute an opportunity for the development and training of the national industry; It follows from the above that there are three central dimensions in the defense industry, such as: the structure and organization of the sector, the industrial defense policy as a public policy, and the productive profile and orientation of the industry [4], the detail is shown in Fig. 1.

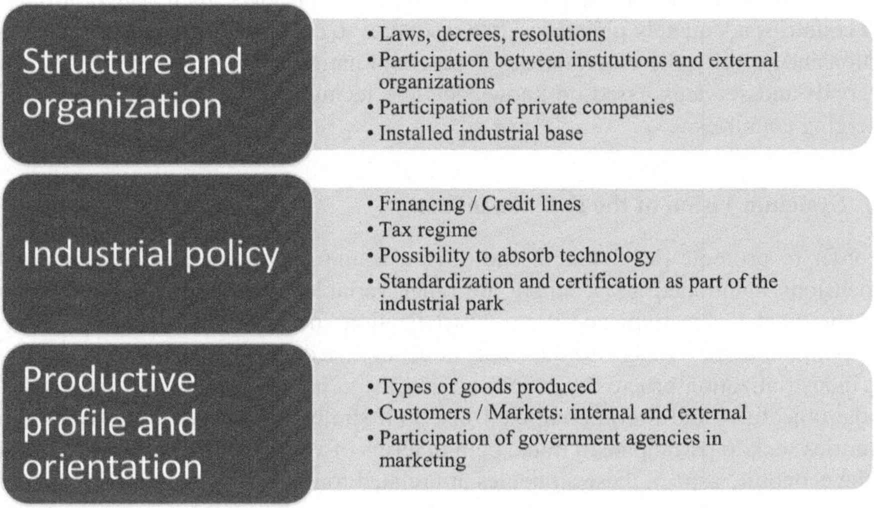

Fig. 1. Analytical scheme of the defense industry by dimensions Source: Own elaboration based on Da Ponte [4].

In Fig. 1 the components of the three dimensions are detailed, where the first refers to the set of organizations, whether state or private (companies in the productive sector and universities) and government institutions, whether political or military, and to Research Centers. and Development, who participate in one or more of the stages of research, development and innovation, production, commercialization and repair and/or maintenance of defense products and the mechanisms they use for it; The second dimension deals with the set of actions that the State deploys in order to develop the defense industry through the defense policy that supports it and details the instruments and tools through which these actions are executed and, the third dimension refers to the productive profile and the internal, external or mixed orientation, this dimension has two edges, one economic and the other economic-industrial and geopolitical; the economic edge responds to the need to achieve sustainability in sectorial development, for which it is necessary to promote commercialization both in the internal and external market and develop a minimum industrial scale; The second edge corresponds to an economic facet related to the type of production and if it is dual in nature and another geopolitical one that has to do with the type of armaments and programs that are supported on the basis of the strategic priorities of both, civilians and the military [4].

From the foregoing, the need to maintain permanent technological monitoring and surveillance is evident as a good practice in organizations dedicated to the Defense industry [8]. Another aspect that stands out in the literature as key to generating synergies between the defense industry and the development of strategic capabilities is knowledge management, as a tool that allows knowledge to be shared among the different agents that are part of the innovation ecosystem for the defense [9].

In this way, Parra, Játiva and Vásquez [10] indicate that the principles of defense economics have varied depending on the complex conditions that currently present themselves for security at the national and international level. Increasingly, it is vital that countries adequately plan the scarce resources to cover their needs in terms of protection and security, therefore, the development of national industries for the generation of goods and services based on knowledge and technology becomes an aspiration of emerging countries.

1.2 Systemic Vision of the Defense Industry

In order to promote development and have a frame of reference, in addition to the dimensions mentioned above, there are other variables that must also be delimited, such as: a) the objectives and the industrialization strategy, which refers to the role that politicians they assign to the State in economic development, the way in which the industrialization process is conceived and the economic rationality that guides it; b) the objectives and the international insertion strategy deal with the way in which countries seek to participate in trade, both in terms of volume and industrialized goods; In the economic aspect, these strategies are related to industrialization and with regard to international relations, it should be noted that military power is decisive both in the strategic and geopolitical fields, therefore, the defense industry occupies a significant place in the economy of a country, so that, the importance that the political, business and military elite give to production for defense, constitutes an indicator of the weight that this factor has in the industrial development model, and, c) the degree of interagency articulation that is located at an execution level, but, acquires vital importance in those countries where informal institutions predominate, since its purpose is to articulate the actors so that they operate with total linkage within the industry of the defense, which makes it possible to connect defense policy with industrialization objectives and strategy, the details are presented in Fig. 2.

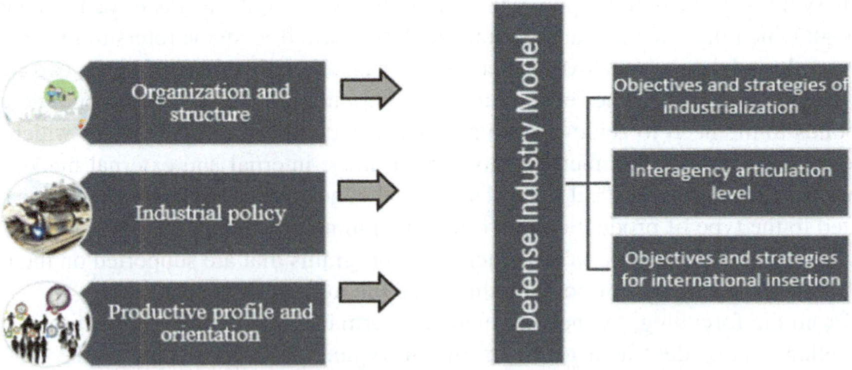

Fig. 2. Systemic representation of the defense industry Source: Own elaboration based on Da Ponte [4].

At the global level, there are many experiences of interaction between industrial policies and regional development, as well as their consequences, which are often contradictory or complementary [11]. In this sense, Maye [12] points out that, thanks to globalization, countries show how the downward pressure to compete and/or save costs in international markets motivates producers and consumers to obtain cheaper goods and services and However, in the defense industry sector, there are other elements, such as: security, concerns and politics, that can overshadow economic logic when it comes to making purchasing decisions, which in turn Instead, it prevents the defense industry in America from using the most profitable supply chains and manufacturing centers through increased transnational cooperation on defense acquisitions, instead, the defense market in Europe is consolidated. The new world order generated a decrease in the demand for military equipment, which at the same time leads to an increase in production and development costs.

The principles of defense economics have varied depending on the complex conditions that currently exist for national and international security. It is increasingly vital that countries adequately plan the scarce resources to cover their needs in terms of protection and security, therefore, the development of national industries for the generation of goods and services based on knowledge and technology becomes an aspiration of emerging countries [10].

The objective of this work is to develop a comparative analysis between South American countries with similarities to Ecuador, in terms of the development of their defense industry systems.

2 Materials and Methods

This research is theoretical and bibliographic, for which bibliographic sources obtained from Redalyc, Scopus, Google Scholar, among others, were used. The information analysis was carried out using theoretical research methods such as historical and systemic; for which the analysis and synthesis techniques and content analysis were applied.

3 Results

When analyzing the Defense industry and the development of strategic capabilities, it is recognized that more and more investment in the Defense sector must be optimized, prioritizing the areas with the greatest capacity for innovation and productivity, observing strategic criteria such as the existence of a base based on research-development, the existence of both public and private sources of financing and the characteristics of human personnel [7].

Another aspect that stands out in the literature as the key to generating synergies between the defense industry and the development of strategic capabilities is knowledge management, as a tool that allows knowledge sharing among the different agents that are part of the innovation ecosystem for defense [9]. Likewise, the need to maintain permanent technological monitoring and surveillance is evident as a good practice in companies dedicated to the defense industry [8].

The global violence of the 20th and 21st centuries has caused innumerable human lives, in a scenario in which the armed forces had the greatest role, where the leaders of the first and second world never came to face each other in the framework of a atomic war; on the other hand, in the third world they maneuvered through internal lines using subversive organizations on the one hand and traditional armed forces on the other. In this vein, it is known that, until the mid-twentieth century, governments exercised censorship on information in order to obtain recruits for military service, which left dissidents a marginal role. At the end of the 1960s, this scenario changed due to the loss of the war by the United States against North Vietnam, which is why public opinion pointed to the military institutions as the main responsible for the defeat, which generated that compulsory military service is terminated in the United States and military technology has given priority over that of civilian origin; this influence was generalized by most of the third world countries [13].

In the South American countries, the greatest war conflict was the Chaco War (1932–1935), where it is estimated that there were 90 thousand casualties; Another conflict with fewer casualties was the Falklands War in 1982, where 635 Argentine men died; There was also the conflict of the border war between Ecuador and Peru from the end of January to the end of February 1995, but the official death toll is unknown since the version of each country varies, thus, Ecuador reported the death of 33 soldiers, but the Association of former Cenepa Combatants indicates that there were 131 killed in battle.

This was the last conflict between South American countries, in addition, the probability of a conflict between countries in this region is minimal, therefore, the South American military institutions have the mission of confronting internal conflicts, for example, the Colombian State is it faces approximately six million victims of forced displacement and other war crimes, such as: threatened, disappeared, kidnapped, homicides, forced recruitment of children, antipersonnel mines, torture, sexual violence and the fight against drugs [13]. Under this scenario, the comparative analysis of the defense system in South American countries is justified, where the factors under study are: the country, the territorial extension, the population, the number of troops in the Armed Forces, the budget allocated to defense and the Gross Domestic Product of the country. Details are set out in Table 1.

From the data analyzed in Table 1, it is identified that Brazil is the country that has the largest budget for defense, exhibits the highest weighted weight in relation to the budget (40.93%), which is understandable given that it is the country with the largest budget. largest in South America and has the largest number of military personnel, followed by Venezuela with a lower percentage (17.54%), but has a number of Armed Forces personnel very similar to that of Brazil, whose territorial extension is greater by a proportion of more than nine times the size of Venezuela and the population is approximately seven times higher, however, the percentage of Brazil's budget is only 2.33 times that of Venezuela; a similar situation occurs between Brazil and Colombia.

Table 1. Defense comparison in South America.

Country	TE	Popul	TAF	GDP	DB	BW
Argentina	2.780.400	43.712.000	79.845	437.856.000.000	4.287.426.700	9%
Bolivia	1.098.580	10.904.000	34.078	33.983.000.000	568.421.520	1%
Brazil	8.514.880	209.486.000	366.614	1.534.782.000.000	19.978.247.480	41%
Chile	756.100	18.276.000	67.683	235.419.000.000	4.571.174.008	9%
Colombia	1.141.750	48.650.000	265.050	253.240.000.000	4.916.946.842	10%
Ecuador	256.370	16.385.000	41.403	94.014.000.000	2.510.507.785	5%
Paraguay	406.750	6.723.000	16.087	26.804.000.000	357.354.910	1%
Peru	1.285.220	31.776.000	78.296	178.643.000.000	2.237.685.498	5%
Uruguay	176.220	3.443.000	22.316	53.145.000.000	770.840.944	2%
Venezuela	912.050	30.936.000	365.315	185.611.000.000	8.549.765.946	18%

Note: Territorial expansion in km^2 = TE, Population = Popul., Troops Armed Forces = TAF, Gross Domestic Product = GDP, Defense budget = DB, Defense budget weighted weight = BW. Source: Own elaboration based on Donadio and Kussrow [13].

Likewise, in relation to the defense systems and legislation in South America, Table 2 shows a detail of both aspects and it is noted that they are similar, with small variations in the name of the entities, positions or entities. laws and regulations, but in essence no major difference has been identified in the concepts, but Argentina, Ecuador, Uruguay and Venezuela are the countries that have the least number of laws and regulations focused on national security and defense; on the contrary, Peru and Colombia are the countries that have the highest number of these types of laws.

Table 2. Defense systems and legislation in South America

Country	Defense system	Systems and concepts of national legislation
Argentina	President, in a functional advisory relationship: the National Defense Council - CODENA and the Crisis Committee. In command relationship: the Ministry of Defense, the Joint Chiefs of Staff of the Armed Forces, the General Staff of the Navy and the General Staff of the Air Force	National Defense Law, Internal Security Law, Ministries Law, Armed Forces Restructuring Law, National Intelligence Law

(continued)

Table 2. (*continued*)

Country	Defense system	Systems and concepts of national legislation
Bolivia	President, with a functional advisory relationship: the Supreme Council of National Defense and the Council for Border Development and Security. In command relationship: Ministry of Defense, Command in Chief of the Armed Forces, General Staff of the Armed Forces, General Commands of the Navy, Army and Air Force	Organic Law of the Armed Forces, Law of organization of the Executive Power, Law that establishes mechanisms of institutional articulation for the execution of comprehensive development policy and border security, Law of the National Citizen Security System, Law of firearms control, ammunition, explosives and other related materials, comprehensive law to guarantee women a life free of violence, law on security and defense of airspace
Brazil	President, in a functional advisory relationship: the Defense Military Council and the National Defense Council. In command relationships: the Ministry of Defense, the Joint Chiefs of Staff of the Armed Forces, the Navy Command, the Army Command and the Aeronautical Command	Law that establishes rules for the departure of Brazilian troops abroad, Law on the organization and operation of the National Defense Council, Law that determines the cases in which foreign forces can transit through the national territory or remain in it temporarily, Law of the Brazilian intelligence system, Law on the organization of the Presidency of the Republic and the Ministries, Law on mobilization, Law that regulates the provisions of the Federal Constitution on terrorism
Chile	President, in a functional advisory relationship: the National Security Council-COSENA and the Board of Commanders-in-Chief. In command relationship: the Ministry of National Defense, Joint Chiefs of Staff, Command in Chief of the Army, Command in Chief of the Air Force and the Command in Chief of the Navy	Mobilization Law, Law on the participation of Chilean troops in peace operations, State Intelligence Law, Law of the Organic Statute of the Ministry of National Defense, Law of the Ministry of the Interior and Public Security, Law of the Ministry of Women and Equity of genre

(*continued*)

Table 2. (*continued*)

Country	Defense system	Systems and concepts of national legislation
Colombia	President, in a functional advisory relationship: the National Security Council. In command relationship: the Ministry of National Defense, the General Command of the Military Forces, the Army Command, the Air Force Command and the Navy Command	Merger of the National Security Council, Superior Council of National Defense; Administrative career law, benefits in favor of relatives of people who died during the provision of compulsory military service, Organization and operation of entities of the national order, Extraordinary powers to issue norms of the military forces and national police, Structure of the Ministry of Defense National, Statute that regulates the civilian personnel administration regime of the Ministry of National Defense, Legal nature of the Nueva Granada Military University, Reincorporation of members of armed groups outside the law, Special administrative career for non-uniformed public employees, Acquisition of goods and services destined to the defense and national security, Special career system, Regulation of the reserved expenses law, Attention, assistance and comprehensive reparation to the victims of the internal armed conflict, Law of risk and disaster management
Ecuador	The President and in a functional advisory relationship: the State and Public Safety Council. In command relationship: Ministry of Security Coordination, Ministry of National Defense, Joint Command of the Armed Forces, General Command of the Land Force, General Command of the Naval Force and General Command of the Air Force	Organic Law of National Defense, Law of Public and State Security
Paraguay	President and in a functional advisory relationship: National Defense Council. In command relationship: Ministry of National Defense, Military Forces Command, Joint Chiefs of Staff, Army Command, Navy Command and Air Force Command	Law of national defense and internal security, Law that establishes the border security zone, Law that creates the National Emergency Secretariat, Law that creates the National Intelligence System, Law of surveillance and protection of Paraguayan airspace

(*continued*)

Table 2. (*continued*)

Country	Defense system	Systems and concepts of national legislation
Peru	President and in command relationship: Ministry of Defense, National Security Council, Secretary of National Security and Defense that has under its command the National Intelligence Directorate, Joint Command of the Armed Forces, General Command of the Army, of the Navy of War and Air Force	Law on states of exception, Legislative Decrees: of self-defense committees of the population and with regulations for the Armed Forces, Law of requirements for the entry of foreign troops into the territory, Law of national mobilization, Law of the national security and defense system, Organic Law of the Executive Power, Legislative Decree, Law of organization and operation of the Ministry of Defense, Law of the National System of Disaster Risk Management, Delegation of power to legislate in the interior sector and national defense, Legislative Decree that regulates the System of National Defense, Legislative Decree that creates the Secretariat of National Security and Defense, Strengthening and Modernization of the National Intelligence System-SINA, Law of control, surveillance and defense of the national airspace

Source: Own elaboration based on Donadio and Kussrow [13]

As can be seen in the analysis carried out in Table 2, the South American countries close to Ecuador have an institutional and regulatory system similar to that of the Andean country in relation to the National Defense system.

Table 3 presents the analysis carried out on influential aspects in the defense industry model in countries with similarities to Ecuador, such as Argentina, Brazil, Chile and Colombia, as a result of the application of the content analysis technique based on the consultation of various bibliographic sources.

The comparative analysis carried out allows us to know the experiences of nearby countries with similarities to Ecuador, with a view to strengthening the national development of the defense industry. As evidenced by the countries analyzed have a greater development of R + D + i in the defense sector than Ecuador, despite the fact that from the point of view of public policy, institutional system and regulations there are no notable differences, although there are significant differences from the defense spending point of view.

Table 3. Defense industry in countries similar to Ecuador

Country	Public policy	Defense development (R + D + i)
Argentina	The main mission of the National Defense System of the Argentine Republic is to ward off and repel, through the use of the military instrument of National Defense, external military state aggressions in order to guarantee and safeguard the sovereignty, independence and self-determination of the nation	On the world stage, knowledge has been identified as a strategic factor. This has changed the nature of military affairs and soft power has emerged as a decisive element in military operations. Therefore, the technological strategy constitutes one of the pillars of Production for Defense and Research and Development (ID). From the particular restructuring and modernization policies, a persistent need to have technological development capacity emerges, in particular in relation to those systems and materials whose international availability is normally conditional and that in crisis situations are of doubtful or difficult to obtain
Brazil	The National Defense Strategy: * Deter the concentration of hostile forces on the Brazilian borders and airspace. * Organize the Armed Forces under the aegis of the trinomial monitoring/control, mobility and presence. * Develop strategic mobility. * Deepen the link between the technological and operational aspects of mobility. * Strengthen the three sectors of strategic importance: space, cybernetics and nuclear energy. * Unify the operations of the three Forces. * Reposition the troops of the three Forces. * Concentrate the presence of Army, Navy and Air Force units on the borders	The Ministry of Defense, in coordination with the Ministry of Science and Technology, will update the Science, Technology and Innovation Policy for National Defense and the resulting regulatory instruments. To meet the objectives of this Policy, the adaptation of the existing organizational structures that act in the area of Defense Science and Technology must occur. The aforementioned documents will contemplate: - measures for the maximization and optimization of research efforts in civil and military scientific and technological institutions, for the development of cutting-edge technologies for the defense system, with the definition of integrated efforts of researchers of the three Forces, especially for priority areas and their respective technologies of interest

(*continued*)

Table 3. (*continued*)

Country	Public policy	Defense development (R + D + i)
Chile	The Political Constitution of the Republic contains purposes related to the Defense of the homeland and the protection of the population. Apart from the constitutional bases, they have aspirations that encourage some of their public institutions, such as: the maintenance of international relations related to national development, the cooperation needs and the international projection of the country, but also the promotion and maintenance of values. contained in the Charter of the United Nations or the achievement of interests common to all humanity	Science, technology and innovation play a central role in the military field and in international security. At present, these dimensions are still relevant, but others are added that need to be taken into account. On the one hand, there has been a process of rapid advances - nanotechnology, information and communications, artificial intelligence, etc. - that are altering the social, political, economic and cultural spheres. On the other hand, access to these advances is not only available to states, but, in many cases, also to individuals, groups, corporations and, in general, different non-state actors
Colombia	According to the Strategic Plan for the Defense and Security Sector 2018–2022, it is based on: * Defense and Security Policy. * Framework Policy for Coexistence and Citizen Security. * International Cooperation Policy. * Public Security Policy. * Policy of Contribution to the protection of Natural Resources and the environment. * Contribution to the development of the country. * Institutional Management and Support	The new systems are part of a recently developed program: Soldado del Futuro Colombiano-SFC, an initiative whose main objective is the well-being of the soldier, infant or police, through the analysis of the different aspects and environments in which they fulfill their mission and how, through technology, the material they have can be improved and provide them with better resources, which will result in higher levels of effectiveness and efficiency when combating and preventing alterations to security and public order

(*continued*)

Table 3. (*continued*)

Country	Public policy	Defense development (R + D + i)
Ecuador	According to the Institutional Strategic Defense Plan 2017–2021, within the legal framework, it is based on: * Constitution of Ecuador. * Organic Law of National Defense 2007. * Law of Public and State Security. * Organic Code of Planning and Public Finance. * Presidential Decree No. 940 of November 16, 2011. * Organic Statute of Organizational Management by Processes of the Ministry of National Defense. * National Development Plan "All a life" 2017–2021. * Intersectoral Security Coordination Agenda 2017–2021	In the field of research and development, it can be seen that the actors that make up the Defense R + D + i system are not efficiently integrated. There is no adequate articulation in the use of available resources and a higher level of use of specialized human talent is lacking. The challenge in this area is to ensure that the Institutes coordinate their actions and resources within the framework of MIDENA's leadership through the under Secretariat for Development Support

Source: Own elaboration

4 Conclusions

The properly planned defense industry can contribute in a sustainable way to develop autonomous strategic capabilities, based on innovation and modernization of forces.

The importance in the defense of the countries analyzed in this work is to fulfill their international obligation to protect and enforce human rights, in recent years the State and important instructions have been adopted, the laudable purpose of which is to guarantee that people dedicated to The defense of these rights can carry out their activities in conditions of tranquility and security, within a framework of tolerance and respect for their work, without falling into the risk of being victims of attacks on their life, personal integrity, and individual freedom, their privacy or honor, or the object of intimidating threats.

The analysis of the Defense industrial sector shows that it is a market that works imperfectly to supply the products and services that this market requires, which can lead to relevant inefficiencies that can only be solved through a set of complementary measures between The ones that must be mentioned are various coordination mechanisms between the actors, in which the Administration plays a prominent role.

For the countries analyzed in this paper, R + D + i (research, development and innovation) represents the expenditure made by many companies or the State in research to acquire knowledge that can be applied to improve products or productive processes for defense of each of the sovereignties, whose development depends on factors such as the institutional and regulatory system, public policy and the allocated budget, among others, with notable differences found only in this last aspect between Ecuador and the countries analyzed.

References

1. INFOBAE: Defense industry (2015). https://www.infobae.com/2015/10/20/1763648-la-industria-la-defensa
2. Adams, J., Chiang, E., Starkey, K.: Industry-university cooperative research centers. J. Technol. Transf. **26**, 73–86 (2001). https://doi.org/10.1023/A:1007836328722
3. Etzkowitz, H.: Networks of innovation: science, technology and development in the triple helix. Int. J. Technol. Manage. Sustain. Dev. **1**(1), 7–20 (2000). https://doi.org/10.1386/ijtm.1.1.7
4. Da Ponte, A.: Defense Industrial Management Models: Contributions for a Systemic Approach. War High School, Bogotá (2016). http://www.enabed2016.abedef.org/resources/anais/3/1465915175_ARQUIVO_daPONTE_Aureliano_2016_AT1_MODELOSDEGESTIONINDUSTRIALENDEFENSA.pdf
5. Fernández, A., Borsic, Z., Paredes, M.: Defense economy: from the theory to practice in Latin America. Ibérica de Sistemas e Tecnologias de Informacao (E29), pp. 80–691 (2020). http://www.risti.xyz/issues/ristie29.pdf
6. Briones, A.J.: Economy of security and defense. Knoledge transfer and innovation associated with the defense industry. Spanish Inst. Strat. Stud. Mag. (3) 5–8, (2013). https://revista.ieee.es/article/download/342/575/
7. Blanco, M.: Defense activity as a source of wealth. Strat. Notebooks (175), 217–259, (2015). https://dialnet.unirioja.es/servlet/articulo?codigo=5261892
8. Escribano, F., Martínez, A., Vicente, S.: Surveillance and technological foresight in Spanish Defense companies. Strat. Think. Secur. Maga. **1**(2), 77–86 (2016). http://uajournals.com/ojs/index.php/cisdejournal/article/view/184/141
9. Briones, J.: Evolution of knowledge management, transfer and innovation in the defense industry. Strat. Notebooks (194), 134–168 (2017). https://dialnet.unirioja.es/servlet/articulo?codigo=6356929
10. Parra, H., Játiva, A., Vásquez, F.: Defense industry and development of strategic capacities in Ecuador. Int. J. Econ. Law World **XVII**, 1–12 (2019). http://www.revistainternacionaldelmundoeconomicoydelderecho.net/?p=971
11. Markusen, A., Costigan, S., eds.: Arming the Future: A Defense Industry for the 21st Century. Council Foreign Relations Press, Washington (1999). https://www.jstor.org/stable/26466841?seq=7#metadata_info_tab_contents
12. Maye, D.L.: Autarky or Interdependence: U.S. vs. European security and defense industries in a globalized market. (E.-R. A. University, Ed.) J. Strat. Secur. **10**(2), 33–47 (2017). https://doi.org/10.5038/1944-0472.10.2.1597
13. Donadio, M., Kussrow, S.: Comparative Atlas of Defense in Latin America and Caribbean. RESDAL, Buenos Aires (2016). https://www.resdal.org/assets/atlas-2016-esp-completo.pdf

Behavior of Reinforced Concrete Column Specimens Under Blast Loading Produced by Pentolite Charge

Nestor Mejía[1(✉)], Enrique Morales[1], Ricardo Durán[1], Rodrigo Mejía[1,2], and Esteban Vásconez[1]

[1] Department of Earth Sciences and Construction, Universidad de las Fuerzas Armadas ESPE, Av. General Rumiñahui S/N, 171103 Sangolquí, Ecuador
ndmejia1@espe.edu.ec

[2] The Ecuadorian Army Corps of Engineers, Av Rodrigo de Chávez, 170111 Quito, Ecuador

Abstract. This study presents an experimental and analytical investigation of the mechanical behavior of reinforced concrete specimens under the impact of explosive charges (Pentolite 50/50). During the experimental studies, 14 explosive charge configurations were made. These were shaped by the interaction of variables such as: weight of the explosive, metallic reinforcements and geometrical configuration of the explosive charges (conical, cylindrical and dihedral). The study found that conical-dihedral configuration with metallic reinforcement caused greater damage to the material. With this same explosive configuration, 11 destructive tests were carried out on reinforced concrete specimens, each measuring 30 × 30 cm in cross section and 50 cm in length with a compressive strength of 210 kg/cm^2. A further 6 destructive tests were carried out on specimens with the same dimensions but with a compression resistance of 280 kg/cm^2. The interaction between the energy released from the explosion and the splinters of the reinforcing steel caused fragmentation of the concrete and deformation of the reinforcing steel. However, the fracturing of the reinforcing steel depends on the amount of the explosive charge. The results of these models are presented in terms of deformation and fracture, which were backed up by the data obtained in the experimental tests, in order to establish a degree of reliability of the analytical models.

Keywords: Pentolite · Reinforced concrete · Steel bars · Blast loads · Deformation

1 Introduction

Ecuador lies within the Ring of Fire that circles the Pacific Ocean, an area in which around 90% of all the world's earthquakes occur and 80% of the strongest. This is reflected in the report of United States Geological Survey (USGS) [1], which established that over the last 70 years Ecuador has suffered 13 earthquakes greater than 7 on the Richter scale. After the last earthquake on April 16, 2016, an official report by the Ecuadorian government determined that 44,813 structures were affected, 5,327

M. Botto-Tobar et al. (Eds.): CIT 2021, LNNS 405, pp. 367–376, 2022.
https://doi.org/10.1007/978-3-030-96043-8_28

of which had demolition order. In order to generate building implosions with national personnel and resources, the present study was proposed to research the mechanical behavior of reinforced concrete subjected to explosive charges.

Many studies have already been carried out to explore the final state of a material subjected to explosive charges. Buczkowski et al. [2] defined the detonation velocity in the explosion of charges of ammonium nitrate based fertilizers and fuels, concluding that a small amount of inert inorganic substances and fuels produce explosive properties. Others have developed thermochemical computer code, which are computational tools that can predict detonation properties based on physical and chemical principles [3].

One of these principles used in the development of thermochemical computer code is the JWL (The Jones-Wilkins-Lee) equation of state. This was used by Pui et al. [4], who carried out experimental test and numerical modeling to describe the blast wave of Sprängdeg m/46 (consisting of 86% PETN and 14% mineral oil) when in touch with a certain material. However he concluded that the results obtained by these two methods deviate because the blast wave was too sensitive to be measured with high accuracy.

The shape of the explosive is just as important as its weight or chemical composition. Geometric conditions can cause variations in the explosive blast pressure profiles. Simoens et al. [5] carried out trials to quantify the geometric effects of cylindrical and spherical explosives of ammonium nitrate, TNT, C4 and emulsion explosives. Regarding the cylindrical loads, the perfect length-diameter ratio was established for the most effective explosion with the highest damage effect. For the spherical loads, the C4 produced the peak overpressure, while the effect was similar for the TNT and the emulsion explosive.

Hryciow et al. [6] developed numerical simulations in LS-DYNA software for cylindrical and spherical loads with 3 different sizes of TNT. The purpose of this study is to determine the pressure field of each shape and their influence on the explosive charge. The study concluded that spherical charges have symmetrical effects and cylindrical charges have asymmetric effects in the pressure profile. Olive et al. [7] measured the pressure-time-distance relationship in a burst of cylindrical charge of RDX (cyclonite) in order to study the effects of the geometric shape on the explosion. They concluded that the effect of the charge shape is significant for scaled distances less than $20ft/\sqrt[3]{lb}$. However, the lack of information about the performance of each of the explosives makes it hard to establish their behavior with exactitude, which is why Diaz et al. [8] applied the TNT equivalent model, to calculate detonation properties of all explosives without considering charge shape.

Some studies have already proposed another approximation model based on the one-degree system of freedom (SDOF). To convert a structure to an SDOF system you need equivalence factors like the load, weight and resistance of the material. These are based on equal energy of the load, of the resistance as well as the equal kinetic energy for the real system and the SDOF system. Yasseri et al. [9] used SDOF to obtain the Force-Impulse (F-I) diagrams, which are necessary to assess the damage of a specific structure caused by the peak load. They also developed Pressure-Impulse (P-I) diagrams, essential for assessing the design and damage of a structure under a blast load.

Figuli et al. [10] studied the blast wave propagation of POLONIT explosives and its dynamic response on a structure by SDOF and experimental testing. The results of the study achieved a 50% reliability.

This literature points to a lack of information about the mechanical behavior and the final state of reinforced concrete under a blast caused by a Pentolite (50/50) explosive. Therefore, the following study has been proposed. The main objective is to design the most effective explosive charge (specific that will cause the most damage to reinforced concrete specimens in terms of fragmented concrete and deformation or fracture of longitudinal and transverse steel bars through of seventeen experimental tests.

2 Design of Experiments

2.1 Specimen Specifications

The reinforced concrete specimens were designed to resist the compressive structural stress under normative parameters [11] with a cross section of 30 × 30 cm and a total length of 500 mm.

Table 1. Specimen design for 21 MPa (28 MPa)

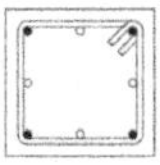	4 steel bars Ø 1.6 cm	Cross Section	
	4 steel bars Ø 1.8 cm	Total	
Area (cm^2)	18.22	900	
	[11]	**Calculated**	**Condition**
	1%<Steel Ratio <3%	Steel Ratio 2%	Approved
[12]	A_{sh}	**Design**	
Equation	(cm^2)	Approved	
21-4	2.03 (2.70)	1 stirrup Ø 1.2 (1.4) cm @10 cm	
21-5	1.08 (1.44)		

3 Design of the Explosive Charge

The explosive used was Pentolite 50/50 ($C_{2.33}H_{2.37}N_{1.29}O_{3.22}$), which is a mixture of 50% pentaerythritol tetranitrate (PETN) and 50% trinitrotoluene (TNT) [13]. There were 14 trials with different combinations, in which the weight, geometry and location of the charge varied. As a result, the conical-dihedral configuration with metallic reinforcement charge was the one that generates the most destructive effect on the specimen. In order to relate explosive the charge with mechanical properties of the specimen and its effect on it, nine types of conical-dihedral with metallic reinforcements were made with different weights.

(a) Elements (b) Conic Dihedral Charge

Fig. 1. Design of charge explosive

Table 2. Explosive characteristics

Parameter	Unity	Specification
Density	g/cm^3	1.6
Gas Volume	l/kg	777
Heat of Explosion	kcal/kg	1300
Velocity of Detonation	m/s	7418
Presion of Detonation	kbar	220
Resistance to water	year	1
Temperature of Explosion	^{o}K	4497

4 Experimental Set-Up

Using the previous nine explosive charge configurations, 11 tests were carried out on specimens with a $f_c = 210$ kg/cm^2 and 6 tests on specimens with a $f_c = 280$ kg/cm^2. Measurements of weight and resistance were taken of each one of the specimens, which were then placed on simple reinforced concrete supports that were separated by 30 cm and embedded in the ground. The explosive charge was then placed in the center of the top face. Recording equipment was also installed to record photographic evidence of the experiment. Finally, the staff moved to a safe area that had already been delimited, and the explosive charge was activated by an electrical impulse.

(a) Support conditions (b) Top face configuration

Fig. 2. Experimental set-up

4.1 Data Acquisition

After the experimental tests, the results were collected by personal inspection, photographic evidence and direct measurement of each specimen. First, personal and photographic inspections were carried out to identify each of the elements that form part of the specimens. Using scaled photographs and a CAD program, the final state of the steel bars were defined, focusing on deformations and whether or not fractures existed. Subsequently, the areas of defragmentation and cracking of the concrete were determined in order to define the residual weight.

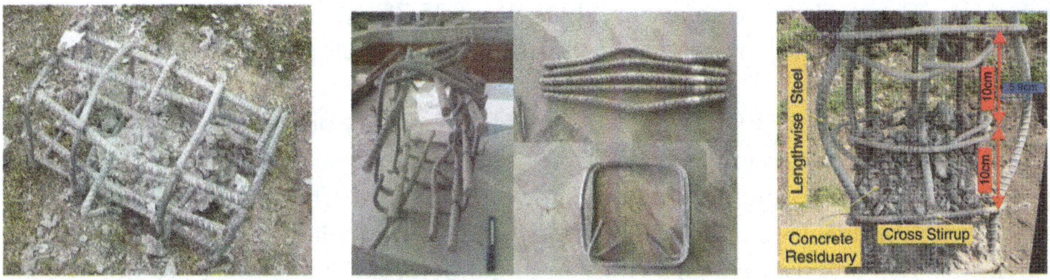

(a) Square column after test (b) Longitudinal bars and stirrups (c) Determining dimensions

Fig. 3. Analysis of specimens

Table 3. Test specification

Charge #	Explosive mass (g)	f'c (MPa)	Charge #	Explosive mass (g)	f'c (MPa)
1	450	21	10	1050	21
2	450	21	11	1050	21
3	660	21	12	1050	21
4	750	21	13	1050	28
5	750	21	14	1050	28
6	760	28	15	1010	21
7	900	21	16	1010	28
8	990	21	17	1350	28
9	990	28	–	–	–

5 Results and Discussion

5.1 Design of Explosive Charge

Figure 4a shows a specimen of f_c = 210 kg/cm^2 subjected to the energy produced by a conical-cylindrical shaped load of 1050 g. After the explosion, what remained of the specimen was 13% residual concrete, as shown in (Fig. (4b)). The steel rod, with an initial length of 20.75 cm, in the 2-1 position suffered a maximum deformation of 5.70 cm. The stirrups present global cracks and deformations. Figure 4c shows a specimen of f_c = 210 kg/cm^2 subjected to the energy produced by a conical-cylindrical shaped load of 1050 g. After the explosion, what remained of the specimen was 73% residual concrete, showing few cracks in the concrete and no destructive effect on the steel rods, as shown in (Fig. (4d)). Figure 4e shows a specimen of f_c = 210 kg/cm^2 subjected to the energy produced by a conical-dihedral shaped load with metallic reinforcement of 900 g. After the explosion, what remained of the specimen was 12% residual concrete, as shown in (Fig. (4f)). The steel rod, with an initial length of 25.75 cm, in the 2-1 position showed a maximum deformation of 6.3 cm. The stirrups present deformations and considerable cracks in all areas.

A specimen of f_c = 210 kg/cm^2 subjected to the energy produced by a load of 760 g of the dihedral type is shown in (Fig. 5a). After the explosion, what remained of the specimen was 21% residual concrete is shown in (Fig. 5b). The steel rod, with an initial length of 21.62 cm, located on the position 1–2 suffered the maximum deformation, which was 1.97 cm. The stirrups show minimal deformation, so they are not affected by the action of the explosive charge. A specimen of f_c = 210 kg/cm^2 subjected to the energy produced by a load of 1010 g of the dihedral conic type is shown in (Fig. (6a)). After the explosion, what remained of the specimen was 12% residual concrete, as shown in (Fig. (6b)). The steel rod, with an initial length of 20.08 cm, located on the position 1–2 suffered the maximum deformation, which was 3.15 cm. The stirrups show considerable deformations on all sides, however the steel armor maintained its shape. Figure 6c shows a specimen of f_c = 210 kg/cm^2 subjected to the energy produced by a

750 g cylindrical load tamped in sand. After the explosion, what remained of the specimen was 18% residual concrete, as shown in (Fig. (6d)). The steel rod, with an initial length of 22.37 cm, in the 2-1 position suffered a maximum deformation of 3.3 cm. The stirrups present lateral deformations, however the steel armor maintains its shape.

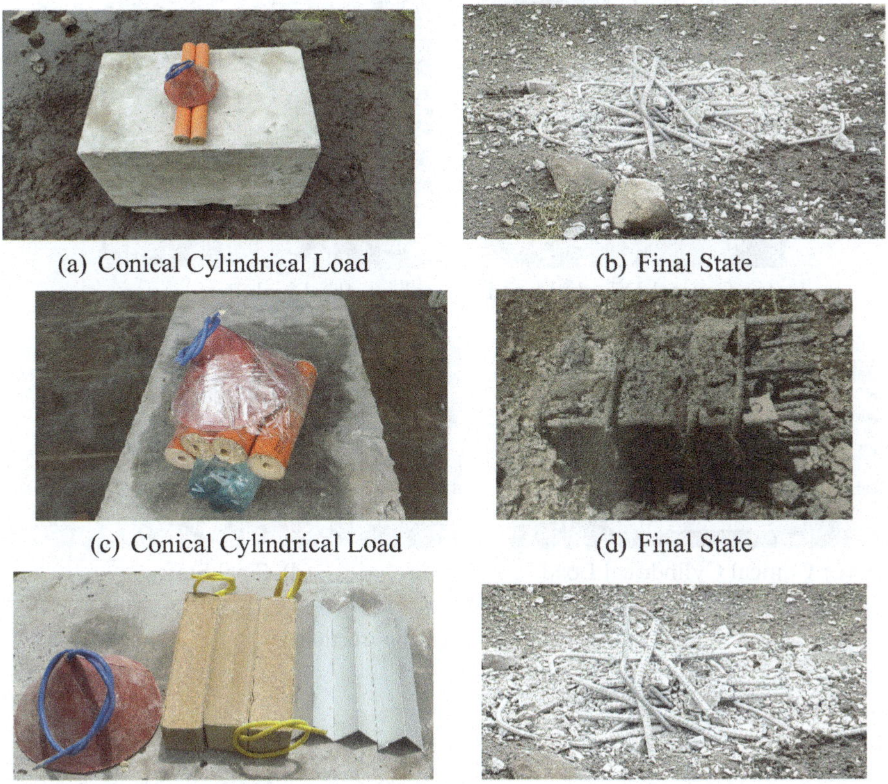

(a) Conical Cylindrical Load (b) Final State

(c) Conical Cylindrical Load (d) Final State

(e) Conical-dihedral shape with metal (f) Final State

Fig. 4. Specimen construction

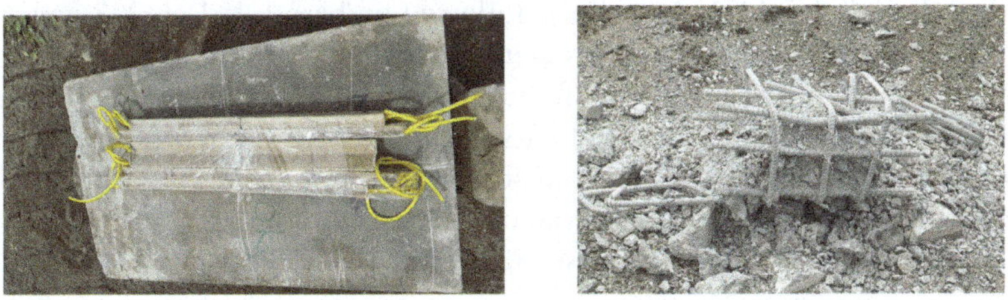

(a) Dihedral load (b) Final State

Fig. 5. Specimen construction

Therefore, after carrying out the 14 tests, the conical-dihedral shaped load with metallic reinforcement is the one that produces the greatest destructive effect on the specimens.

(a) Conical-Dihedral Load

(b) Final State

(c) Conical Cylindrical Load

(d) Final State

Fig. 6. Specimen construction

5.2 Specimens

A total of 11 tests were carried out on specimens of $f_c = 210\,\text{kg/cm}^2$ and 6 tests on specimens $f_c = 280\ \text{kg/cm}^2$. The bar located in point 1–2 has the greatest deformation in all the trials. This is because the explosive charge has a metal reinforcement that becomes a projectile and breaks the concrete; the energy generated by the explosion is absorbed directly by the 1–2 steel bar, followed by bars 1-1, 1–3, 2–3 and 2-1. The size of the concrete fragments, we can determine that the configuration of the explosive charge is enough to break more than 80% of the concrete. Depending on the weight of the explosive, the damage on the steel bars is variable; however, with weights greater than 990 g of explosives, bar 1–2, and in some cases bars 1-1 and 1–3, reached deformations greater than 6 cm and considerable cracks. Therefore, we can conclude that in real-scale, the specimen would be considered collapsed.

Table 4. Steel bars displacement

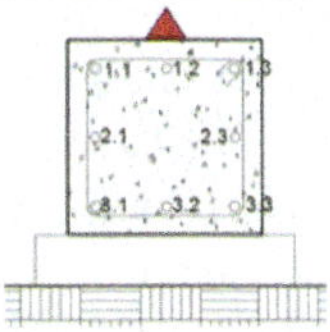

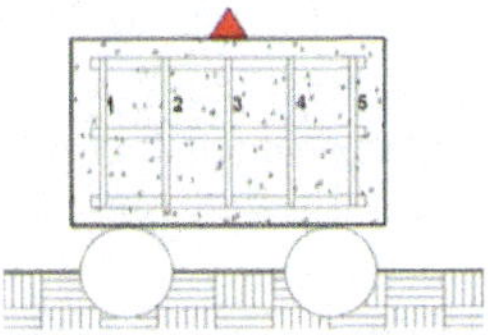

Specimen	Steel Bars Configuration								Residual Concrete
Weight (g)	1.1	1.2	1.3	2.1	2.3	3.1	3.2	3.3	(%)
1 (760)	2.14	2.18	2.06	0.89	1.10	-	-	-	39
2 (750)	2.3	3.41	3.42	1.64	1.65	-	-	-	23
3 (660)	2.31	2.76	2.27	0.88	0.82	-	-	-	21
4 (1010)	2.05	3.38	2.73	1.62	1.56	-	-	-	12
5 (990)	3.46	4.37	4.04	3.06	1.72	1.17	-	2.86	11
6 (750)	1.84	3.52	2.55	1.79	1.43	-	-	-	18
7 (1050)	4.92	5.90	5.20	4.62	3.2	0.26	2.30	1.68	13
8 (1050)	-	-	-	-	-	-	-	-	73
9 (1050)	-	-	-	-	-	-	-	-	76
10 (1050)	5.24	8.05	7.05	4.79	5.62	-	-	-	11
11 (990)	3.78	6.98	3.68	3.99	1.94	-	-	-	12
12 (1050)	4.15	6.48	3.06	3.94	4.20	-	-	-	11
13 (1110)	2.30	3.85	2.64	2.33	1.63	-	-	-	15
14 (1350)	4.08	5.38	4.76	3.00	3.45	-	-	-	13
15 (450)	2.17	3.06	3.00	2.49	2.59	-	-	-	30
16 (450)	2.97	3.42	1.36	2.93	0.15	-	-	-	15
17 (900)	3.10	1.68	2.52	1.18	0.77	-	-	-	36

6 Conclusions

After carrying out 14 tests to design the most optimum explosive charge, several variables were considered such as: the geometry, energy released, weight and stand-off. There were some geometric combinations of explosive charge that created a high fragmentation of the concrete, but with little effect in the steel rods, which is why they were excluded from the study. After observing the effect of each load, it was established that the load with the conical-dihedral shape with metal reinforcement is the one that produces the greatest deformation in the reinforcing steel.

The data collection was carried out post-test, so it was only possible to obtain information of concrete fragmentation and deformations of the steel rods. Direct measurements could not be obtained during the test due to the fact that on the day of the tests, there were no accelerometers or pressure sensors with a nominal capacity greater than that produced by the detonation of the explosive charge. Since our devices were not sufficient in measuring the real pressure values of the blast, the SDOF model could not reach a high level of confidence.

Due to the complexity and sensitivity of the research, it is necessary to complement and develop in more detail the research methodology. This is why today we are carrying out real-scale bench testing. These are being recorded in various ways, using: Phantom technology, a camera that can record up to 1 million frames per second; 3D scanners with Go Scan technology; thermochemical computer code that analyzes the properties of the explosive charge; direct measurement devices and computational modeling by commercial software.

References

1. USGS: Earthquake History in ECUADOR (2018). https://earthquake.usgs.gov/. Accessed 22 Jan 2018
2. Buczkowski, D., Zygmunt, B.: Detonation properties of mixtures of ammonium nitrate based fertilizers and fuels. Central Eur. J. Energet. Mater. **8**(2), 99–106 (2011)
3. Keshavarz, M.H., Oftadeh, M.: New method for estimating the heat of formation of CHNO explosives in crystalline state. High Temperat. High Pressur. **35**(4), 499 (2004)
4. Pui, Y.H.: Numerical and experimental studies of blast loading, vol. 26 (1995)
5. Simoens, B., Lefebvre, M.: Influence of the shape of an explosive charge: quantification of the modification of the pressure field. Central Eur. J. Energetic Mater. **12** (2015)
6. Hryciów, Z., Borkowski, W., Rybak, P., Wysocki, J.: Influence of the shape of the explosive charge on blast profile. J. KONES **21**(4), 169–176 (2014). Powertrain and Transport
7. Olive, G.: DISTRIBUTED BY: national technical information service. Ocean Eng. (1974)
8. Alonso, F.D., et al.: Characteristic overpressure-impulse-distance curves for the detonation of explosives, pyrotechnics or unstable substances. J. Loss Prevent. Process Ind. **19**(6), 724–728 (2006)
9. Yasseri, S.: R 541 Iso-damage diagrams for blast resistant design (2015)
10. Figuli, L., Bedon, C., Zvaková, Z., Jangl, Š, Kavický, V.: Dynamic analysis of a blast loaded steel structure. Proc. Eng. **199**, 2463–2469 (2017)
11. NEC-SE-DS: Cargas Sísmicas Diseño Sismo Resistente 1 (2015)
12. ACI 318-14: ACI 318-14 - building code requirements for structural concrete (2014)
13. Dobratz, B.M.: Properties of chemical explosives and explosive simulants. Lawrence Livermore National Laboratory, vol. 1, pp. 1–334 (1972)

The Social Explosion Which is Advancing in South America: The Cases of Chile and Bolivia

Luis Recalde(✉), Klever Bravo, and Luis Villa

Universidad de Las Fuerzas Armadas - ESPE, EC 170603 Sangolquí, Quito, Ecuador
{llrecalde,kabravo,lovilla}@espe.edu.ec

Abstract. This article aims to analyze the social protests carried out in Chile and Bolivia, South American countries, taking as a prelude the social outbreak unleashed in October 2019 in Ecuador. Although these protests occurred simultaneously, the reasons were different: in Chile, the reaction of the poor classes to the socio-economic polarity was clearly seen. This country is recognized as an economic and productive model in Latin America; while in Bolivia, the cause of the internal social conflict was the fraud in October 2019 elections, and the rejection of the indefinite reelection of Evo Morales and his system of government clinging to Socialism of the 21st Century. Even though the causes of the social protests were different, the consequences were similar: deaths, injuries, detainees, damage to public and private property, and a high burden of violence.

Keywords: Social unrest · Poverty · Molecular revolution · Fraud · Outbreak · Violence

1 Introduction

In recent years, in several South American countries, social mobilizations have multiplied aggressively, which maintain their own dynamics and are consistent with the times and political events, showing enormous social discontent in the subregion [1]. With different socio-economic roots, an amalgam of actors and groups is visualized that for several decades have been proclaiming the great social inequality and discrimination that characterizes the "hard nuclei" of poverty, vulnerability, violence, and social exclusion which makes them persist and reproduce [2].

The different protests and mobilizations carried out in the subregion since October 2019, have demonstrated the willingness of social actors to influence, often at a local and sometimes at national level, on decision-making processes [3]. It is important to establish that these proclamations of nonconformity seek to delegitimize representative democracy as we know it. It tries to revolutionize the form of citizen participation with "renewed" doctrines and ideological foundations such as Socialism of the 21st Century, which seeks -mainly- a new form of representation and centralization of power, especially in Bolivia.

M. Botto-Tobar et al. (Eds.): CIT 2021, LNNS 405, pp. 377–388, 2022.
https://doi.org/10.1007/978-3-030-96043-8_29

In the case of Chile, it will be established that the October 2019 events are not different from what has been happening in Ecuador, Bolivia and Colombia: a deep social inequality that is becoming more and more profound since Chile is one of the 15 most unequal countries in the world. The pensions and low wages that a middle class perceives have reflected poverty and a polarized economy. [4] which was exported to the region [5]; not so in Bolivia. There, the export of gas and other natural resources has allowed it to conceive the "Bolivian economic miracle" that has been mentioned so much, and that has been well managed by the Minister of Economy and current president, Luis Arce.

Considering also that the growing criminalization of the student movement ended up unleashing one of the largest social protests in that country. It has been complemented by the massive use of new digital technologies. They have been a breeding ground for the social explosion since the new generations - formed in this new "emerging middle class" - are the ones that currently lead the protests, both in the streets and on Social Networks [5]. Thus, achieving a historic milestone to promote a Constitutional Reform that reconfigures the Chilean Democratic Institutionalism.

2 Materials and Methods

This article is part of a large research project at the University of the Armed Forces ESPE-Ecuador, on the study of social protests carried out in October 2019 in Ecuador, and which were replicated in Chile, Bolivia and Colombia. As it is a social research work, the required materials have not gone beyond printed and digital sources. They make it possible to obtain the required information, such as; digital publications in social media, printed books, and other scientific articles.

Regarding the methodology, this research work starts from an analysis of the information obtained through descriptive, analytical and narrative methods about the disturbances of public order that occurred in Chile and Bolivia in October 2019. This way, it was possible to identify the causes and consequences of social protests, as well as the main actors. In this context, an analysis about the radical strategies that have been used in Latin America by anarchist social groups whose objective is to disturb the state of "normality" of society, and its democratic institutions through chaos and vandalism, causing the decontextualization of the guiding principles of a harmonious and civilized coexistence is carried out [6].

3 Results and Discussions

3.1 The Evolution of Social Protests in South America

The incidents caused by protests and social unrest in South America have considerably cracked the weak participatory democratic systems, represented by parties and political movements with very little credibility. The participatory democratic system is based on a fair, plural, and with greater social inclusion society [7].

From this perspective, it could be established with great clarity that there are structural causes in common to subregion countries regarding to the social discontent that has grown in recent years. First, the rise of middle classes and the emergence of a wide swath of population vulnerable to the economic recession. Second, a generalized "malaise in democracy" that takes shape in a broad crisis of political representation, citizen disaffection, and mistrust in institutions and political elites [8].

It is important to understand and comprehend that the popular uprisings in subregion countries represent a social response to the increase in income inequality that is clearly reflected in poverty and extreme poverty. It actually decreased in Latin America in the past twenty years. The Gini coefficient average (simple) in the region fell from 0.56 in 2000 to 0.51 in 2017 [9]. However, according to the criteria of Stefan Peters, an academic at Justus Liebig University, "... although the Gini coefficient is an important measure, other indicators must be considered, such as wealth, land ownership and education, which show that elites have managed to maintain their privileges. In each of these factors, Latin America is the most unequal region, but if we consider everything together, it is much more" [10].

There are also other no less relevant factors such as gender inequalities, ethnic and racial inequalities, inequalities between rural and urban areas and between different regions of each country [2]. As a consequence, there is greater social stratification, residential segregation [11], and conflict, which can lead to situations of political and social violence [12, 13].

3.2 Chile: Social Discontent and the Outbreak of a Society Full of Frustrations

After three decades of an apparent successful economic model, inequality, social injustice, socioeconomic burdens and, above all, the lack of attention of a discredited political class that has been eroding democratic coexistence [14], led the Chilean people to a colossal protest at the end of October 2019.

It is important to highlight that the historical germ of this social outbreak was developed in several previous demonstrations led by students, feminist, environmental movements, indigenous peoples and ethnic groups: Mochilazo (2001), the Revolution of the Penguins (2006), March of the Umbrellas (2011); Niunamenos (2016); Ola Feminista or Mayo Feminista (2018). These movements evidenced the ineffectiveness of the political class [15], but mainly due to the huge gap in inequality, wage precariousness, level of indebtedness, and poverty that had been increasing with the passing of the years.

According to the Social Panorama of Latin America report, prepared by the Economic Commission for Latin America and the Caribbean (CEPAL), the wealthiest 1% of the country was left with 26.5% of the wealth in 2017 while 50% of lower-income households accessed only 2.1% of the country's net wealth [16]. It is possible to assert that a voracious neoliberal economy has generated one of the most unequal countries in the world [5]. Based on this panorama, it can be clearly established that the gaps in the distribution of wealth have generated the greatest number of social conflicts. Considering that according to the research carried out in this field and published in the Annual Conflict Observatory Report 2018; if it could reduce inequality from 0.49 to 0.43 (GINI), it could mean a reduction of about 18% in conflicts at the national level [17].

Another source of conflict is job insecurity, characterized by the inability of the State to establish policies that equitably regulate the needs of workers and the aspirations of employers. According to COES Conflict Observatory, based on the 1,789 protest actions that took place in 2017, 621 (34.7%) correspond to labor protests [17]. Considering that the minimum wage in Chile is 301,000 pesos (US$423). On the other hand, the National Institute of Statistics of Chile (70% less than 20 countries that in different years reached the same PIB per capita [18]), half of the workers in that country receives a salary equal to or less than 400,000 pesos (US$562) per month [16].

As a consequence of the precarious salary analyzed, another cause of the social explosion arises; the level of indebtedness of the poor and middle class in Chile. Low minimum and average wages, and the absence of a moderately acceptable social security for forty years, forced 63% of Chilean households to take out consumer loans to afford the basic needs [16]. It has originated a phenomenon called "Salary extension" that helps to understand how debt ceased to be an emergency option and became day-to-day. So daily that Chileans refer to their credit line as part of their income [19]. This leads 35% of households to high or excessive indebtedness especially in the lower income strata. According to researcher Emmanuelle Barozet "Debt is nothing more than an obligatory and complex response to low wages. It serves to cover basic expenses, to buy merchandise in the supermarket, to buy medicine, to help a relative; In other words, it is not a debt to indulge in luxury" [20].

The diagnosis of the poverty situation in Chile, according to the Social Development Report 2019, was established based on two large dimensions: income poverty and multidimensional poverty (education, health, work and social security, housing and environment, and networks and social cohesion). Considering the first dimension; in 2017, 8.6% of the population is in poverty situation, and 2.3% in extreme poverty. While 20.7% of the population is in a multidimensional poverty [21]. These statistical data, despite being favorable in relation to previous years, poverty is still a relevant factor for social discontent and anger. It was believed that the successful economic model had allowed the emergence of a new middle class or emerging middle class with a better quality of life, but what really germinated was only the upper layer of a motley group of poor people [22]. It needs an urgent review of the current neoliberal economic model that resulted in a mirage.

3.3 The Emerging Power of Social Movements and Their Leading Role in the Popular Outbreak in Chile

The Black October revealed the structural flaws of the Chilean model, which for many was nothing more than a fiction and had been hidden for a long time by a myopic political class and an economic elite far from the reality that the common citizen lives. It was sponsored by a rigid constitutional architecture incapable of processing institutionally the citizen demands [23]. They have been the best breeding ground for the emergence of social movements claiming historical aspirations of a precarious and impoverished middle class by the neoliberal economic model established since several decades ago. Middle class that accumulates enormous frustration around health, pensions, education and employment inequalities added to other inequities linked to socio-structural categories, such as class, gender and ethnicity, and strong regional and subnational inequalities [23]. They fully coincide with the theoretical framework about the causes claimed by social movements, and may be of three types: programmatic, identity and positional [24]. The following table provides a chronological report of the events that led to a violent social explosion in Chile in October 2019 (Table 1).

It is important to note that gradually protests and disturbances of social movements have become increasingly violent. It is necessary to consider that conflict emphasizes that the gaps between expectations and reality tend to be behind social conflicts and even revolutions [24]. Despite the fact that the correlation between the average intensity of cases of civil unrest <<civil unrest>> and the PBI per inhabitant is negative while the correlation between the intensity of civil unrest and the decrease in the inequality is positive. The Chilean crisis in 2019 is an exceptional case that goes beyond these two correlations. Since the relatively high PBI per inhabitant and the reduction in inequality between 2000 and 2018 do not explain the duration and intensity of the crisis [25], so it could be configured that the crisis was an amalgam of outraged citizens, organized crime, insurrection, and terrorism that led to acts of violence never before seen in the Southern Cone.

The scenario established in October 2019 fully agrees with Gerald Raunig's conceptualization that the revolutionary machine arises from the concatenation of resistance, insurrection and constituent power [26]. These three variables had been silently germinating during the crisis in Chile, and they took shape and led to chaos, destruction and anarchy. Events that could be clearly configured under the strategy that Félix Guattari and Gilles Deleuze called "Dissipated molecular revolution," which consists of multiple, small, and headless dissipated boils in society. They are far from those of the last century. There, it was possible to determine who the enemy is, now it is not known who articulates them, or how one does it [27]. It advocates establishing the disappearance of a formal leadership, so it is impossible to identify leaders. Otherwise, collectives are identified.

The strategy of the dissipated molecular revolution consists of dividing the units of the public force to achieve decentralize and weaken their actions, confront them without the principle of force, but rather under the scheme of isolated force and occupied zone [6]. It produced violent actions against the public force and acts of vandalism to destroy public and private infrastructure. As a result of more than 1,400 million dollars in destruction, looting and decreased sales. As a corollary of the social outbreak in Chile, there is the third component of the revolutionary machinery defined by Gerald Raunig as the "constituent power." That forced the political elites to call elections for a Constituent Assembly to design a new Constitution that includes the legitimate aspirations of the Chilean people, and that is consistent with the reality of a globalized world.

3.4 Social Unrest in Bolivia

Evo Morales Ayma, an indigenous person of Uru-Aymara descent, assumed his first term in January 2006. His binomial was Álvaro García Linera, intellectual, representative of the indigenous coca-growing insurgency, former member of the Tupac Katari Guerrilla Army, EGTK, economist and good planner [28].

At that time, Morales' political party, Movimiento al Socialismo, MAS, reached 53.7% of the votes, favoring its international relations with Presidents Fidel Castro and Hugo Chávez, Cuba and Venezuela respectively [29]. Hence the political alignment has given the much-bellowed "turn to the left."

Table 1. Chronological compilation of events that have occurred as of October 2019.

Day	Event
October 18th	- Protests over the increase in the price of the metro ticket at rush hours lead to clashes between protesters and police - The infrastructure of the metro and several public transport buses were destroyed - Around midnight, President Sebastián Piñera declares a state of emergency and a curfew
October 19th	- Hooded protesters confront riot police and soldiers in different parts of the capital and the rest of the country - The President of the Republic annuls the increase in the price of the metro and calls for dialogue
October 20th	- President Piñera affirms that Chile is "at war against a powerful enemy," and calls for unity and to condemn "violence and crime" - The tragic death toll is 11 deaths in the violent protests in the streets - The curfew is extended to 16 regions of the country
October 21st	- Demonstrations resumed shouting "military out" - Suspension of classes in almost all schools and universities in the capital
October 22nd	- The main unions and social movements of the country call for a general strike - Sebastián Piñera asks "forgiveness" and announces a package of measures, raising the lowest pensions, freezing electricity rates
October 24th	- Michelle Bachelet, United Nations High Commissioner for Human Rights, announces that she will send a UN verification mission to examine the human rights situation in Chile
October 25th	- More than 1.2 million people march in Santiago in one of the largest mobilizations in the history of Chile in the last 30 years
October 27th	- Piñera announces the lifting of the state of emergency that was in force for eight days and takes the military off the streets - The death toll reaches 20 people, five of them from shots fired by the forces of order
October 27th	- The president changes eight of his 24 ministers, including his chief of staff - A UN mission investigates allegations of human rights violations during the wave of violence
October 30th	- Sebastián Piñera announces the suspension of the APEC and COP25 summits, of which Chile was to host. Given this, Greta - Thunberg cancels her trip across the Atlantic - President Piñera rules out resigning
November 1st	- Hundreds of women dressed in black take to the streets in front of La Moneda in the so-called-march of mourning-, to pay tribute to the fallen

(continued)

Table 1. (*continued*)

Day	Event
November 2nd	- The government announces a project that cancels a tax cut for the richest businessmen in the country
November 5th	- The 7th Guarantee Court of Santiago declares admissible the complaint against President Sebastián Piñera for alleged crimes against humanity that occurred in the state of emergency - Piñera gives the first international interview to the BBC network, so he justifies and supports the actions of the Chilean police
November 6th	- 12 members of the Special Forces and 2 police officers are formalized for crimes of torture committed against two detainees, one of them an under 18
November 7th	- The president summons the National Security Council (Cosena) in La Moneda. The decision is criticized by public opinion - The first president hardens the tone and announces a series of projects to increase measures to control of public order
November 10th	- Government announces implementation of a Constituent Congress for a new constitution and a ratification plebiscite. Discard the holding of a Constituent Assembly
November 11th	- Constitutional Commission of the Chamber of Deputies approves national plebiscite on a new Constitution
November 13th	- Government presents three proposals to overcome the crisis, an agreement for peace, one for justice and one for the Constitution. - - One of the immediate measures to mitigate the violence consists of the reintegration of former agents into the Carabineros and the Investigative Police
November 15th	- Chile's political sectors managed to establish an agreement classified as "historic" - The Government and the opposition came together to hold a plebiscite, in which citizens will decide whether or not they want a new Constitution
November 18th	- The executive secretary of the Inter-American Commission on Human Rights (CIDH) begins the investigation in Chile
November 21st	- Amnesty International concludes its investigative mission in Chile recognizing that military and police agents carried out "generalized attacks against the population, using unnecessary and excessive force"
December 5th	- Piñera calls a plebiscite on April 26, 2020, an instance in which citizens can decide whether or not to draft a new constitution

Since his electoral victory in December 2005, Evo Morales considered important the issue of Bolivian economic improvement through the nationalization of hydrocarbons. Thus, the increase in the exploitation of natural gas was notorious. It raises from 8.9 million cubic meters in 2000 to 37.9 million cubic meters in 2007. Added to this were tax collections and remittances as an example of the famous "economic miracle." On the other hand, Morales imposed a new distribution of land on the indigenous peasant communities under the seal of Suma Qamaña or Living Well whose backdrop was the beginning of decolonization. Of course, there was a socialist backdrop. That is, the presence of Cuba in the illiteracy eradication programs with the name "Yes, I can"; as well as the presence of Cuban doctors in health programs [30].

The 2009 Constitution allowed him to change the historical name of the Republic of Bolivia to the Plurinational State of Bolivia. Among other things, even the social initiatives imposed by Cuba and Venezuela were more than enough reasons for their political identity with the revolutionary left to increase tensions with the United States of America. An extent that in 2008, Morales expelled from Bolivia to US Ambassador Philip Goldberg, for his alleged participation in the attempted coup of that year. That motivated a more active participation in the Union of South American nations, UNASUR, and in the Bolivarian Alliance for the Peoples of Our America, ALBA [30].

Given his indigenous leadership, it was easy for Evo Morales to win the next elections. In other words, his subsequent victories were "overwhelming": 2009, with 64.2%; 2014, with 61.3% and 2019, with 47% compared to Carlos Mesa, who reached 36.5%. It canceled a second round for exceeding the second with more than 10%.

Taking the attributions of a permanent winner at the polls, Morales called a referendum on February 21st, 2016, with the central question on indefinite presidential reelection. The Bolivian people answered "no" with 51.3% and the "yes" with 48.7%. This produced a reeling within the spectrum of "chavization of the Bolivian Government," due to the construction of the museum inspired by President Evo Morales, a kind of cult of the person. On the other hand, the opposition made permanent complaints about the rumors that he had a son who, supposedly never was born, or the proclamation of his ex-partner to a position of "senior management" that attributed large contracts with the Chinese. For all this, the protests did not wait. Some universities, the residents of Potosí, and the occasional coca-growing community joined. In other words, the political and moral resource of the MAS, "was eroding" [31].

With Morales' political landscape in decline, at November 27th, 2017, when the Plurinational Constitutional Court gave Evo Morales the green light to run in the general elections of October 20th, 2019, under the argument that the Pact of San José de Costa Rica was above the Bolivian Constitution, attributing that Morales' candidacy was a human right.

At the end of the general elections in October 2019, Morales surpassed his main opponent, Carlos Mesa, with a margin of 10.5 points. That is, Morales reached 47.07% and Mesa 36.52%. In a divided Bolivia, social protest came. The civic strikes and blockades ignited in a short time, forcing Morales and García Linera to take refuge in Chapare, the starting point of their political biography. There, its inhabitants, coca growers, interrupted all access roads with the slogan of protecting their leader. However, the use of their political bases instead of the public force, made the crisis increase. There were serious clashes between civilians, involving three deaths from the opposition which was led by Luis Fernando Camacho, president of the Santa Cruz Civic Committee. Next, the protesters proceeded to burn the houses of the ministers, who began to resign from their positions. Seeing this upheaval, the mining sectors and the Bolivian Central Workers, COB, called for the resignation of President Morales [31].

More than the request of the Central Workers, Morales' resignation was at the "suggestion" of General Williams Kaliman Romero, Chief of the Armed Forces, who publicly made such a recommendation in order to "pacify the country," and maintain stability. Kaliman's textual message was: "After analyzing the internal conflict situation, we suggest that the President of the State resign his presidential mandate, allowing the pacification and maintenance of stability for the good of our Bolivia" [32].

For this, it should be remembered that during the 14 years of Government, very few times did he resort to the Armed Forces for the solution of internal conflicts. His relationship with this Institution gained strength from the ideological side, beginning with the war cry of "homeland or death, we will win." It was coined by the legendary subversive Ernesto Che Guevara. Likewise, it is worth remembering that the President-Armed Forces relations were also sustained in the diplomatic appointments of the military and their families, since they were considered "soldiers of the change process" [33].

On the other hand, in the cities of La Paz, Sucre, Oruro and Potosí, the Police joined the protests, abandoning the protection of the Government Palace and the seat of the Legislative Power. A spokesperson for this Institution issued the following message: "The Police are not due to any political party; they are due to the people." With this, the police riots did not wait. The Oruro Police made a symbolic takeover of the governor's office while in other barracks, the agents waved Bolivian flags on the terraces in support of the protests. Evo Morales persisted in asking his followers to stay on fight for the "defense of democracy" [34].

According to the audit report of the Organization of American States, OAS, a body specialized in evaluating elections, after its analysis, it stated that "statistically, the profit of Evo Morales with the margin of 10% is improbable. There are even physical records with alterations and forged signatures. Added to which is the manipulation of the computer system and disrespect for the chain of custody of the records" [35].

Regarding the Final Report on Electoral Integrity, a 95-page document prepared by the two bodies responsible for auditing the OAS: The Strengthening of Democracy and the Department for Electoral Cooperation and Observation Secretary, the report was published on December 5th 2019, disclosing the irregularities in five findings:

1) "Flawed" system for the transmission of results and computations
2) Manipulations, falsifications and adulterations of electoral records in six departments
3) Poor chain of custody
4) Computational records do not provide confidence
5) Highly unlikely trend of the last 5% of the scrutiny shown [36].

Morales' resignation raised a stir with some phrases such as: "there has been a civic, political and police coup", "my sin is being an indigenous person, a union leader and a coca grower." These phrases, among others, were echoed in Latin American countries, especially in Mexico, where the secretary of the Ministry of Foreign Affairs offered

Morales political asylum since 20 officials from the Executive and Executive Legislative powers were welcomed at the Mexican embassy in La Paz.

With the resignation of the Bolivian president, other authorities directly linked to MAS also resigned: Adriana Salvatierra, president of the Senate, along with seven senators; Víctor Borda, president of the Chamber of Deputies, along with six congressmen; Javier Zavaleta, Minister of Defense and the Ministers of Education, Hydrocarbons, Mining, Sports; various vice ministers and ambassadors; Yuri Calderón, general commander of the National Police; various mayors and governors… [37].

After 21 days of internal unrest in the Andean country, the results were painful: 36 dead and 804 injured, according to the Ombudsman's Office.

Of the 36 deceased, 18 lost their lives in the massacres that occurred on November 15th and 19th in the towns of Sacaba and Senkata, either due to the impact of a bullet or inhalation of tear gas. There were even deaths from blows by mobs such as Juan José Mamani in Cochabamba. In order to "heal in good health", the Government issued Supreme Decree No. 4100 which economic compensation was established for the relatives of the deceased and injured during the confrontations between the security forces, opposition groups and the followers of Evo Morales [38].

With the resignation of Morales, Jeanine Áñez, second vice president of the Senate assumed the presidency. This assumption of power was endorsed by the Constitutional Court, the Armed Forces and the Bolivian right party. The highland country returned to the polls on Sunday, October 18th, 2020, where the winner was the MAS candidate, Luis Arce. The victory over Carlos Mesa was with 20 points, thus ratifying the supremacy of this party for a long time [39].

4 Conclusions

The neoliberal economic model implemented for three decades in Chile, despite being described as successful and that on many occasions it has been taken as an example for the rest of Latin American countries, has produced huge gaps of inequality, poverty and discrimination. It leads to extremes such as being among the 15 most unequal countries in the world, a dying middle class, and the impediment to a better level of human development.

It should be considered that the Chilean social movements of October 2019 have reacted as a response to the imperturbability and impudence of a discredited political class that has failed to respond to the pressing needs of its constituents. It is necessary taking into account that its democratic system survives within of an outdated and orthodox constitutional model. A scenario that has given rise to demands from collectives from across the social spectrum through protest, vandalism, destruction, and chaos. For example, the new model of protesters' reaction lacks a protest model peaceful as happened in Ecuador and Bolivia.

Regarding the political and social unrest in the Plurinational State of Bolivia, it is clear that they were caused by fraud in the October 2019 elections. It was ratified by the OAS report. This impasse led to serious confrontations between followers of Evo Morales and opposition groups resulting in 36 deaths, 804 injuries, and the resignation of the coca grower leader. However, the subsequent triumph of Arce in the 2020 elections achieved the tremendous return of Morales. Attached to this, came the Socialism of the 21st Century and its typical: revenge and continuity in power for a long time.

References

1. Recalde, L., Criollo, M.A., Santos, M.D.: Social mobilizations as a silent instrument for seizing power: the ecuadorian case. In: Rocha, Á., Paredes-Calderón, M., Guarda, T. (eds.) MICRADS 2020. SIST, vol. 181, pp. 529–543. Springer, Singapore (2020). https://doi.org/10.1007/978-981-15-4875-8_46
2. CEPAL: La matriz de la desigualdad social en América Latina. Santiago: Comisión Económica para América Latina y el Caribe (CEPAL), División de Publicaciones y Servicios Web, publicaciones@cepal.org. (2016)
3. García-Linera, Á.: Crisis del Estado y Poder Popular. New Left, pp. 66–77 (2006)
4. Vargas, V.M.: El Tiempo. Recuperado el 11 de 03 de 2021, de (2019). https://www.eltiempo.com/mundo/latinoamerica/las-causas-de-las-protestas-en-chile-427450
5. Cuadra, Á.: Protesta social en Chile, 2019–2020: fracaso de un modelo económico. Textos y Contextos (2020)
6. Pazmiño, M.: Las cenizas de la desestabilización se reavivan. La República (2020)
7. Democracia participativa: Significados.com. Recuperado el 09 de 03 de 2021, de (2018). https://www.significados.com/democracia-participativa/
8. Fundación Carolina.: Crisis políticas y protestas sociales en América Latina. Artículos, Boletín, Editoriales. Obtenido de (2019). https://www.fundacioncarolina.es/crisis-politicas-y-protestas-sociales-en-america-latina/
9. Ferreira, F., Schoch, M.: Inequality and social unrest in Latin America: the Tocqueville Paradox revisited. WordBank.org. Recuperado el 27 de 2 de 2021, de (2020). https://blogs.worldbank.org/developmenttalk/inequality-and-social-unrest-latin-america-tocqueville-paradox-revisited
10. Dannemann, V.: América Latina: Radiografía de la desigualdad. DW.com (2021)
11. Naciones Unidas: pág. 70 (2013)
12. Trucco, D., Ullmann, H.: Juventud: realidades y retos para un desarrollo con igualdad. CEPAL (septiembre de 2015)
13. Escotto, T.: Las juventudes centroamericanas en contextos de inseguridad y violencia: realidades y retos para su inclusión social. CEPAL (2015)
14. Fundación Jaime Guzmán: En La insurrección chilena (pág. 126). Santiago (2020)
15. Jiménez-Yánez, C.: #Chiledespertó: Causas del estallido social en Chile. Revista Mexicana de Sociología, pp. 949–957 (2020)
16. Paúl, F.: Protestas en Chile: 4 claves para entender la furia y el estallido social en el país sudamericano. BBC News Mundo, 7. Recuperado el 27 de enero de 2021, de (2019). https://www.bbc.com/mundo/noticias-america-latina-50115798
17. COES: Informe Anual Observatorio de Conflictos 2018. Santiago (2018)
18. Matus, M.: Desigualdad: la grieta que fractura la sociedad chilena. En "U. Chile", Chile Despertó: Lecturas desde la Historia del estallido social de octubre, p. 109. Santiago (2019)
19. Ramos, M.: Por qué estamos más endeudados que nunca en nuestra historia. CIPER. Recuperado el 12 de marzo de 2021, de (2020). https://www.ciperchile.cl/2020/04/26/por-que-estamos-mas-endeudados-que-nunca-en-nuestra-historia/
20. Marchetti-Michels, P.: "Democratización del crédito": El análisis del endeudamiento en Chile y sus efectos en el malestar social. Emol. Recuperado el 16 de marzo de 2021, de (2019). https://www.emol.com/noticias/Economia/2019/11/01/965944/Endeudamiento-chilenos-detonante-crisis-social.html

21. Ministerio de Desarrollo Social y Familia: Informe de Desarrollo Social. Santiago de Chile. Recuperado el 16 de marzo de 2021, de (2019). https://www.desarrollosocialyfamilia.gob.cl/storage/docs/Informe_de_Desarrollo_Social_2019.pdf
22. Candina-Polomer, A.: La clase media que no era: ira social y pobreza en Chile. En Chile despertó: Lecturas desde la Historia del estallido social de octubre. Santiago: Universidad de Chile (2019)
23. Castiglioni, R.: El ocaso del «modelo chileno»? Nueva Sociedad. Recuperado el 15 de marzo de 2021, de. https://www.nuso.org/articulo/el-ocaso-del-modelo-chileno/
24. Tilly, C.: "Los Movimientos Sociales, 1768–2008: Desde sus Orígenes a Facebook". Barcelona.: Editorial Crítica (2019)
25. Schmidt-Hebbel, K.: La crisis de Chile: causas, efectos y salidas. El Mercurio. Recuperado el 9 de marzo de 2021, de (2019). https://negocios.udd.cl/noticias/2019/11/la-crisis-de-chile-causas-efectos-y-salidas/
26. Raunig, G.: Revoluciones moleculares y prácticas artísticas transversales. Madrid, España. Recuperado el 11 de marzo de 2021 (2007)
27. Carrasco, D.: 18 de octubre: El inicio de una Revolución Molecular. En F. J. Guzmán, La Insurrección Chilena: Desde la mirada de la Fundación Jaime Guzmán (2019)
28. Fierro, R.: "Voto nulo e historias funestas". El Comercio (1 de abril de 2021)
29. De Mesa, J., Gisbert, T.: Historia de Bolivia. La Paz: editorial Gisbert (2008)
30. Uharte, L.: Una década del Gobierno del MAS en Bolivia: un balance global. Revista Barataria No. 22(22) (2017). Obtenido de file:///C:/Users/ROSITA/AppData/Local/Temp/Dialnet-UnaDecadaDelGobiernoDelMASEnBolivia-6279014.pdf
31. Stefanoni, P.: Bolivia después de Evo. Fundación Carolina (2019). https://www.fundacioncarolina.es/wp-content/uploads/2019/11/AC-29.pdf
32. El Universo.: Luis Arce asumió la presidencia de Bolivia; Fuerzas Armadas le entregó el bastón de mando. (8 de noviembre de 2020)
33. Miranda, B.: Renuncia de Evo Morales: por qué fue tan decisivo el respaldo de las Fuerzas Armadas en Bolivia. BBC News Mundo. Obtenido de, 11 de noviembre de 2019. https://www.bbc.com/mundo/noticias-america-latina-50382253
34. BBC News Mundo: Motín de policías en Bolivia: agentes en varias ciudades se declaran en rebeldía contra el gobierno de Morales, quien denuncia un golpe de Estado, 9 de noviembre de 2019
35. BBC: News Mundo, Redacción: Evo Morales renuncia a la presidencia de Bolivia y denuncia un golpe de estado, 10 de noviembre de 2019
36. OEA: Fortalecimiento de la Democracia, Departamento para la Cooperación y Observación Electoral. Obtenido de (2019). http://www.oas.org/es/sap/deco/Informe-Bolivia-2019/0.1%20Informe%20Final%20-%20Analisis%20de%20Integridad%20Electoral%20Bolivia%202019%20(OSG).pdf
37. CNN Español: Lo último sobre la renuncia de Evo Morales a la presidencia y la crisis en Bolivia. Obtenido de, 11 de noviembre de 2019. https://cnnespanol.cnn.com/2019/11/11/lo-ultimo-sobre-la-renuncia-de-evo-morales-a-la-presidencia-y-la-crisis-en-bolivia
38. CIDH, comunicado de prensa: CIDH presenta sus observaciones preliminares tras la visita a Bolivia y urge una investigación internacional para las graves violaciones a los derechos humanos ocurridos en el marco del proceso electoral desde octubre de 2019. Obtenido de, 10 de diciembre de 2019). https://www.oas.org/es/cidh/prensa/comunicados/2019/321.asp/
39. Burbano de Lara, F.: El triunfo del MAS. El Universo, 20 de octubre de 2020

National Strike in Colombia 2019–2020: Between "Rumba", Protest and Cohesion

Kléver Antonio Bravo(✉), Fernando Alvear, and Roberto Jiménez

Universidad de las Fuerzas Armadas ESPE, 170603 Sangolquí, Quito, Ecuador
{kabravo,faalvear,rxjimenez2}@espe.edu.ec

Abstract. This article aims to analyze and describe the national strike in Colombia that took place between November 21st, 2019 and the first weeks of February 2020, taking into account the causes, actors and consequences of these social protests. The national strike "for life and peace" had reasons of an eminently social nature; having as main actors the urban youth of the middle class and trade unions, who demanded improvements in labor and tax reforms; as well as, better quality education and the rejection of corruption, especially due to the regional scandal caused by the company Odebrecht. Tens of thousands of protesters were able to express their common needs through batucadas, while groups of infiltrators looted and vandalized, actions that were repressed by public force. Finally, it could be said that this episode in Colombian history allowed us to visualize the needs and rights of a national group, whose only stage were the streets of the main cities of Colombia.

Keywords: Protesters · Social protest · Public force · Rights · Government

1 Introduction

The social protests that took place in Colombia had two characteristics that made them look different from those that previously occurred in that country. First, the preceding social protests carried out in Ecuador, Bolivia and Chile, where it was evident that social organizations had engaged in violent actions as never seen before in those countries, which suggested that Colombia's situation was not going to be much different. Second, the attitude of the Colombian protester, as the perception was that, in previous years, protests were self-repressed by Colombian citizenry due to the lengthy internal conflict in which they were mired (due to) with the guerrillas (or drug traffickers) and the real possibility of extremist groups infiltrating the protests, causing a higher rate of violence that the Colombian people could no longer bear.

Iván Duque's government took advantage of the nuance that social protests were acquiring and took actions to tone them down. In the same way, they managed to create the perception that these disturbances, far from having an approach of rejection to the government, had been arranged from Venezuela with the intention to destabilize the Government through infiltrators or Chavista agents, and their role of terrorists; just as it happened in Ecuador. Another strategy used by the Government was to create a

M. Botto-Tobar et al. (Eds.): CIT 2021, LNNS 405, pp. 389–396, 2022.
https://doi.org/10.1007/978-3-030-96043-8_30

campaign of fear participating in the protests and supporting those who did not joined them. These measures took effect partly because the protests in Colombia were the least violent compared to those in other South American countries such as Ecuador, Chile and Bolivia. Although, for many, the effect was contrary since the problem was overstated.

2 Results

Social protests in Colombia – and Latin America- have always been subject of study; starting from the fact that a social protest is a collective action deployed by a group of people seeking a common goal. And that a social protest is also linked to a social movement, where a human group pursues a shift [1].

Could it be a coincidence that protests across South America were a chain phenomenon from Ecuador to Colombia, passing through Chile and Bolivia? It's possible. However, the case of Colombia has its peculiarity since, there, "social protests have been stigmatized for many years" [2].

The social protests that took place in Ecuador, in October 2019, following President Moreno's announcement about the rise of the price of fuel, were joined by criminal acts occurred mainly in Quito, the capital city: the destruction of the Historic Center, setting on fire three buildings of the Comptroller General and a negative balance of 1,330 detainees and 1,507 injured, including 435 policemen [3].

One of the main actors of the social protests in Colombia, November 2019, was the National Police, as an element of "repression", which was supported by the Military Forces as "assistance to the Police", better known as "Army Policing", whose legal background is in the Citizen Security Law (2011), and is the one that discusses the "repressive management" of social protest, according to several analysts [4].

The national strike by "life and peace" began on November 21st, 2019. The name given to this national protest could suggest that the intensity and violence seen in Chile was not going to be seen in Colombia. The social groups who called for the demonstrations were fully identified as student representatives and workers' confederations, taking into account that these protests took place on 550 of the 1,222 municipalities, considering also that in the country's main cities such as Bogotá, Medellín and Bucaramanga almost 450,000 people (out of a population of almost eleven million inhabitants) mobilized, without dismissing the response that the call had throughout the whole population [5].

Given that Colombia is not a country inclined to strikes and mobilizations, we note a brief history: in 1977 there was a strike called by teachers, students and workers against President Alfonso López Michelsen; in 2008, an indigenous strike against President Álvaro Uribe; and, in 2013, the agrarian strike against the government of Juan Manuel Santos [6].

One of the main precedents of protests dates back to 2016, when the Congress finally approved the peace process between the Colombian State and the FARC. This agreement was signed in the face of strong opposition from a significant sector of the population that argued that a treaty to which the population did not agree, had been legalized, as pronounced in the planned plebiscite on October 2, 2016. Finally, the government of Juan Manuel Santos saw the urgent need to close the outstanding issues with the FARC, making a sort of cost - benefit calculation to end a conflict of more than sixty years, in the face of the rejection by a considerable sector of the Colombian society; also, because the Nobel Peace Prize's dream was not being put to rest.

Among the sectors of opposition to the signing of the peace agreement was former President Álvaro Uribe who maintained that the country was being handed over to the Revolutionary Armed Forces of Colombia (FARC) since this terrorist group had been granted unacceptable demands, which would represent a high cost for the State to fulfill this agreement [7].

This background, added to the general perception that the gap between rich and poor had increased, matured the idea on the leadership of social groups to carry out an effective action, added to the fact that historically, in Colombia social protests never had a massive reception, because the strike organizers were frequently associated to guerrilla groups. However, much remains to be deciphered from the role of armed groups in Colombian social protests [8].

It should be considered that since 2013 there has been a significant conflict with drug trafficking cartels, this is the most serious problem that Colombian society currently has, having an intensity of four (4), on a scale of 1 to 5, since its trend remains stable due to the number of cartels present (Caparrapos, the Army of National Liberation, ELN, FARC, the Popular Liberation Army, EPL, Los Rastrojos...) and the conflict existing between them, in front of governmental entities and with Colombian society in general.

Something worrying is that in recent years, new players in conflict have emerged or reemerged with a new flag of ideals. In 2017, FARC and PLA dissidents had effectively returned to the armed struggle along with their ideological indoctrination, seeking to recover armed dominance in various areas of Colombian territory, also trying to take over the drug trade, which is their main source of financing for these illegal groups. Taking into account that, in 2017, the FARC "officially" disappeared, but on the same year a number of unidentified armed groups made their appearance identified as FARC dissidents, who decided to continue their own armed struggle led by new actors who cannot join Colombia civil society and decided to create their own fronts to continue their terrorists and criminal practices, linked to drug trafficking and other transnational crimes.

Another aspect which deserves to be taken into account is the emergence of social organizations; that, taking advantage of the dissolution of the most important guerrillas such as FARC, have found the free route to carry out their protests without the government authorities being able to link them with the guerrillas. Thus, we now have the Catatumbo Peasant Association since 2013 and the Antioquia Artisan Miners since 2017. These organizations have stepped up to be considered in the list of organizations that cause social conflict and would have the capacity to create problems for the government in turn, due to its convening, economic and increasing political power. These attributes become important since the aforementioned organizations are aware of their convening capacity and their real chance of creating chaos at any given time. Which could force the executive branch to sit at the negotiating table to address their social demands, particular and class interests.

3 The National Strike "for Life and Peace"

Protests of 2013 and 2017 ended with a series of promises made by president Santos' government that – ultimately - were not kept, partly because President Duque did not have the will to do so and largely because he did not have the resources to fulfill them [9]. In the same way, promises made to the peasant sector in 2013, one of the many petitions proposing the implementation of input prices reduction, and protection of domestic production against imports prompted by the Free Trade, adding smuggling and the extension of the coffee subsidy [10], were not honored by President Duque either. All this was the trigger which generated the November 2019 national strike.

Here, it is appropriate to mention the role of social media: Facebook, WhatsApp and Twitter, elements of an extremely successful convocation for massive presence of protesters adding the impossibility to physically locate their anonymous promoters. Such has been the success of these social networks, that cohesions were effective, especially for solid and violent mobilizations in Colombia or anywhere in the world [2].

Among the causes that led to the protests in November 2019, are the unfulfilled offers done in previous years; however, it is important to see in context what demands were made by the different sectors in Colombian society for the strike to have had such popularity throughout all national territory.

Faced with this list of causes that led to the protests, the Government took some obvious and uncreative actions such as delegitimizing the call for a strike by making calculations on what the monetary damage would be for the Colombian State. This fact was totally ignored by those demonstrators, who with threats still wanted to start negotiations, to which the Government was reluctant. Partly because they did not want to fall into the unionists strategy and also because they were aware that there was no room for maneuver or any resources to address the demands of the protesters. This seemed to be aggravated if those reforms to earn the revenue the state needed, were not applied. In addition, the demands from the protesters needed to be addressed [11].

It should be understood that this strike was not entirely shouting and repression. It also showed that Colombian people can combine protest with a festive atmosphere, as it was seen since day one, on November 21st 2019, better known as 21N. This panorama was put together by young people with *batucadas* and chants. About the batucadas, repetitive musical rhythms with wind and percussion instruments, Central places in Bogotá, Cali, Medellín, Tunja, Ibagué, Manizales, Pereira, Armenia, Popayán and other cities were taken, with the purpose of combining the cry of a collective need and a festive spirit, something typical of Colombian spirit, because it was necessary to "have a good time" at that strike-party [12].

During the day, the batucadas and at night, the casseroles. Indeed, at the end of the day, young people and adults made great sounds with casseroles - without any rhythm -, to maintain the air of protest through noise against repression of the Mobile Anti-riot Squad, ESMAD, a police force created by the end of the twentieth century "as a force duly trained in the use of non-lethal weapons and in the proper handling of marches, strikes and, in general, everything related to attending crowds" [13].

Protests are transformed into a war of attrition, where the government was the one who had the urgency to reach a favorable solution, facing another battle of attrition where the main side was in no hurry to reach an understanding. Their only concern was to keep having the masses on the streets perpetrating the demonstrations. This slowly becomes a rope closing around the President's neck, challenging his popularity amongst the population.

It should be noted that the confrontations between the masses and security forces did not wait. The youths organized the famous "first line", a human barrier with prefabricated blue shields, destined to protect the rest of the group of protesters that was later reinforced with the "indigenous guard" who traveled to Bogotá, and was hosted by the National University of Colombia [12]. This is the reason why batucadas and casseroles did not come alone. President Iván Duque ordered the Army's intervention and a curfew in Bogotá starting at 9 pm, given that looting and vandalism did their thing: 76 stations and 79 buses were destroyed within the first days of the demonstrations, with an initial balance of three dead and 250 injured people [14].

On the other hand, the role of mass media, especially television, where the vast majority of the population could comfortably watch from their homes how events unfold and the intensity and violence generated, often magnified by the very nature of communicators to portrait news as "shocking", provoked into the citizenry the perception that the country is in a high degree of upheaval, creating an exaggerated perception of the intensity of the demonstrations and the violence caused.

As noted in previous lines, in Colombia there is the right to protest under articles 20 and 37 of its Constitution: *"Art. 20. Everyone is guaranteed the freedom to express and disseminate their thoughts and opinions, to report and receive truthful and impartial information, and to found mass media. These are free and have social responsibility. Art. 37. Any part of the people can meet and demonstrate publicly and peacefully. Only the law may expressly establish the cases in which the exercise of this right may be limited"*, while some part of the Colombian society maintains that the Citizen Security Law has a repressive character against social protest [10].

As for phrases, or "rallying cries", Chile coined continuous phrases such as "we are not at war, we are united", "it is not 30 pesos, with 30 years", and a more classic one: "Chile woke up." In line with the phrases, Colombia also coined a weak and insignificant phrase: "Colombia woke up". Considering this country to be more triggered by music, there was a phrase that did catch on the population: "A song for Colombia", a phrase that managed to bring together 40 different groups of Colombian artists who performed in various concerts, which allowed the ratification of the concept - or the idea - that the protests that started on November 21st were peaceful [15].

An unfortunate event was undoubtedly the death of young Dilan Cruz, a 20-year-old Colombian high school student who died on November 28th, 2019 during the protests due to a head trauma and a gunshot wound inflicted by the members of the Mobile Antiriot Squad, ESMAD, becoming a symbol of the demonstrations [8]. This event became a rallying cry for all sectors that were part of the protests and for Colombian society in general, since its peculiarity was that the young man was killed by an impact gun, when it is assumed that The Public Force should have only non-lethal material, so, many questions raised from this situation: Why did members of the Public Force to carry lethal weapons during the demonstrations? What was the situation in which the member of the Esmad, was forced to use his firearm?

4 Colombian Opinion Regarding 2019 National Strike

To begin talks between the government and the strike leaders, the latter established a list of 13 petitions (at the end the petitions increased to 104), one of which was "Dissolve ESMAD and purge the Police." This issue took more strength after the death of young Dilan Cruz.

At the end of demonstrations, Ipsos Grupo SA carried out the following analysis of Colombia 2019 protests, where Colombian citizen's opinion regarding the national strike which lasted nearly three months.

From the results obtained by this survey, it can be seen that the vast majority of the population considers that the issues being addressed do not offer a solution in the short term. It is thought that the changes taking place in the country will be positive at the end, they will help to improve economic and social situation since 51% of the population believed that "*It is the beginning of a long process with changes in the future*", to which the government will have to make enormous efforts to comply with the agreement, taking into consideration that State revenues will be reduced and the few existing ones are already committed [11].

On the other hand, there are divided opinions regarding Colombia's image in face of the rest of the world. Many (42%) think that "*It will affect negatively but it will be over soon*", 29% think it will not affect the country, while 24% consider that the country's image abroad will be seriously affected, there is no marked difference in public opinion. However, it does not appear to be a point of general concern.

Regarding public opinion about fading away ESMAD, 61% of those surveyed do not agree with the elimination of the Mobile Anti-Riot Squads and only 31% agree with the disappearance of this entity dedicated exclusively to the control of disturbances of public order, which suggests that people agree with having a police unit specialized in quelling disturbances.

In something that the citizenry fully agree on, is in the decision taken by the mayors of Cali and Bogotá when decreeing the curfew for each of their cities, the social situation led them to make these exceptional decisions due to the circumstances existing at that time. In that sense Cali's Secretary of Security, Andrés Villamizar said: "we are going to take all the measures, as the mayor said, against the vandals, the misfits and the violent ones, who have been trying to loot warehouses in Cali, who have been blocking". This functionary of State also noted that the 20,000 protesters who marched through the city center had done it peacefully, but - taking advantage of this circumstance - vandals and criminals vandalized sectors of the city [16].

In late December, tensions were decreasing due to the advent of Christmas, but as soon as January 2020 entered, demonstrations took an upturn with the relentless middle-class urban youth [5]. By the end of that month, the balance was six dead and 769 wounded. Of these, 379 Police officers and 390 civilians. Regarding the dead, it should be noted that one of them was a military nurse who committed suicide after suffering the pressure from his colleagues because of his position in favor of the protesters. Another deceased was a young victim of the explosion of a "papabomba" that he was carrying.

This national strike "for life and peace" was put to an end by covid 19, as it seemed that dialogues between representatives of the Civic Committee of the National Strike and President Duque were planned until March 15, 2020. By that day, the Congress had passed the long-awaited tax reform on late last December [17], while the world was beginning to falter due to a pandemic that was not anticipated.

5 Discussion

All of the aforementioned is aggravated by the existence of a media war between two completely opposing speeches, on one hand the Government, which declared that the progressive use of force was linked to the level of violence in the demonstrations. For their part, social leaders argued that there was a strong repression by law enforcement authorities.

Both the criminalization of social protest and disobedience to the rules of coexistence and respect for private property, were concepts used in the agreement of each of the parties. Once again, the press, especially television, played a very important role since their comments and images often responded to business interests, *ratings*, pressure from one or both parties, or the threat of extremist groups that also sought to give an image of downfall instead of owning the lead.

One thing in which a strong opinion is not established is in determining whether the demonstrations were peaceful or violent. Three out of five Colombians consider that the protests have been mostly peaceful, while the rest think otherwise. Unlike the historical protests in this country, October 2019 protests were marked with criminal acts and vandalism, these violent acts were the common denominator throughout the region, but if we compare them with the protests that occurred in Ecuador, Bolivia and Chile; it can be seen that those that occurred in Colombia were the least violent of all, despite the fact that there was one death as a direct action of ESMAD in these days of protest. The intensity and impetus shown in the streets by the protesters did not reach the levels of violence shown on neighboring countries.

6 Conclusions

The social conflict in Colombia comes from a long time ago, being recognized across the region and around the world, for many years, by the fact that the guerrillas have been the main concern of every government in turn. Pointing out as a turning point for Colombia’s politics when peace was finally signed with the FARC in 2016. It is that moment when social groups saw the opportunity to make their demands heard which had been silenced for many years due to the fear of being “accused by the government to act in favor of the guerrillas” and also due to fear of being “used by the guerrillas to divert attention and state resources in another direction. This is how these protests were the highlight in response to many years of neglect by the government into hearing the needs of different sectors of the population, especially peasants and students.

This national strike was also motivated by the failure to comply with the commitments made in the government of Juan Manuel Santos in 2013 and 2017 related to peasants and students respectively, added to the economic measures that were underway by the government of Iván Duque and the appearance of new demands of social groups. All this created the perfect environment to achieve national mobilization that coincided with the violent protests that occurred in other countries of the South American sub region.

As predicted, the November 2019 national strike was characterized by unusual violence demonstrated by protesters, as was the case in Ecuador, Bolivia and Chile. Infiltrated personnel were present among the protesters who wanted to cause chaos and create a hostile environment; ideal to drive public opinion against national and local authorities, at the same time it was becoming more difficult to find an agreement that helps depose the strike made by social leaders, seeing that citizens begin to exert pressure to find a solution to the strike and get back to exercise their daily activities, which only could be done in peace.

References

1. Casilimas, J.: Management of social protest: between the doctrine of national security and the proposal of citizen security. Universidad Externado de Colombia, Bogotá (2017)
2. García, J.K.: The police in the post-conflict: impact analysis of the new police code and the mobile anti-riot squad in Colombia. Logos Sci. Technol. Mag. **11**(3). https://doi.org/10.22335/rlct.vll3.986
3. Ministry of Government, bulletin of 15 October 2019. https://www.ministeriodegobierno.gob.ec/1330-detenidos-y-1507-heridos-fue-el-resultado-de-las-paralizaciones-en-el-Ecuador/
4. Barbosa, E.: Military containment of the protest. In: Revista Brújula, Semillero de Investigación, vol. 8 (2020). https://doi.org/10.21830/23460628.44
5. Hylton, F.: Something is happening in Colombia (2020). https://www.nuso.org/articulo/algo-esta-pasando-en-colombia/
6. Pardo, D.: National strike in Colombia: 3 unprecedented factors that made November 21 a historic day, in BBC News Mundo, 22 November 2019. https://www.bbc.com/mundo/noticias-america-latina-50520302
7. Tovar, L.: The fracture: a reading of recent social mobilizations in Colombia (2020). http://www.historiaypresente.com/hyp/wp-content/uploads/2020/02/Luchas_Sociales.pdf
8. Barrera, V, Hoyos, C.: Violent or messy. analysis of the repertoires of social protest in Colombia. Polit. Anal. Mag. (98), Bogotá (2020)
9. Buendía, G.: Fundación Carolina (2019). https://www.fundacioncarolina.es/wp-content/uploads/2020/01/AC-1.20.pdf
10. Cruz, E.: The right to social protest in Colombia. Juridical Thought 01/2016 (2015)
11. Rodríguez, E.: Colombia 2020: social mobilization as an opportunity and a reflection of change (2020). https://www.fundacioncarolina.es/wp-content/uploads/2020/01/AC-1.20.pdf
12. Aguilar-Forero, N.: The four Co's of youth collective action: the case of the national strike in Colombia. Polit. Anal. Mag. **33**(98), 26–43 (2020). Bogotá
13. Rozo, W.: Esmad, lethal weapon, Cinep/PPP (2013)
14. BBC News Mundo: Protests in Colombia: Iván Duque takes the Army to the streets and calls for a national conversation, 23 November 2019. https://www.bbc.com/mundo/noticias-america-latina-50511211
15. Week: The artists listened: the concert on December 8 will not be in the Simón Bolívar park, December 8, 2019. https://www.semana.com/musica/articulo/concierto-del-8-de-dicDiciembre-no-sera-en-el-parque-simon-bolivar/79562/
16. El Comercio. https://www.elcomercio.com/actualidad/toque-queda-cali-vandalismo-paro.html
17. Torrado, S., Oquendo, C.: The protests against the government of Iván Duque are reactivated in Colombia. El País, 22 January 2020. https://elpais.com/internacional/2020/01/22/colombia/1579648383_033676.html

Causes and Variables of Forest Fires, a Brief Review of the Ecuadorian Case

Henry Cruz(✉), Santiago Jácome, Tatiana Gualotuña, Diego Marcillo, and Efraín R. Fonseca C.

Universidad de las Fuerzas Armadas - ESPE, Sangolquí, Ecuador
hocruz@espe.edu.ec

Abstract. Forest fires means a potential threat to life on the planet since their effects can be catastrophic. Nowadays, the frequency of FF has increased, and the control capacity for it has decreased. The causes are diverse; however, the human factor is crucial in the generation, prevention, control, and mitigation of the impact of FF. This study presents the latest main FF events and the factors, causes, and variables involved in a general and global way. Additionally, has been carried out a study regarding the leading FF cases in Ecuador, it has been identified a sustained increase of burned hectares from 2015 to 2020, being 2020 the year with the highest incidence with 27,904.91 Ha. It has also been found that the increase or decrease of hectares burned depends on the fluctuations and relationships between wind speed, ambient temperature, humidity, and the characteristics of the geographic area where the events occur. It has also been identified that in the period 2018–2020 the Inter Andean region, presents most of the forest fires with a high level of damage.

Keywords: Forest fires · Causes · Factors · Variables · Ecuadorian case

1 First Section

Forest fires (FF) are becoming more and more frequent worldwide, and their consequences are increasing. There are multiple causes of forest fires, ranging from the burning of weeds to favor agriculture to ignition forests due to lightings. However, climate change and forest degradation, and fragmentation have resulted in conditions more prone to FF globally. Although in areas with hot and dry conditions, FF is more likely, human action is the leading cause than natural factors. On the other hand, there is a correlation between the increasing number of fires and increasingly warmer ocean temperatures due to climate change [1, 2].

There is an increased probability of burning large areas of forests in South America and Africa and particularly in mid-latitudes and extra-tropics, in the coming years [3]. Indeed, forests degraded by logging and fragmented by deforestation are more susceptible to destruction by fire.

Likewise, during 2019, 789,729.2 Ha have been burned in Europe, Middle East, and North Africa, according to the European Forest Fire Information System (EFFIS) report,

M. Botto-Tobar et al. (Eds.): CIT 2021, LNNS 405, pp. 397–410, 2022.
https://doi.org/10.1007/978-3-030-96043-8_31

which shows an increase in areas of forests and vegetation burned in such areas [4]. A large contingent of media and security forces have been deployed where the fire has devastated property and caused many victims, particularly in Mediterranean countries and regions such as Greece, Spain, southern France, and the Italian island of Sardinia.

In addition, around 10,000 people were evacuated due to uncontrolled fires in western Canada in 2019. This year, more than 4.5 million FF of more than one square kilometer were counted, concerning 2018, where forest fires were increased by a total of 400,000, reaffirming the growth trend of these events [5, 6].

FF affect areas about 350 million hectares each year, causing damage to property, livelihoods, but most seriously, loss of human life. Wildfires also contribute to global warming, air pollution, desertification, and loss of biodiversity. Prevention is one of the most practical measures, and effective surveillance can help with early warnings, hand in hand with decision-making to intervene and measure the fire impact [7]. Several methodologies detect and establish FF classification maps [8, 9] and architectures and variables that allow FF detection and early warning [10].

This study is a comprehensive analysis of forest fires, which is structured as follows: Sect. 2 shows the research methodology, including a review of the main events reported in recent years and an analysis of the most frequent variables involved in the process. Section 3 shows the conditions and main variables that cause FF and an analysis of the situation of forest fires in Ecuador in recent years, which defines the different variables and behavior patterns. Section 4 details a discussion of the results obtained. Finally, Sect. 5 reports the main conclusions and future work.

2 Methodological Framework

2.1 Method

The present work was developed based on a deductive model of information search and analysis, based on the following activities:

- Identify recent FF events around the world.
- Define the most common variables that FF generate.
- Learn about FF prediction systems.
- Analyze the FF events in Ecuador in recent years.
- Establish the geographic sectors in Ecuador with the highest incidence of FF.
- Analyze the most common variables that affect the generation of FF in Ecuador.

2.2 Recent Forest Fire Events Around the World

In 2020, there were several large fires worldwide, with developing countries being the most vulnerable to the destructive impact of fire, the aftermath of the loss of human life and property, and the destruction of natural resources. Among the main recorded forest fire events, the following are the most representative:

Dozens of fires on the west coast of the United States have burned about 1.2 million hectares in California and another 647,500 in Oregon and Washington in September.

These events left at least 34 fatalities, devastated small towns, and destroyed at least 8,000 homes and other property [11].

In El Pantanal, Brazil, extreme events occurred in July; a time considered the fire season. The burned area was 19,000 square kilometers, which represents 10% of the forest in the area. The region is experiencing a historic drought that, according to experts, is caused by a climatic imbalance. The place depends on the water that comes from the forests of the north of Matto Grosso, in the Amazon region of Brazil, where it rains less and less due to deforestation and the expansion of agricultural lands [11].

Australia has been destroyed by wildfires never seen in it until the end of January 2021. The start of the fires is associated with high record temperatures reached and months of severe droughts. More than 10.6 million hectares burned, dozens of people died, thousands of homes devastated, livelihoods and buildings destroyed, and more than 1.25 billion animals died directly or indirectly due to the fires [12].

2.3 Common Variables that Affect the Generation of Forest Fires

FF are a recurring event started by humans or by natural phenomena; however, the reality indicates that most fires are caused by humans, whether due to negligence, economic interests, wild use of fire in agriculture, generation of grasslands, illegal land clearing, or intentional fires [13]. The increase of FF due to provoked actions is tangible and has become a severe threat to society in such a way that certain governments around the world consider these actions as acts of terrorism [14]. On the other hand, the increase in population density implies a higher risk of fires due to the greater demand for land and other natural resources.

Evaluating the environmental conditions is considered preponderant to establish the origin, cause, and variables of the FF, such as elevation, temperature, and winds, which allows, among other things, to establish the most vulnerable areas [15]. It is necessary to evaluate the effects of different catalytic factors of fire occurrence patterns to identify the variables associated with the purpose of timely warning, acting, and mitigating these events [16]. Human action associated with environmental conditions contributes to an increase in the incidence of FF. Ecologists have recognized the relationship between site productivity and factors such as environmental conditions and the structure of the forest for this type of event to increase [17, 18]. Indeed, a study on the causes of FF carried out from 1996 to 2010 was determined that agricultural activity and the renewal of pastures are the leading causes [19].

The forest fire prevention policy is related to the development of control variables for human action and governments and the control of meteorological variables that influence fires [18]. Within the environmental conditions, there are multiple factors that directly and indirectly affect the start of fires. These factors are related to vegetation type, precipitation levels, topographic relief, biophysical factors in the different seasons of the year, maximum and minimum environmental temperatures, rainfall, wind speed, solar radiation, water vapor pressure, atmospheric pressure, and vegetation type. On the other hand, human intervention factors such as population density, pollution, lack of protection of natural reserves, and gross domestic product (GDP) are based on the primary economy. Controlling factors for monthly rainfall, daytime temperature, land use form, land cover, elevation, distance to populated areas, and distance to unpaved roads are also considered factors.

Environmental conditions have two groups: atmospheric and climate conditions, and the other related to geographic, soil, and existing vegetation conditions. In the first group, the standard variables that lead to FF are low humidity levels, high temperatures, high atmospheric pressure, and lightning. For [20], the ambient temperature is proportional to the ignition, while the level of precipitation is inversely proportional. Similarly, in [21], the authors established four preponderant factors to determine the probability of occurrence of an FF event: relative humidity, daily precipitation level, daily average temperature, and maximum wind speed. The latest uncontrolled FF events in California show a typical pattern of atmospheric manipulation due to aerial spraying, environmental contamination, moisture absorption, and reduced drought tolerance by plant species [22]. In hot regions with subtropical influences, the leading causes of FF are high air temperatures and high wind speeds; on the contrary, high rainfall and air pressure reduce the number of burned areas [23].

Another essential aspect of being considered is that the highlands could contribute to the generation of a FF. Indeed, forests on higher ground have a higher probability of fire events [15]. The type of vegetation is another element that encourages the ignition of the forest. An example of this is the frequent burning of dry coniferous forests, this type of vegetation associated with climatic variations (humidity and cold) contributes to low-severity fires, which allows the renewal of the forest [24]. Table 1 summarizes the primary factors and causes that generate FF.

Table 1. Factors and causes that generate the forest fires.

Factors	Causes	Description
Anthropogenic (caused by humans)	- Soil preparation for cultivation through precarious activities	Pasture renovation, soil use change
	- Pyro-maniacal actions and unscrupulous acts of people (arson)	Visits of people to natural areas and forests
	- Lack of knowledge and information about the controlled use of fire	Misinformation
Natural phenomena	- The vegetation's dryness (moisture content and amount of living vegetation)	About phenomena, environmental, atmospheric, and characteristic of the soils type and vegetation
	- Changes in climatic variables that influence the spread and fire intensity ("el Niño" phenomenon)	

(*continued*)

Table 1. (*continued*)

Factors	Causes	Description
	- The main variables involved in forest fires are humidity, precipitation level, temperature, atmospheric pressure, altitude, wind speed, and lightning	The main factors contributing to increasing burned areas are high air temperature and wind speed; however, high rainfall and air pressure reduce the number of burned areas
Other factors	- The lack of Early warning systems (fire intelligence) or its accessibility	Factors of artificial origin generated and/or created by humans
	- Misuse of fuels	
	- The proximity of populations (people, towns, cities, factories)	
	- Electricity supply networks	

2.4 Forest Fires Prediction Systems

FF have become one of the prominent disturbances of terrestrial ecosystems worldwide, so it is vital to establish patterns, models and evaluate factors that cause FF for their prevention and the sustainable management of biodiversity and the environment [20]. Predictive models undoubtedly allow the establishment of actions that help prevent, predict the spread, issue alarms, plan contingencies, as well as mitigate the impact of FF; therefore, it is necessary to comprehensively understand all the physical, chemical, and environmental phenomena that occur before, during and after of FF [25].

Predictive systems through Bayesian networks classify the causes of FF into two groups: The first corresponding to the human factor, where arson, smoking, hunting, picnic fire, shepherd fire, stubble burning, and natural causes such as lightning strike, power lines. Similarly, the community established the factors that increase the possibility of uncontrolled FF. A study [26] carried out from 2008 to 2018 considered the factors temperature, relative humidity, tree species, distance from the road, wind speed, distance from agricultural land, amount of burnt area, and month and distance from the settlement.

There are also predictive systems based on the analysis of year, climate, land use, and ease of access to sensitive areas. These aspects allow modeling of probabilities of occurrence [20]. In [21], the authors show relative humidity levels, and the precipitation levels generate a greater probability of FF, finding that it is higher, even than the environmental temperature and the wind [27]. In the prediction of FF, aspects that influence the generation and spread of FF must be considered, as shown in Table 2.

We believe that management systems must be developed to prevent FF. These systems will make it possible to strengthen administrative capacities for FF prevention, hand in hand with adequate announcement and education to achieve greater awareness in compliance with policies. In a complementary way, we believe that it is essential to have an FF map of the events and risks.

Table 2. Aspects to consider in predicting forest fires.

Considerations for prediction	Features
Dryness evaluation and vegetation's climate	Consider statistical parameters of local drought
Active fire detection and monitoring	Generate early warnings and prediction of fire behavior
Prediction of the most frequent climatic variables for forest fire generation	Establish the frequency of active forest fire variables and their thresholds
Establish the relationship between the physical and chemical processes of forest fire combustion with the interactions of climate, time, fuel, vegetation type, and topography	Establish results and weights of the combination of variables in the prediction of forest fire
Integration and processing of data in fire information systems with other relevant information	Combine, for example, vegetation cover and securities at risk
Capabilities of modeling the occurrence and behavior of fires	Generation of predictive maps
Dissemination of meteorological information in real time such as temperature, humidity, wind speed and others; with the purpose of feeding predictive models	Implement advanced technologies based on remote sensing data

2.5 Some Metrics to Calculate the Possibility of a Forest Fire

The variables of ambient temperature, wind speed, humidity percentage, drought index, among others can be used to calculate indicators that allow to establish both the possibility of a forest fire occurring and the danger of it [28].

The McArthur Forest Fire Danger Index (FFDI) is an empirical model that analyses fire behaviour in forests, evaluating some external variables. Although fire behaviour is different depending on vegetation types, most of the basic fire hazard factors are common to all of them. The inputs required to run the model are air temperature, relative humidity, wind speed, precipitation, time since last precipitation, and Keetch-Byram Drought Index (KBDI) [29].

The FFDI incorporates a drought factor (DF) and the KBDI. The DF drought factor is an estimator of the moisture content of dry fuels. It is calculated by combining estimates of the effects of direct smoking by recent significant rains and smoking from below, which depend on the moisture content of the soil [30].

In KBDI the water balance is expressed as soil moisture deficiency, which is the amount of water that the soil can contain in its capillaries against the effects of gravity. Low soil moisture deficiency means very low or no restriction for evaporation from the soil or transpiration by plants, and very low infiltration capacity in case of heavy rains. A high soil moisture deficiency means that there is very little water available for evaporation by the soil or transpiration by plants. KBDI ranges from 0 (no drought) to 800 (extreme drought) [31].

FFDI may indicate high to extreme fire hazard for very hot summer days after a few days without precipitation. However, after the first rains of the wet season, the flammability of the fuel decreases considerably due to the presence of a high proportion of primary tissues with high water content (foliage), typical of a new vegetative period. The FFDI hazard scale in the forest ranges from low-moderate (0–11) to catastrophic (+100).

3 Analysis of the Most Common Variables of Forest Fires in Ecuador

In Ecuador, most FF are caused by human activity (anthropogenic) since fire is considered a work tool for soil preparation, renovation of pastures, and the change of soil use. Such practices are carried out inappropriately, which is why they are the leading cause of FF. Likewise, pyrotechnic activities are hand in hand with evil actions by people who visit natural areas and forests where they make uncontrolled use of fire, perhaps due to ignorance.

From 2015 to 2020, around 141,139.91 Ha were burned in Ecuador [32], being 2020 the year with the highest incidence, 27,904.91 Ha burned. This year there was an increase of 5,235.91 Ha burned, which represents 18.71% compared to 2019. From January 1, 2018, to January 31, 2021, 8,972 forest fire events were registered and were burned 78,680.64 Ha. The most affected provinces were Loja with 18,671.13 Ha, Pichincha with 10,520.13 Ha, Azuay with 8,101.56 Ha, and Guayas with 6,786.84 Ha [33].

The dry forests of the inter-Andean region are the most affected by fire, with a high probability of spread and a moderate level of the Haines index. In 2020 this index increased in the central inter-Andean zone in Cotopaxi, Tungurahua, and Chimborazo.

From June to September in the inter-Andean zone, the most significant number of FF incidents were generated, which is undoubtedly related to the dry air, the temperature difference, the different levels of the atmosphere of the Central Sierra, and the low-level humidity in the indoor atmosphere. Seasons with the highest humidity were presented from February to May and October to November. Such a situation is due to the interaction of the Pacific Ocean and the Amazon [34]. Table 3 shows the places where the most destructive FF events have occurred from 2018 to 2020. The information provided by the Secretaría Nacional de Gestión de Riesgos y Emergencias del Ecuador (SNGR) has been considered to carry out data analytics and statistical calculations to establish trends and patterns of behavior [35].

3.1 Variable Wind Speed

The wind is a factor that directly affects the intensity, direction, and speed of fire spread and, therefore, represents a fundamental element to be considered in the fight against forest fires. The main effects of the wind on the fire are the following:

- Desiccation of vegetation, which makes it more prone to ignition.
- Boost the fire by adding oxygen to the combustion.
- Propagation of flames towards elements that are not burning, causing their ignition.

- Displacement of sparks or firecrackers to areas not yet compromised, which causes secondary sources of fire.
- Unpredictable changes in fire progress related to variations in wind speed and direction.

In general terms, the higher the wind speed, the greater the intensity and speed of fire propagation. However, changes in wind direction are not always harmful since a change in the opposite direction to the fire's advance can contribute to its extinction.

After a METAR analysis [36] was possible to verify that the wind has activated and caused contamination by fire in the forests in different parts of Ecuador's geography. The inter-Andean zone made up of Pichincha, Cotopaxi, Chimborazo, Tungurahua, was the most affected. In the Austral region, the provinces of Loja and Azuay show high damage rates from the same factor. On the Ecuadorian coast, the province of Guayas was also seriously affected.

The data analysis showed that, in the areas where the most destructive forest fire events have occurred in Ecuador, the average wind speed for 2018 was 33.18 km/h, for 2019 it was 28.15 km/h, and for 2020 it was 26.19 km/h; that is, the range between 26.19 km/h to 33.18 km/h represents a threshold of favorable conditions for contamination by fire. Table 3 exhibits the chronological wind speed data.

3.2 Variable Ambient Temperature

The environmental temperature is a parameter that defines the generation of forest fire because it stimulates the production of warm air currents, which reduce the level of humidity and increase the drought index [15, 17, 20, 23]. We carried out an analysis of the 2018–2020 period, which revealed that the annual average temperature for 2018 was 21.4 °C. In 2019 this average fell to 18.76 °C. For 2020 there was an average of 21.56 °C. Precisely the increase and decrease of forest fire events in Ecuador directly correlate with the fluctuation of the environmental temperature. There is a decrease of 16.13% of burned hectares, while between the period 2019 to 2020, there is a 17% increase [33]. Table 3 presents the characterization of the environmental temperature data.

3.3 Variable Ambient Humidity (Water Vapor)

Humidity is the amount of water vapor present in the air. The dryness of the air indicates the presence of less water vapor; that is, the air contains little humidity, and in these circumstances, there will be few clouds. On the contrary, if the air contains a lot of water vapor (humidity), the clouds fill up, fog forms, and there may even be precipitation. On the other hand, the amount of water vapor that an air mass contains depends on the temperature; if the air temperature rises, it contains more moisture. So, the higher the air temperature, the greater the amount of water vapor [37, 38].

According to [39, 40], the highest probability of fires, and the risk of the fire start, depends on the relative humidity. The probability of ignition depends on the fuel (biomass)'s humidity, whose humidity is balanced with that of the air. The relative humidity is the relationship between the water vapor in the air and the maximum amount of vapor it could contain. Relative humidity depends on the temperature and the amount of water vapor in the air. If the temperature is high, the capacity of the air to contain water vapor increases, then the relative humidity decreases. Consequently, higher temperature and/or low amount of vapor in the air, then low relative humidity, and high probability of fire. Table 3 presents the humidity percentages obtained in the 2018–2020 period.

Table 3. Most significant forest fire events in Ecuador in 2018–2020.

Common places destroyed for forest fires	Date	Wind speed (Km/h)	Temp °C	Hum level
Pichincha/Cayambe/Cangahua	29/09/2018	33.5	24.2	72%
Pichincha/Cayambe/Cerro Pisambilla	29/09/2018	33.5	24.2	72%
Pichincha/Mejía/Cerro Atacazo	29/09/2018	33.5	24.5	72%
Azuay/Nabón/Sector Las Nieves	09/09/2018	24.1	19.9	83%
Azuay/Girón/Pichanillas	05/09/2018	27.7	20.6	88%
Chimborazo/Riobamba/La Chorrera	02/09/2018	37.1	19.9	33%
Loja/Loja/Vilcabamba	02/09/2018	37.1	19.9	77%
Loja/Loja/Vilcabamba	08/09/2018	38.9	18	64%
Guayas/Guayaquil/Juan Gómez Rendón	10/11/2019	20.5	19.9	59%
Guayas/Guayaquil/Pascuales/Monte Sinaí	09/11/2019	20.5	19.9	59%
Guayas/Guayaquil/Pascuales/Socio Vivienda 3	09/11/2019	20.5	14.8	59%
Guayas/Guayaquil/Chongón/Km 22 Vía a la Costa	09/11/2019	20.5	19.9	59%
Azuay/Cuenca/Sayausi/Dos Chorreras	05/11/2019	46.4	20.2	89%
Manabí/Chone/Chone/Vía Colorado, atrás de Puerto Arturo	01/11/2019	24.1	20.2	64%
Guayas/Guayaquil/Juan Gómez Rendón (Progreso)	01/11/2019	24.1	20.2	58%
Santa Elena/Santa Elena/Simón Bolívar (Julio Moreno)	30/09/2019	24.1	21	44%
Loja/Calvas/Cariamanga, cabecera cantonal/Animuro	26/10/2019	31.7	16.9	47%
Guayas/Guayaquil/Juan Gómez Rendón/vía La Costa	20/10/2019	33.5	15.8	41%
Carchi/Bolívar/Monte Olivo/sector Monte Olivo	19/10/2019	20.5	16	82%
Pichincha/Quito/Jipijapa/Simón Bolívar de las Palmeras	14/10/2019	37.1	18	61%

(continued)

Table 3. (*continued*)

Common places destroyed for forest fires	Date	Wind speed (Km/h)	Temp °C	Hum level
Manabí/Pedernales/Vía Cojimíes frente al Hotel Punta Azul	15/10/2019	35.3	17.2	86%
Manabí/Pedernales/Pedernales, Cabecera Cantonal	01/10/2019	31.7	21	80%
Loja/Calvas/El Lucero/San Roque	26/09/2019	31.7	20.4	27%
Pichincha/Quito/El Condado/Mena del Hierro, vía a Nono	08/11/2020	35.2	17	51%
Loja/Saraguro/San Antonio De Qumbe (Cumbe)/Marculoma	06/11/2020	33.3	21	95%
Azuay/Camilo Ponce Enriquez/Moras	05/11/2020	20.4	22	93%
Azuay/Paute/San Cristóbal (Carlos Ordóñez Lazo)/Cantagallo	05/11/2020	20.4	22	93%
Pichincha/Cayambe/Cangahua/Comunidades Moras Chupa	02/11/2020	25.9	22	56%
Cotopaxi/Sigchos/Sigchos, Cabecera Cantonal/Yacuchaqui	03/11/2020	22.2	22	35%
Tungurahua/Santiago de Píllaro/San José de Poaló	02/11/2020	25.9	22	27%
Azuay/Cuenca/San Joaquín/Soldados	01/11/2020	25.9	23	72%
El Oro/Chilla/Chilla/Chilola	01/11/2020	25.9	23	66%

Source: compendium of various sources [10, 11, 28, 29]

Table 4 shows the climatic variables in Ecuador in the period 2018–2020 and in Fig. 1 the behaviour curve of the variables in this period.

Table 4. Climatic variables in Ecuador in the period 2018–2020.

Year	Wind speed (Km/h)	Temp (°C)	Humidity (%)
2018	33.18	21.4	77.63
2019	28.15	18.76	56.07
2020	26.12	21.56	65.33
Average	29.15	20.57	66.34

The highest average wind speed (33.18 km/h) occurs in 2018, while the highest average ambient temperature is generated in 2020 (21.56 °C), on another side, the lowest humidity level occurs in 2019 (56.07%).

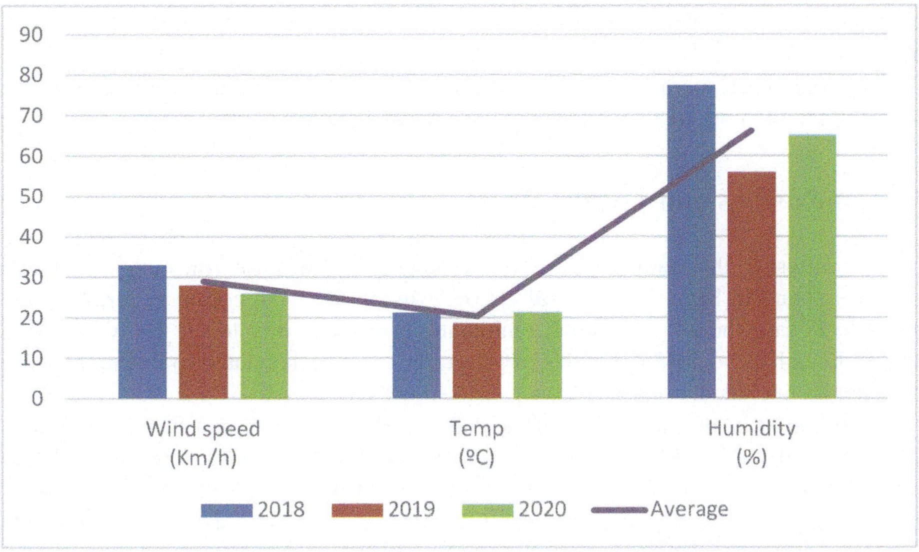

Fig. 1. Climate variables behaviour in the period 2018–2020

4 Discussion

The analysis carried out shows that the causes of FF are of natural origin and coming from humans. FF have increased their frequency of occurrence, and each time, they turn out to be more destructive and harmful to the environment and the stability of biodiversity. In fact, in the period 2015 to 2020, Ecuador a peak of growth of burned hectares in 2020 of 27.904,91.

In the year 2018, the majority of the most important forest fire events have occurred in the Ecuadorian Sierra, in the year 2019 the pattern changes being the Ecuadorian coast the most affected, however, in 2020 again, the Ecuadorian Sierra, especially the Inter Andean region, presents most of the forest fires with a high level of damage.

On the other hand, the places most affected by FF in Ecuador are in the inter-Andean region, which is associated with the confluence of several factors such as the decrease in air pressure and environmental temperature. In this region, the ignition and spread of FF are favored with variables of wind speed greater than 25 km/h, ambient temperatures greater than 20 °C, and humidity with a low levels' tendency.

In the province of Loja, there is a low level of humidity tending to 0%; high wind speeds greater than 30 km/h and ambient temperatures greater than 16 °C. In the province of Guayas, the value of the variables that promote FF is found in the thresholds of 20 to 25 km/h of wind speed, 19 °C to 22°C of ambient temperature, and relative humidity tending to low levels.

In 2018 the wind speed is higher compared to 2019 and 2020. Around 27,614.51 Ha were burned in 2018.

The 2019 has the lowest average ambient temperature in relation 2018 and 2020 (18.76 °C) as well as the lowest humidity level 56.07%. This year, a smaller number of hectares of forest were burned compared to 2018 and 2020, with a total of 23,161.22 hectares.

The ambient temperature in 2020 was the highest (21.56 °C), this year more hectares were burned than in 2018 and 2019 with an approximate of 27,904.91 Ha.

5 Conclusions and Future Work

In general terms, the leading causes of FF have a relationship with human action or omission, hand in hand with natural action. Among the most common variables that promote FF are the environmental temperature, the humidity level, the wind speed, and, in Ecuador, the different levels of geographical altitude related to the atmospheric pressure are included.

From the analysis carried out, it has been possible to verify that there is a trend of sustainable growth of FF in Ecuador and the world and an increase in burned areas. Ecuador has three risk zones identified to be exterminated by FF: The inter-Andean region, including Pichincha, Cotopaxi, Chimborazo, and Tungurahua provinces; the Austral region, with the provinces of Azuay and Loja; and finally, the coastal region, mainly with the provinces of Guayas and Manabí. In the period 2018–2020 the Inter Andean region, presents most of the forest fires with a high level of damage.

Furthermore, it has been verified that in years where the ambient temperature is higher (2018 and 2020) the risk of burning a greater number of hectares of forests also increases.

There is a trend towards the generation of FF in the indicated geographical areas of Ecuador when there is a correlation of wind speed, environmental temperature, and humidity level. Such a situation generally happens at higher wind speed values of 20 km/h, temperature 20 °C, and humidity levels will vary between regions with values tending to 0%. Likewise, the conditions will vary depending on the altitude where the forests and vegetation are found, hand in hand with the atmospheric pressure values. In a general way, in Ecuador exist a sustained increase of forest fires events and areas burned between 2015–2020 period.

The future work is oriented to develop an automatic system for surveillance and monitoring of forests with capabilities to predict FF.

Acknowledgments. This work was partially supported by the PIS-04 "Detección temprana de incendios forestales basado en aprendizaje automático y en algoritmos de clasificación aplicados al espectro visible" research project of the Universidad de las Fuerzas Armadas ESPE, Ecuador and partially supported by the Work Group "Internet de las Cosas y Ciudades Inteligentes" of the Corporación Ecuatoriana para el Desarrollo de la Investigación y la Academia (CEDIA), Ecuador.

References

1. Ganteaume, A., Camia, A., Jappiot, M., San-Miguel-Ayanz, J., Long-Fournel, M., Lampin, C.: A review of the main driving factors of forest fire ignition over Europe. Environ. Manag. **51**(3), 651–662 (2013). https://doi.org/10.1007/s00267-012-9961-z

2. World Wildlife Fund. https://www.wwf.org.ec/. Accessed 07 Nov 2020
3. Santoro, A., Venturi, M., Piras, F., Fiore, B., Corrieri, F., Agnoletti, M.: Forest area changes in cinque terre national park in the last 80 years. Consequences on landslides and forest fire risks. Land **10**(3), 293 (2021)
4. The European Forest Fire Information System (EFFIS). https://effis-gwis-cms.s3-eu-west-1.amazonaws.com/effis/reports-and-publications/annual-fire-reports/2019_Fire_Report_HighRes_final_PTcorrection/Annual_Report_2019_final_topdf_2.pdf. Accessed 04 Feb 2021
5. Da Silva, A.S., Justino, F., Setzer, A.W., Avila-Diaz, A.: Vegetation fire activity and the Potential Fire Index (PFIv2) performance in the last two decades (2001–2016). Int. J. Climatol. **41**, 78–92 (2021)
6. Global Forest Watch International Organization. https://www.globalforestwatch.org. Accessed 14 Feb 2021
7. Food and Agriculture Organization. http://www.fao.org/news/story/es/item/29145/icode. Accessed 14 Dec 2020
8. Cruz, H., Eckert, M., Meneses, J., Martínez, J.F.: Efficient forest fire detection index for application in unmanned aerial systems (UASs). Sensors **16**(6), 893 (2016)
9. Cruz, H., Gualotuña, T., Pinillos, M., Marcillo, D., Jácome, S., Fonseca C., E.R.: Machine learning and color treatment for the forest fire and smoke detection systems and algorithms, a recent literature review. In: Botto-Tobar, M., Cruz, H., Díaz Cadena, A. (eds.) CIT 2020. AISC, vol. 1326, pp. 109–120. Springer, Cham (2021). https://doi.org/10.1007/978-3-030-68080-0_8
10. Fonseca C., E.R., Marcillo, D., Jácome-Guerrero, S.P., Gualotuña, T., Cruz, H.: Identifying technological alternatives focused on early alert or detection of forest fires: results derived from an empirical study. In: Botto-Tobar, M., Cruz, H., Díaz Cadena, A. (eds.) CIT 2020. AISC, vol. 1326, pp. 354–368. Springer, Cham (2021). https://doi.org/10.1007/978-3-030-68080-0_27
11. France vingt-quatre. https://www.france24.com. Accessed 24 Nov 2020
12. World Wildlife Found International Organization. https://www.worldwildlife.org/. Accessed 09 Jan 2021
13. World Health Organization. https://www.who.int/hac/techguidance/ems/vegetation_fires/es/. Accessed 14 Mar 2021
14. Besenyő, J.: Inferno terror: forest fires as the new form of terrorism. Terror. Polit. Violence **31**(6), 1229–1241 (2019)
15. Zúñiga-Vásquez, J.M., Pompa-García, M.: The occurrence of forest fires in Mexico presents an altitudinal tendency: a geospatial analysis. Nat. Hazards **96**(1), 213–224 (2018). https://doi.org/10.1007/s11069-018-3537-z
16. Muñoz-Robles, C., Santana-Arias, R.: Puntos de calor en la Sierra Madre Oriental de San Luis Potosí: patrones espaciales y factores asociados. Madera y bosques **24**(1), 41–44 (2018)
17. Ruffault, J., Mouillot, F.: Contribution of human and biophysical factors to the spatial distribution of forest fire ignitions and large wildfires in a French Mediterranean region. Int. J. Wildland Fire **26**, 498–508 (2017)
18. Xiong, Q., et al.: Fire from policy, human interventions, or biophysical factors? Temporal–spatial patterns of forest fire in southwestern China. For. Ecol. Manag. **474**, 118381 (2020)
19. Lourenço, L., Fernandes, S., Nunes, A., Gonçalves, A.B., Vieira, A.: Determination of forest fire causes in Portugal (1966–2010) **4**(3), 171–175 (2013)
20. Vacchiano, G., Foderi, C., Berretti, R., Marchi, E., Motta, R.: Modeling anthropogenic and natural fire ignitions in an inner-alpine valley. Nat. Hazards Earth Syst. Sci. **18**(3), 935–948 (2018)
21. Lu, J., Zhou, T., Li, B., Wu, C.: Scale analysis and correlation study of wildfire and the meteorological factors that influence it. Math. Probl. Eng. **2018**, 1–10 (2018)

22. Herndon, J.M., Whiteside, M.: Incendios Salvajes en California: el papel de la geoingeniería y de la manipulación atmosférica encubierta **17**(3), 1–18 (2018)
23. Rasul, A., Ibrahim, G.R.F., Hameed, H.M., Tansey, K.: A trend of increasing burned areas in Iraq from 2001 to 2019. Environ. Dev. Sustain. **23**(4), 5739–5755 (2020). https://doi.org/10.1007/s10668-020-00842-7
24. Parks, S.A., Dobrowski, S.Z., Panunto, M.H.: What drives low-severity fire in the southwestern USA? Forests **9**(4), 165–169 (2018)
25. Sullivan, A.L.: Improving operational models of fire behaviour. In: Proceedings of the 18th World IMACS Congress and MODSIM 2009 International Congress on Modelling and Simulation. Modelling and Simulation Society of Australia and New Zealand and International Association for Mathematics and Computers in Simulation, pp. 2377–2383 (2009)
26. Sevinc, V., Kucuk, O., Goltas, M.: A Bayesian network model for prediction and analysis of possible forest fire causes. For. Ecol. Manag. **457**, 117723 (2020)
27. Taufik, M., Veldhuizen, A.A., Wösten, J.H.M., van Lanen, H.A.J.: Exploration of the importance of physical properties of Indonesian peatlands to assess critical groundwater table depths, associated drought and fire hazard. Geoderma **347**, 160–169 (2019)
28. Gannon, C.S., Steinberg, N.C.: A global assessment of wildfire potential under climate change utilizing Keetch-Byram drought index and land cover classifications. Environ. Res. Commun. **3**(3), 035002, 1–15 (2021)
29. Dowdy, A.J., Mills, G.A., Finkele, K., de Groot, W.: Index sensitivity analysis applied to the Canadian forest fire weather index and the McArthur forest fire danger index. Meteorol. Appl. **17**(3), 298–312 (2010)
30. Griffiths, D.: Improved formula for the drought factor in McArthur's Forest Fire Danger Meter. Aust. For. **62**(2), 202–206 (1999)
31. Heim, R., Jr.: A review of twentieth-century drought indices used in the United States. Bull. Am. Meteorol. Soc. **83**(8), 1149–1166 (2002)
32. El Comercio. https://www.elcomercio.com/tendencias/incendios-forestales-fuego-casitagua-pichincha.html. Accessed 14 Jan 2020
33. Reporte Anual de la Secretaria Nacional de Gestión de Riesgos y Emergencias del Ecuador. https://www.gestionderiesgos.gob.ec/wp-content/uploads/2021/01/Informe-de-Situacion-No-031-Cierre-2020-Incendios-Forestales.pdf. Accessed 28 Mar 2021
34. Pazmiño, D.: Peligro de incendios forestales asociado a factores climáticos en Ecuador. FIGEMPA: Investigación y Desarrollo **1**(1), 10–18 (2019)
35. Secretaria Nacional de Gestión de Riesgos y Emergencias del Ecuador. https://www.gestionderiesgos.gob.ec. Accessed 02 Feb 2021
36. Cedar Lake Ventures, Inc. https://es.weatherspark.com. Accessed 14 Jan 2021
37. Pausas, J.G.: Incendios forestales. Los Libros de la Catarata (2020)
38. Villers, L.: Incendios forestales. Ciencias **81**, 234–244 (2006)
39. Abatzoglou, J.T., Williams, A.P.: Impact of anthropogenic climate change on wildfire across western US forests. Proc. Natl. Acad. Sci. **113**(42), 11770–11775 (2016)
40. Peña-Fernández, E., Valenzuela-Palma, L.: Incremento de los incendios forestales en bosques naturales y plantaciones forestales en Chile. In: Memorias del segundo simposio internacional sobre políticas, planificación y economía de los programas de protección contra incendios forestales: Una visión global, pp. 595–612 (2008)

Robust System for Crowd Counting and Estimation Using Multi-column Filters on Unmanned Platforms

Henry Cruz[1,4(✉)], Darwin Merizalde[1,2], Juan Amón[4], Manolo Paredes Calderón[1,3], Marco Calderón[4], and Pablo Albán[2]

[1] Universidad de las Fuerzas Armadas ESPE, Sangolquí, Ecuador
hocruz@espe.edu.ec
[2] Centro de Investigación de Aplicaciones Militares-CICTE, Quito, Ecuador
[3] Escuela de Formación de Soldados, Ambato, Ecuador
[4] Ejército Ecuatoriano, Quito, Ecuador

Abstract. The automatic estimation combined with the use of sensors is increasingly useful for various applications that improve the quality of life of human beings and their environment. These developments become more relevant if they are used in complex scenarios, such as those generated from images and videos obtained from aerial images, from optical sensors installed in unmanned systems. In this research, an algorithm with a Multicolumn Gaussian Filter was implemented, which, from images and videos obtained in real-time, with flights at altitudes above 100 m, can estimate the number of people that make up conglomerate crowds in certain sectors. A particularity of the contribution of this research is the ability to perform estimation in real-time, with a consequent multicast transmission, whose low-cost architecture allows the information to be transmitted to geographically distant locations with minimal latency. The results obtained in the research, generate high reliability, to estimation, detection, and counting of people in complex scenarios. The calculation of the error in the training phase shows 6.37% MAE error in estimation accuracy and 15.33% MSE error, these measures verify robustness in the estimate. Also, the solution implemented shows an accuracy rate of 92.39% in the classification, in addition, the field tests using UAVs show high performance in real application situations. Finally, a compensation factor has been introduced, which has made it possible to avoid dispersion in the estimation of the automatic crowd count in real employment situations.

Keywords: Automatic crowd counting · Gaussian multicolumn filter · Artificial intelligence · Unmanned aerial systems · Convolutional neural networks · Multicast transmission · Computer vision

1 Introduction

The multidisciplinary approach to security makes aspects related to monitoring, surveillance, and control necessary to guarantee comprehensive security. In this context, the use

M. Botto-Tobar et al. (Eds.): CIT 2021, LNNS 405, pp. 411–426, 2022.
https://doi.org/10.1007/978-3-030-96043-8_32

of information and communication technologies are essential to improve capabilities, especially surveillance and reconnaissance.

On the other hand, surveillance systems through optical sensors have gone from being passive to acting as active, this in consideration of the existing interaction in the taking of signals, signals pre-processing, segmentation, generation of descriptors, classification, recognition, and interpretation of the scene. Precisely, optical sensors interact with computer vision to generate virtual information in real time.

There are several applications that computer vision has generated, among the most important are the detection, monitoring, recognition, location and counting of people. Indeed, a variant of the people counting represents the counting in dense scenarios or also called crowd counting, this application is key in surveillance and recognition, since it allows estimating the number of people in a scene, site, or place, in such a way that the security forces can foresee any type of incident and thereby guarantee the safety of both people and infrastructure, as well as neutralize acts of violence and destruction, if applicable.

The main works that are oriented to the counting of crowds or counting in crowded scenarios have as a principle the classification based on the extraction of characteristics of the images or video frames. Precisely, in [1] the information sources are based on the total of the edge pixels, edge orientation and Minkowaski dimension as input for the classification.

Other works use automatic classification as the way to obtain more effective results considering the Convolutional Neural Networks (CNN), to classify pixels using the Gaussian distribution as a source of information, and establishing switched classification and training levels, which allows a verification of characteristics in different schemes and recognition from different perspectives [2].

Also, regression models allow to establish local and global estimates; precisely the extraction of multilevel characteristics in combination with multi-output regression models, allows the localization, recognition of the pixels and the learning of the characteristics of a certain group of pixels that correspond to the number of people [3].

In effect, the predictive models have as generic processes the taking of characteristics product of the pre-processing of the samples, to subsequently submit to training procedures through set training and validation set and the comparison with the test set in such a way that the prediction is appropriate. Precisely, one of the ways to get an estimate of the number of people in dense or crowded scenarios is through the calculation of density [4].

The most important works use density maps, that is, in the pre-processing, regions of pixels are established that change color and represent the accumulation of energy; based on these variations, even more or fewer people are established. Other methods make comparisons between density maps, establishing the signal peaks and generating a global map. The basis of the classification is through CNN to establish the regions of interest (ROI's) [5]. Following the same scheme in [6], density maps are established by taking the characteristics of the pixels that make up people's heads using RGB-D as a source of information for classification.

On the other hand, other works show that aerial surveillance of crowds is more effective using Unmanned Aerial Systems or Unmanned Aircraft Systems, this due to the flexibility, versatility, and ability to interact with various technologies [7]. Indeed, the combination of the use of UAVs for the extraction of information and interaction with crowd counting algorithms shows that it improves the precision of the detection and the estimation of the crowd count; where having density maps is a requirement to fulfill [8].

In previous work, the capabilities of drones for surveillance and reconnaissance have been verified and improved [9], as well as density maps and classification through CNN to estimate the number of people during real monitoring and surveillance actions, control of violent mobs [10].

This work shows the implementation of a surveillance and reconnaissance platform through drones, developed to count people in crowded settings. The system includes a real-time video streaming component, a data acquisition and processing console, and a crowd detection and counting algorithm. Section 2 shows the materials and the work methodology used, where the different components of the platform are highlighted. In Sect. 3 the results of the experimentation carried out both in the laboratory and in the field are described. In Sect. 4 there has been a discussion of the results obtained and finally in Sect. 5 the main conclusions and future work to be developed are presented.

2 Method Used to Estimation

One of the main objectives of this research is to perform a crowd count with valid precision, from an image or set of arbitrary still images, with an arbitrary camera perspective and crowd density derived from the above conditions. This significant contribution to science and society considers meeting the following challenges:

a. Complex foreground segmentation is essential to improve the accuracy of results, as inaccurate segmentation will have an irreversible negative effect on the final count; however, our algorithm can achieve high precision from an arbitrary image, without segmenting the multitude that make up the image, and without having information on the geometry or movement of the scene.
b. The density and distribution of the crowd vary significantly (or data sets), generating significant occlusions for most people in each image. Therefore, the solution to be reached differs from traditional detection methods.
c. The dynamic fluctuation in the scale (size) of the people, generated by the physical distance, focal length of the camera, sensor, etc., being necessary to use an adaptive algorithm that allows estimating the number of people from the diversity of images who can enter the system.

For the development of the crowd counting system by means of artificial vision, a series of steps has been designed in which the detection of both images and videos is possible, for which they have been structured as follows, how it is shown in the Fig. 1.

It is possible to observe that there is a series of combinations of processes and threads that in the end allow reaching the count using subroutines that allow the call of the neural network for the prediction of the numerical quantity that is required as a result. Figure 2 shows the processes of analysis, classification, and prediction in the counting of people in crowded settings.

Once the use of the different tasks has been developed for an orderly and synchronized processing of work sentences and file management, it is time to call the most important part of the entire algorithm of the system, which is the neural network itself that it will be detailed in one of the following sections.

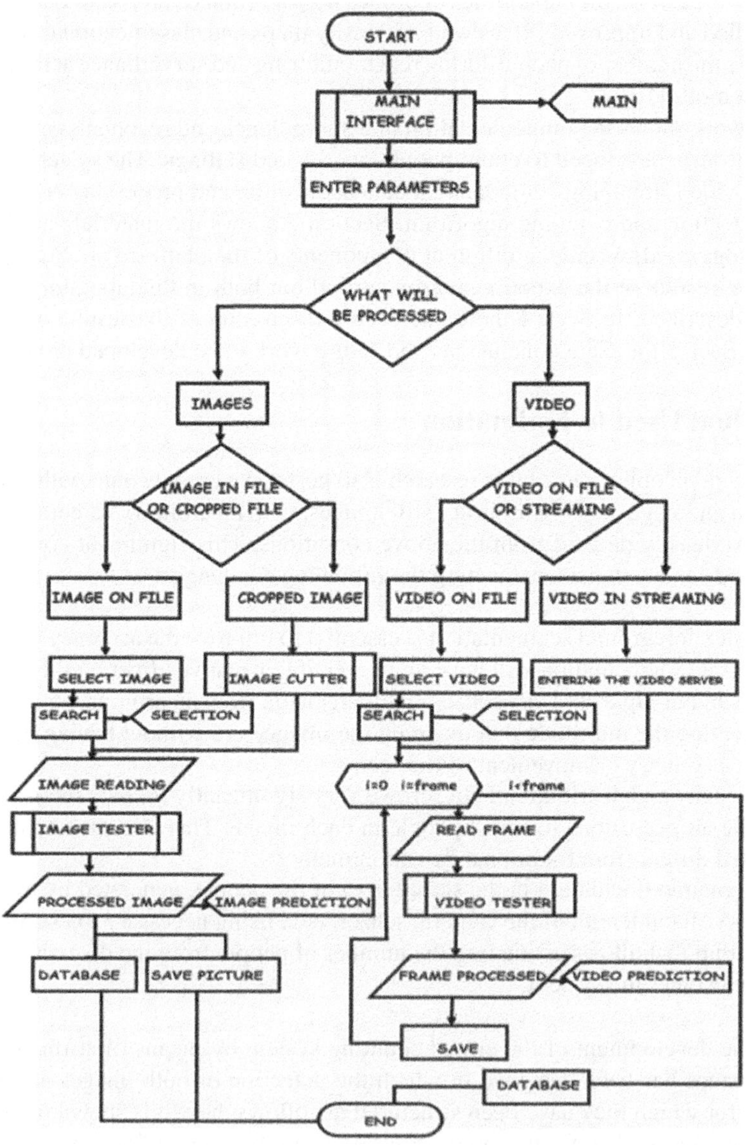

Fig. 1. Outline of routines and subroutines to be executed to get the crowd count.

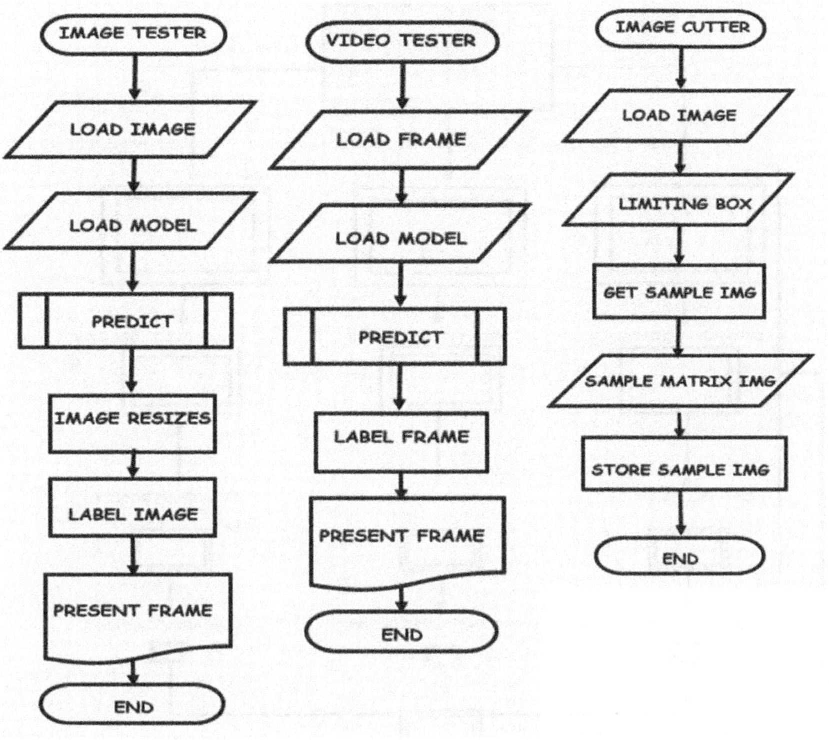

Fig. 2. The information classification and presentation process

2.1 Scheme of the Proposed Algorithm

The present investigation makes an estimate that gives a possible solution to the different and complex conditions that an image can have, it can reach a highly reasonable estimate from the use of a convolutional neural network (CNN).

Therefore, as mentioned [11], it is possible to use a deep, multicolumn convolutional neural network (MCNN), whose arbitrary number of columns can be trained on preprocessed inputs in different paths. Subsequently, the resulting predictions are obtained by averaging each of the individual predictions from all deep neural networks.

The MCNN algorithm proposed in this work is based on multiple convolutions that perform them in parallel and that converge towards an output layer that will allow estimating the number of people within an image in any type of environment in which they have been trained. The layer is composed of three columns of convolutional neurons with filters of different sizes, to find characteristic and important patterns that will allow us to recognize everything that has been taught. The input of the MCNN is the image itself and its output is a crowd density map that represents the mass concentration of people in the scene, whose normalized integral in valid and invalid values is the result of the crowd count. Figure 3 shows the architecture of the neural network and the scheme of the convolutions to be developed.

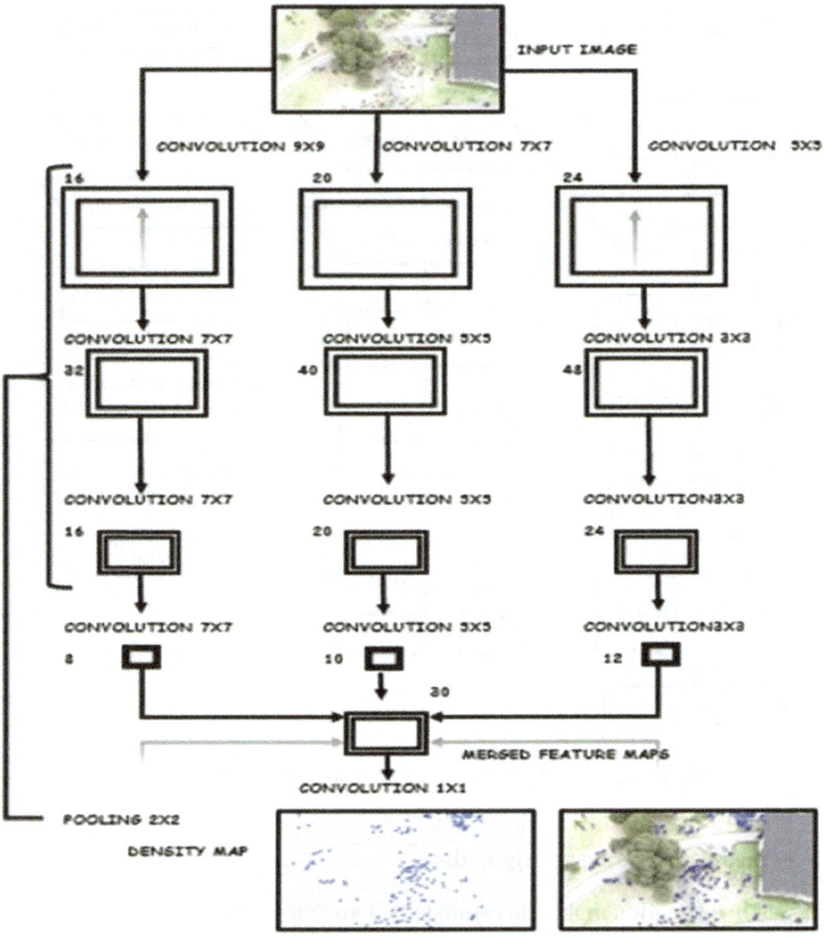

Fig. 3. Architecture of the convolutional neuronal network

In general terms, the proposed architecture has the following characteristics:

A neural network architecture has been considered with a total of three columns, the three correspond to filters of different sizes (large, medium, small) so that the characteristics learned by each CNN convolutional column are adapted to a great variation of characteristics as can be the variations in the size of the regions of interest (ROI's) and changes in lighting, scale, and visual perspective. This allows to affirm that the general network covers the resolution of problems related to the recognition of patterns by angle of view, height of observation, size, and resolution of the images.

The layer that is fully connected has been replaced with a convolution layer whose filter size is 1×1; this change allows that, independently of the type and characteristics of the input image, distortion problems can be solved. The immediate exit from the network is an estimate of the density of the crowd, from which the total count is derived.

2.2 Training System

The training system consists of several phases, initially considering the taking of the information (input), later the preprocessing is carried out considering how the use of Gaussian density maps as a valid option to achieve precision.

The training has been carried out through Convolutional Neural Networks (CNN) in a supervised way, the parameters and rules of learning compliance have been provided, for this in a set of images, markings are made through assignments of points or coordinates in the image matrix of h-x-w, where (h) means the height of the image and the (w) represents the width.

The marking looks for the pixels that make up the head as the referential pattern, a point is established where the person's head is located, this will allow you to deliver an input to the training process where it will be indicated exactly where there is a person in the environment of the image.

With the teaching already completed, these images and location files are passed through a preprocessing phase, and this consists of the combination of the image matrix and the scoring matrix, which, when operationalized, allow obtaining the necessary inputs for the input of the algorithm trainer.

The trainer algorithm is made up of a set of convolutional neural networks, a convolutional neural network is a type of multilayer network that consists of several convolutional layers and alternate pooling (subsampling), and at the end it has a series of full-connected layers as a multilayer perceptron network. The input of a convolutional layer network is usually an image m * m * r, where m is both the height and width of the image and r is the number of channels (as can be seen in the images, we work with both grayscale, so r equals 1, as well as RGB colors whose value of r is 3). Convolutional layers have k filters or kernels whose dimensions are n * n * q, where n and q are chosen by the designer, generally (q) is usually between 5 or 7. Each filter is generated by convolution a map of features or size characteristics (m − n + 1) * (m − n + 1) * p, where p is the number of filters to use. Then each map is subsampled in the pooling layer with the "mean pooling" or "max pooling" operation on contiguous regions of size where p can take values from 2 for small images to, commonly, no more than 9 for large images. Before or after subsampling, a sigmoidal activation function plus a bias is applied for each feature map, which in this case are the heads of each person.

The convolution of the system in continuous time is given by an equation that will allow us to obtain a resulting function required by means of a mathematical operator, this operator is an integral whose input inputs are two functions, one of them is the image to be processed and another is the convolutional displacement filter. The equation in continuous time would be represented as follows.

$$h = \int_{-\infty}^{\infty} f(t) * g(t - \tau) d\tau \tag{1}$$

The interval of the integral will depend on the domain over which each of the functions are defined. Representing this function in a discrete way, we are left with:

$$H = \sum_{-\infty}^{\infty} f(k) * g(n - k) \tag{2}$$

Using the function, it is possible to perform the convolutions where the classification of the characteristics required in our initial teaching process is carried out. In our case, the teaching model in which the image with N number of labeled heads is shown is represented by the following equation:

$$H(x) = \sum_{i=1}^{N} \delta(x - x_i) \tag{3}$$

That being convolved by a Gaussian kernel of type G (x) remains.

$$F(x) = \sum_{i=1}^{N} \delta(x - x_i) * g(x) \tag{4}$$

As a complement to obtain interesting data about the training and its learning behavior of cases with the algorithms already established, the calculation of errors such as Mean Absolute Error (MAE), Mean Square Error (MSE) and the function of Loss based on the following equations. For the loss function it is denoted according to [11] we mention the Euclidean distance measured is between the differences of the estimate of the density map with the markings made in the teaching process.

$$L(\Theta) = \frac{1}{2N} * \sum_{i=1}^{N} |F(x_i, \Theta), F_i|_2^2 \tag{5}$$

Where is a learning parameter of the multi-convolutional model, xi represents the input image to the system, Fire represents the marking, and F (x_i, Θ) represents the generated density map.

Finally, to evaluate and obtain metrics of the model used, we obtain the MAE and MSE of each iteration with the following formula as mentioned [11], with which it is possible to evaluate the learning of the neural network during training.

$$MAE = \frac{1}{N} * \sum_{i=1}^{N} |x_i - \hat{x_i}| \tag{6}$$

$$MSE = \frac{1}{N} * \sqrt{\sum_{i=1}^{N} (x_i - \hat{x_i})} \tag{7}$$

Where N is the number of images to be evaluated, x is the marked number of people, xi represents the estimated number of people by the neural network. In these calculations, the MAE indicates the accuracy of the estimates and the MSE represents the robustness of the estimates [11]. Once the convolution has been carried out, the density maps of the images of our data set can be seen in the results, as can be seen in Fig. 4.

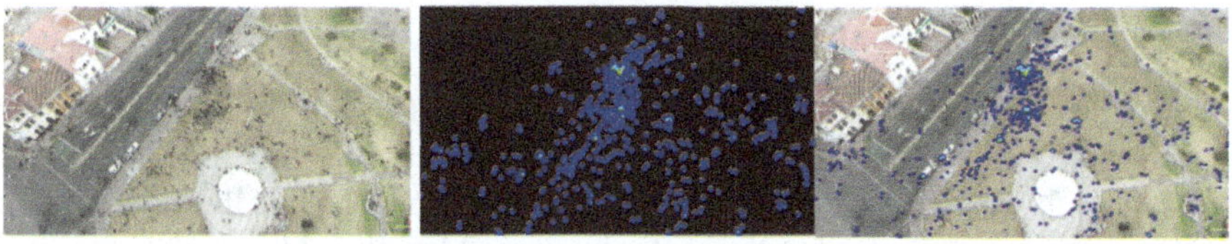

Fig. 4. Result of training with a data set in crowded settings.

2.3 Multi-destination Transmission System

The real-time transmission system is supported by the WLAN network architecture, which allows the radio control of the unmanned vehicle to be interconnected with the local video server-SVL. The radio control connects to the router wirelessly, using transmission frequencies close to 5 GHz. This architecture allows the radio control to retransmit the video that arrives from the unmanned vehicle to the SVL, which has the real time messaging protocol (RTMP) and allows the video to be hosted as a buffer. The architecture is amply illustrated in Fig. 5.

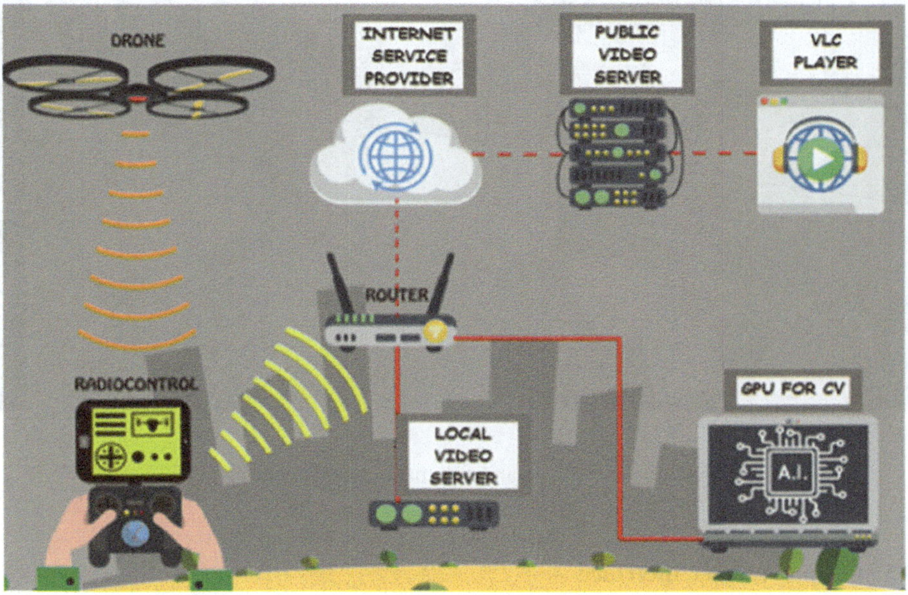

Fig. 5. Network architecture of the transmission system to estimate and count crowds in real time.

2.4 Integration of the Systems

The surveillance and reconnaissance subsystem are made up of the unmanned vehicle and its respective radio control. The configuration proposes that only the radio control is incorporated into the WLAN network through the 5 GHz band, while the video streaming subsystem (SVL) is connected to the router by wire. Also, the video analytics system is wired to the router.

This allows the computer to consume and execute the analytics of the video stored in the video server, and not in the radio control, thus avoiding interference in the normal operation of the radio control, as these interferences can have catastrophic consequences.

Once the required video analytics algorithm is executed, the results are transmitted to another video host server, using an internet service provider. Once hosted there, the end user can view the results using the VLC video player installed on the main computer. Real-time monitoring can be seen in Fig. 6.

Fig. 6. Viewing the estimate and automatic crowd counting in real time

2.5 Graphical User Interface

The system has a friendly graphical interface that allows the monitoring, control, and configuration of options. In effect, the interface design facilitates the selection of the type of information source to process, such as original photography, cropped photography, video, streaming video in real time. Figure 7 shows the interface developed and the various options described. The steps for using the options for processing are as follows:

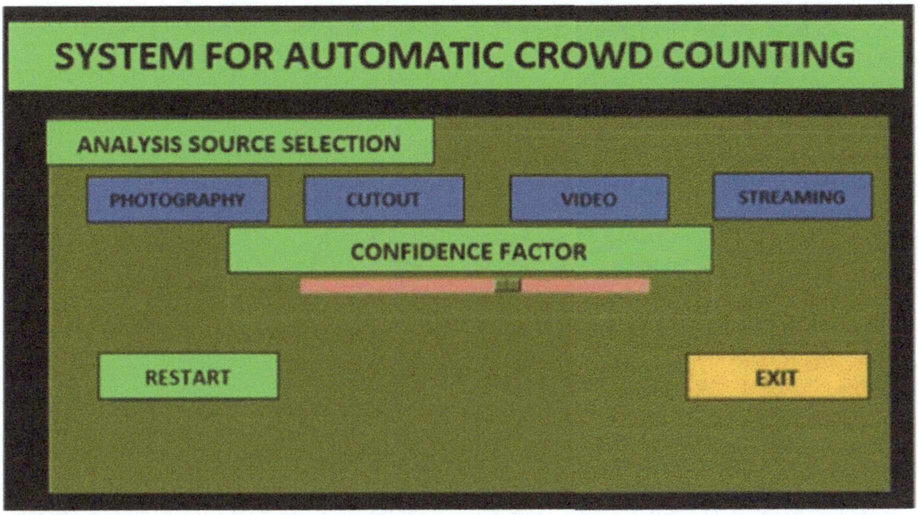

Fig. 7. Graphical user interface for robust algorithm for crowd counting and estimation.

a. Photograph: Image provided by the unmanned vehicle is downloaded to the computer to obtain its EXIF data.
b. Cropping: Image provided by the unmanned vehicle, once downloaded, can be cropped to select specific objectives of interest to the requesting entity or end user. This image also contains the EXIF data.
c. Video: Video provided by the unmanned vehicle and downloaded to the computer that performs the analysis. This file does not contain the EXIF GPS position data.
d. Streaming: Video provided in real time by the radio control of the unmanned vehicle when it is transmitted in the WLAN network to the local video server.

3 Experimental and Field Tests

As results of the tests regarding training, data of the Mean Absolute Error (MAE) and the mean square error (MSE) were obtained, during the training these data have been calculated to be able to make an appreciation of how our estimator is correcting its errors in each iteration, revealing how our neural network is learning and estimating more precisely, these results are represented as follows in the Fig. 8:

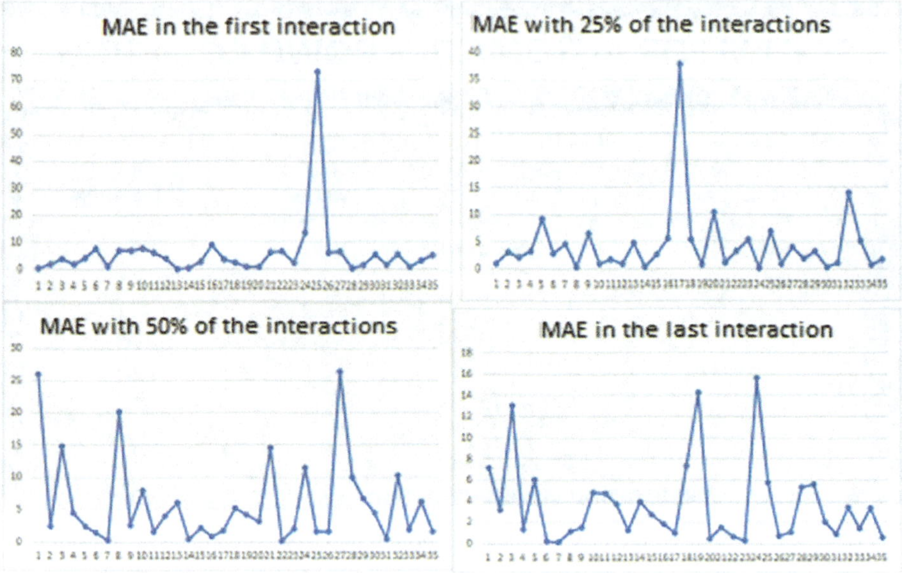

Fig. 8. Comparison and graphic representation of the Mean Absolute Error taking as samples some periods of training.

It can be clearly observed that for each epoch or iteration during the training process, the absolute error tends to decrease considerably. Considering a training of 100 iterations, in the first iteration a peak value of 70 is seen as the most erroneous value, in iteration No. 25 it can be seen that the error has been reduced by 50% being able to see a peak value of 35 error, in iteration No. 50 a peak value of 26 can be observed, which means that the error has been reduced by 25.7413% with respect to the previous data and finally in the last iteration it can see a peak error value of 16, which represents 38.4615% compared to the previous data. In conclusion, the error has been reduced by 77.1429% with respect to the first iteration, resulting in that the greater the number of iterations, it has a greater possibility of reducing the estimation errors, and this would represent a greater amount of time for the realization. Of a good training of the network as well as its computational cost. In Table 1, it can see MAE and MSE results obtained in the training, and in Fig. 9, the Mean Absolute Error and the Mean Square Error behavior are shown.

The operation and performance tests have been carried out under two schemes, one based on laboratory performance and the other based on field performance tests. Laboratory tests have been developed to establish the precision of the classification, in this sense, around 410 samples have been analyzed and the estimated value compared to manual counting, this being the basis of Ground Truth. The values of true positives (TP), false positives (FP), true negatives (TN) and false negatives (FN) have been obtained, based on these inputs the Precision $p = TP/(TP + FP)$ has been calculated, The Recall $r = TP/(TP + FN)$, in addition to the Dice Index $DI = 2\,(pr)/(p + r)$, these indicators provide us with information on the global accuracy of the algorithm in the classification. On the other hand, the Jaccard Index has also been calculated, which indicates the percentage of

Table 1. MAE and MSE results obtained in the training.

Type	Number	MAE	Value	MSE	Value
epoch	0	valid_mae:	6.5461	valid_mse:	15.3260
epoch	1	valid_mae:	6.3080	valid_mse:	15.6522
epoch	2	valid_mae:	6.9400	valid_mse:	15.8280
epoch	3	valid_mae:	6.1693	valid_mse:	14.9179
epoch	4	valid_mae:	5.9028	valid_mse:	14.9455
Average			**6.3732**		**15.3339**

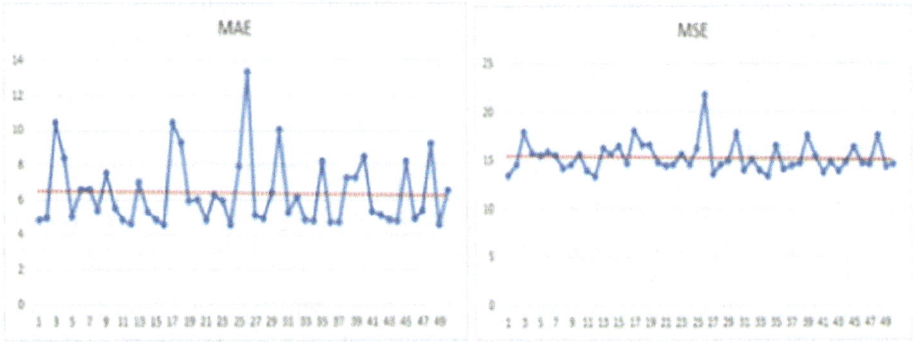

Fig. 9. Graphic representation of the behavior of the Mean Absolute Error and the Mean Square Error during the entire training

similarity between the expected result and the one obtained. Table 2 shows the average result of the samples analyzed in their global context.

Table 2. Average metrics calculated over 140 test images.

Images	r	p	DI	JI
140	91.40%	92.39%	91.84%	88.41%

The resolution of the video frames depends on the specific characteristics of the optical sensors, and it is necessary to consider that adequate focal length capabilities, CMOS type sensors, and adequate horizontal and vertical resolution are available. In addition, the height at which the unmanned vehicle takes the samples must be in accordance with the capabilities and characteristics mentioned above.

The field tests were carried out using DJI drone cameras of the Matrice 210 and Mavic 2 Enterprise Dual versions, during daylight hours, with favorable environmental conditions without rain, with winds below 12 knots, and heights ranging between 75 and 120 m.

On the other hand, from the observation carried out, it was verified that the main problems presented by the system are in reference to the handling of scales, partial occlusions, variation of perspectives and changes in lighting, so that at the level of modeling, processing and estimation in real time conflicts, errors and instability are generated in the system, therefore; In order to achieve better performance rates, a compensation factor has been implemented, it allows to reach an estimate with greater precision, according to the scenario and the quality of the image.

The adjustment factor is an adjustable coefficient that ranges from a value of 2 to 6 in which it allows to operationalize a percentage value that will allow the counting of high and low values of the resulting matrix normalized for counting. This operation is carried out by obtaining the highest value and the minimum value of the resulting matrix and the adjustment factor will allow obtaining the percentage between this maximum and minimum value to be considered in the count, which after normalization will become values of about and zeros, the values of one will represent a valid book value as a person within the scenario.

4 Discussion

The developed system consists of two components, one that generates the prediction and estimation of the number of people in a crowded scenario: and on the other hand, the transmission and reception system in real time.

Indeed, from the measurements carried out, it is verified that the crowd counting system developed has an accuracy rate of 92.39%, which generates a high confidence in the estimate; this result is corroborated with those obtained from recall and DI. In addition, the Jaccard Index shows correlation of results by obtaining an 88.41% similarity between what was obtained and what was desired.

It has been calculated in the training phase, the MAE and MSE, whose values are 6.37 and 15.33 respectively, which means that exist 6.37% of the MAE error in precision in the estimation and while in MSE exist a 15.33% error in the robustness of our estimates, which is very similar to the experimental metrics, which indicates that is a stable crowd counting system.

On the other hand, tests in the field and in real time define that it is necessary to consider external and environmental factors that limit the classification and detection, especially due to problems of occlusions, changes in lighting and change of perspectives, so it is necessary to introduce factors correction to improve results.

5 Conclusions

The crowd counting platform developed has combined two components, one based on the estimation of the number of people and on the other hand, the design and implementation of monitoring and transmission in real time.

The implemented system shows a high accuracy rate of 93.39%, which together generates a high performance, as corroborated by the Jaccard Index of 88.41%. Also, the MAE and MSE calculated are low, this means that the system developed is stable.

The system introduces a correction factor especially for use in real time, which allows to maintain a high level of performance.

The system can count all the people in the frame, for that reason a module was added that allows a specific area to be counted, to concentrate the threat. The system has the ability to consume the existing EXIF data in the photographs taken from the drones, which allows generating virtual information in real time on the location of the targets showing the respective geographic coordinates in DD° MM'SS".

Future work includes developing the classification and prediction system in scenarios, the system can be improved, performing training with night vision and thermographic images.

Acknowledgments. This work was partially supported by the PIS-04 "Detección temprana de incendios forestales basado en aprendizaje automático y en algoritmos de clasificación aplicados al espectro visible" research project of the Universidad de las Fuerzas Armadas ESPE, Ecuador and for Centro de Investigación de Aplicaciones Militares-CICTE.

References

1. Chan, A., Liang, Z., Vasconcelos, N.: Privacy preserving crowd monitoring: counting people without people models or tracking. In: 2008 IEEE Conference on Computer Vision and Pattern Recognition, pp. 1–7 (2008)
2. Sam, D., Surya, S., Babu, R.: Switching convolutional neural network for crowd counting. In: 2017 IEEE Conference on Computer Vision and Pattern Recognition (CVPR), pp. 4031–4039 (2017)
3. Chen, K., Loy, C., Gong, S., Xiang, T.: Feature mining for localized crowd counting. In: BMVC, vol. 1, no. 2, p. 3 (2012)
4. Liu, J., Gao, C., Meng, D., Hauptmann, A.: DecideNet: counting varying density crowds through attention guided detection and density estimation. In: Proceedings of the IEEE Conference on Computer Vision and Pattern Recognition, pp. 5197–5206 (2018)
5. Kang, D., Ma, Z., Chan, A.: Beyond counting: comparisons of density maps for crowd analysis tasks—counting, detection, and tracking. IEEE Trans. Circ. Syst. Video Technol. **29**(5), 1408–1422 (2018)
6. Lian, D., Li, J., Zheng, J., Luo, W., Gao, S.: Density map regression guided detection network for RGB-D crowd counting and localization. In: Proceedings of the IEEE/CVF Conference on Computer Vision and Pattern Recognition, pp. 1821–1830 (2019)
7. Motlagh, N., Bagaa, M., Taleb, T.: UAV-based IoT platform: a crowd surveillance use case. IEEE Commun. Mag. **55**(2), 128–134 (2017)
8. Tzelepi, M., Tefas, A.: Human crowd detection for drone flight safety using convolutional neural networks. In: 2017 25th European Signal Processing Conference (EUSIPCO), pp. 743–747 (2017)
9. Paredes, M., Calderón, M., Riquelme, J., Merizalde, D.: Multipurpose unmanned system: an efficient solution to increase the capabilities of the UAVs. In: 2020 IEEE ANDESCON, Quito, pp. 1–6 (2020)

10. Cruz, H., Reyes Ch., R.P., Pinillos, M.: Automatic counting of people in crowded scenes, with drones that were applied in internal defense operations on October 20, 2019 in Ecuador. In: Rocha, Á., Paredes-Calderón, M., Guarda, T. (eds.) MICRADS 2020. SIST, vol. 181, pp. 111–121. Springer, Singapore (2020). https://doi.org/10.1007/978-981-15-4875-8_10
11. Zhang, Y., Desen, C., Siqin, G., Shenghua, G., Ma, Y.: Single-image crowd counting via multi-column convolutional neural network. In: Proceedings of the IEEE Conference on Computer Vision and Pattern Recognition (CVPR), pp. 1–15 (2016)

Author Index

M. Botto-Tobar et al. (Eds.): CIT 2021, LNNS 405, pp. 427–428, 2022.
https://doi.org/10.1007/978-3-030-96043-8

The manufacturer's authorised representative in the EU is Springer Nature Customer Service Centre GmbH, Europaplatz 3, 69115 Heidelberg, Germany. If you have any concerns regarding our products, please contact ProductSafety@springernature.com

Printed and bound by CPI Group (UK) Ltd, Croydon, CR0 4YY
30/09/2026
02225693-0007